THE THEORY OF COMPUTABILITY

INTERNATIONAL COMPUTER SCIENCE SERIES

SELECTED TITLES IN THE SERIES

An Introduction to Numerical Methods with Pascal *L V Atkinson and P J Harley*
Handbook of Algorithms and Data Structures *G H Gonnet*
Microcomputers in Engineering and Science *J F Craine and G R Martin*
Software Specification Techniques *N Gehani and A D McGettrick* (Eds)
Programming Language Translation: A Practical Approach *P D Terry*
Data Abstraction in Programming Languages *J M Bishop*
System Simulation: Programming Styles and Languages *W Kreutzer*
The Craft of Software Engineering *A Macro and J Buxton*
An Introduction to Programming with Modula-2 *P D Terry*
Cost Estimation for Software Development *B Londeix*
Parallel Programming *R H Perrott*
The Specification of Computer Programs *W M Turski and T S E Maibaum*
Software Development with Ada *I Sommerville and R Morrison*
Logic Programming and Knowledge Engineering *T Amble*
Performance Measurement of Computer Systems *P McKerrow*
Syntax Analysis and Software Tools *K J Gough*
Practical Compiling with Pascal-S *M J Rees and D J Robson*
Concurrent Programming *N Gehani and A D McGettrick* (Eds)
Functional Programming *A Field and P Harrison*
Comparative Programming Languages *L B Wilson and R G Clark*
Software Prototyping, Formal Methods and VDM *S Hekmatpour and D Ince*
Numerical Methods in Practice: Using the NAG Library *T Hopkins and C Phillips*

THE THEORY OF COMPUTABILITY

Programs, Machines, Effectiveness and Feasibility

R. Sommerhalder
S. C. van Westrhenen

University of Delft

Wokingham, England · Reading, Massachusetts · Menlo Park, California
New York · Don Mills, Ontario · Amsterdam · Bonn
Sydney · Singapore · Tokyo · Madrid · San Juan

Cover designed by Crayon Design of Henley-on-Thames and printed by The Riverside Printing Co. (Reading) Ltd.
Typeset by P&R Typesetters Ltd, Farnham.
Printed and bound in Great Britain by Mackays of Chatham PLC, Kent.

First printed 1988.

British Library Cataloguing in Publication Data

Sommerhalder, R. (Rudolph)
The theory of computability: programs, machines, effectiveness and feasibility.—
(International computer science series).
1. Computer systems. Programming.
Algorithms
I. Title II. Westrhenen, S. C. van
(S. Christian van) III. Series
005.1′2′028

ISBN 0-201-14214-7

Library of Congress Cataloging in Publication Data

Sommerhalder, R.
The theory of computability.

(International computer science series)
Includes bibliographies and index.
1. Algorithms. 2. Computational complexity.
3. Computable functions. I. Westrhenen, S.
Christiaan van, 1928– . II. Title. III. Series.
QA9.58.S64 1988 511′.8 88-16809
ISBN 0-201-14214-7

Preface

Much research into the theory of algorithms has been done by mathematicians interested in the foundations of mathematics. The advent of ever bigger digital computers after the Second World War made questions about solvability by algorithmic methods and feasibility of solutions to problems also important for computer scientists.

The main goal of this book is to provide a rigorous mathematical introduction to the theory of algorithms for computer scientists. We are interested in questions like 'What is the nature of a computation?', 'What is an algorithm?', 'What is the computational complexity of a problem?', 'What is an efficient algorithm?' and 'What are the limitations of computation?'

The following points indicate where we deviate from the traditional approach to the theory of computation.

(1) Algorithms are expressed as programs in a programming language SAL (Simple Algorithmic Language), as opposed to the use of computation models like universal register machines or Markov algorithms.

We deviate from this line in the chapters about complexity theory. There we use Turing machines instead of SAL. In our opinion it would be nicer – from a conceptual point of view – to use the language SAL throughout the whole book and to define the classes **P, NP** etc. in terms of SAL programs. However, this would make the book useless as an introduction to the literature about complexity theory, since almost every paper on complexity issues refers to the Turing machine as the underlying model. Furthermore, in the discussion of complexity theory, we generally use high-level descriptions of algorithms which are largely insensitive to the details of the underlying model of computation.

(2) We also introduce programming languages over strings. These reflect more accurately the way digital computers process information, and are therefore adequate for assigning a cost to computations.

(3) In the definition of the recursive functions we use the function operators iteration and exponentiation instead of the classical primitive recursion scheme and the μ-operator. These operations are the formal counterpart of the times statement (a repetition statement with fixed number of repetitions) and the while statement.

(4) Since programs have multiple inputs and outputs, functions accept vector values as arguments and also return vector values.

(5) For functions we introduce the operation 'Cartesian product of functions', and a generalization of function composition for vector-valued functions.

The book is essentially self-contained and is designed to make the theory of algorithms accessible to beginning graduate students in computer science.

The book is organized as follows.

Chapter 1 introduces mathematical concepts, and notations are given that are assumed known in the rest of the book.

Chapter 2 provides an informal introduction to the field and also contains a short historical introduction to the theory of computation.

Chapters 3 and 4 establish the framework for the theory of computability. In Chapter 3 the programming language SAL and its semantics are discussed and the concept of a computable function is defined. In Chapter 4 recursive functions and their relation to SAL computable functions are discussed extensively.

In Chapter 5 the basic tools of recursion theory are developed, such as Gödel numberings, universal programs and universal functions, standard indexing, acceptable indexings, and the s–n–m theorem.

In Chapter 6 the Halting Problem and its variants are discussed and recursively enumerable sets and predicates are introduced. Various unsolvable problems are discussed.

Chapters 7 and 8 provide an introduction to the theory of reducibility – many-to-one and Turing reducibility.

Chapter 9 covers the basic requirements that are needed for a model of computation in order to get practically valid complexity results.

In Chapter 10 an introduction to algorithmic machines working with strings of symbols is given, and the Turing machine is introduced. The Turing machine is used in the remaining chapters of the book as the model of computation.

In Chapter 11 various complexity classes are defined, and the relations between them discussed. Hierarchy theorems, the power of non-determinism, speed-up theorems and the basic results of Blum's machine-independent complexity theory are investigated.

Chapter 12 concentrates on tractable problems, and presents the case for polynomial time as a formalization of practical feasibility. In this chapter the various resource-bounded reducibilities are discussed.

Chapter 13 discusses the classes **P**, **NP** and *co*-**NP**, and presents a number of **NP**-complete problems.

Chapter 14 completes the discussion by presenting results about and complete problems for the classes **L**, **NL**, **P** and *PSPACE*.

Chapter 15 gives guidelines for further study, in particular with respect to issues not treated in detail in the book itself, such as approximating algorithms, probabilistic algorithms, and the impact of parallel computing on feasibility results.

At the end of each chapter a number of exercises and some references to the literature are provided. The references are not intended to be complete. At the end of the book we suggest some entry points to the vast literature on the theory of algorithms.

There are various possible paths through the book. Two are of particular importance.

First, readers mainly interested in recursion theory can restrict their reading to Chapters 2–8, and then make their own selection from Chapters 9–12.

Second, readers interested in the properties of the classes **P** and **NP** in connection with the theory of computation are advised to read Chapters 2–6, to look at the definitions of Cook and Karp reducibility in Chapters 7 and 8, then continue with Chapters 9 and 10, look at the results in Chapter 11 and then continue with Chapters 12–15.

This book grew out of our teaching classes for graduate and undergraduate students at the Technical Universities of Eindhoven and Delft over several years. In our experience the second path can be taught in a two-semester course 'Introduction to the theory of algorithms'. In fact both authors have given such courses over a number of years.

We thank the various students for their criticism of those parts of the manuscript we used as material for our courses in computability theory. We also thank our colleagues Henk Claassen and Hans Tonino of the Technical University of Delft and Peter Asveld of the Technical University of Twente for their criticism of various versions of the manuscript and for supplying problems.

Finally we thank the Faculty of Mathematics and Informatics for the use of computer facilities while preparing the text.

Last but not least we thank the referees of Addison-Wesley for their constructive criticism, and the editorial staff, in particular Simon Plumtree, Steven Troth, and Allison King, for their flexibility throughout the preparation of the book.

R. Sommerhalder
S. C. van Westrhenen

Delft, 1987

Contents

Chapter 1
Mathematical Preliminaries

In this chapter we give a brief survey of mathematical notations and facts used in other parts of the book. The reader is advised quickly to glance over this chapter, and get back to it only when necessary.

Do however, note the following.

(1) We use the word function to mean a partial function. If a total function is intended, this is explicitly stated.

(2) Letters, usually lower-case letters, are used to denote variables and constants. There is no typographical distinction between letters denoting indivisible values, and letters denoting aggregated values. We may write for example

let $a \in N$

and also

let $b \in N^n$

In the latter case, b denotes a vector of n components; the ith component is denoted by b_i. Whether a letter denotes an indivisible or an aggregated value follows from the context.

(3) If $a \in N^n$ and $b \in N^m$, we write (a, b) to denote the $n+m$-vector $(a_1, a_2, \ldots, a_n, b_1, b_2, \ldots, b_m)$.

(4) If two entities A and B are equal, we write $A = B$. If, on the other hand, we define an object A by saying that it is equal to B, we write $A \triangleq B$.

1.1 Sets

The concept 'set' is one of the most fundamental notions of mathematics. The branch of mathematics concerned with the study of sets is called set theory; this discipline was founded by the German mathematician Georg Cantor (1845–1918).

In the theory of sets, the concepts 'set' and 'element of' are primitive; they are not defined in terms of other concepts. Sets are usually denoted by capital letters.

Let A be a set: '$x \in A$' means that x is an element of the set A, and '$x \notin A$' means that x does not belong to the set A. A set is fully specified by the elements it contains, but the way in which these elements are specified is immaterial; there is for example no difference between the set consisting of the elements 2, 3, 5 and 7, and the set of all prime numbers less than 11. In other words, two sets A and B are equal if and only if they contain the same elements.

There exists a special set, denoted by $\varnothing$, which does not contain any element at all.

Let A and B be sets. The notation '$A \subseteq B$' (in words 'A is a subset of B') means that every element of A is also an element of B. The notation '$A \subset B$' means that A is a proper subset of B, i.e. $A \subseteq B$ but there also are elements of B which do not occur in A. If $A \subseteq B$ and $B \subseteq A$, then A and B are equal; this is denoted by $A = B$.

The set denoted by $\{x \mid P(x)\}$, where $P(x)$ is a statement involving x, is the set of all x such that $P(x)$ is true. Finite sets and sometimes infinite sets can be specified by a (partial) enumeration of their elements. For example,

$$\{1, 2, 3, 4\}$$
$$\{1, 3, 5, 7, \ldots\}$$

This partial enumeration is intended to specify the set

$$\{x \mid x \text{ is an odd natural number}\}.$$

In Table 1.1 are specified some sets and the letters which will always be used to denote these sets.

We define the following operations on sets A and B.

(1) The *power set* $P(A)$ of A, $P(A) \triangleq \{x \mid x \subseteq A\}$

(2) The *union* $A \cup B$ of A and B, $A \cup B \triangleq \{x \mid x \in A \text{ or } x \in B\}$

(3) The *intersection* $A \cap B$ of A and B, $A \cap B \triangleq \{x \mid x \in A \text{ and } x \in B\}$

(4) The *difference* $A - B$ of A and B, $A - B \triangleq \{x \mid x \in A \text{ and } x \notin B\}$; this set is also called the complement of B relative to A

(5) The *symmetric difference* $A + B$ of A and B, $A + B \triangleq (A - B) \cup (B - A)$

Table 1.1 Sets and letters denoting these sets.

Letter	Denotes the set of all	
N	natural numbers	$N = \{0, 1, 2, \ldots\}$
N^+	positive natural numbers	$N^+ = \{1, 2, 3, \ldots\}$
Z	integers	$Z = \{0, 1, -1, 2, -2, \ldots\}$
Q	rational numbers	
R	real numbers	
B	truth values	$B = \{\text{TRUE}, \text{FALSE}\}$

(6) The *Cartesian product* $A \times B$ of A and B, $A \times B \triangleq \{(x, y) | x \in A \text{ and } y \in B\}$

$$A \times \varnothing = \varnothing, \text{ and } \varnothing \times B = \varnothing$$

The Cartesian product of n sets $A_1, \ldots, A_n$, $n \geqslant 2$, is defined similarly

$$A_1 \times A_2 \times \cdots \times A_n \triangleq \{(x_1, x_2, \ldots, x_n) | x_i \in A_i, \text{ for all } i, 1 \leqslant i \leqslant n\}.$$

(7) Let A be non-empty. The *powers* A^n of A are defined as follows

(a) $A^0 \triangleq \{\varnothing\}$

(b) $A^1 \triangleq A$

(c) $A^n \triangleq A \times A \times \cdots \times A$ (n factors A), for $n \geqslant 2$.

We can form the union and the intersection of any arbitrary number of sets. Let A be any finite or infinite set of sets.

$$\bigcup_{X \in A} X \triangleq \{x | \text{there is an } X \in A \text{ such that } x \in X\}$$

$$\bigcap_{X \in A} X \triangleq \{x | x \in X \text{ for all } X \in A\}$$

1.2 Functions and the lambda notation

Let A and B be sets.

(1) A *binary relation from A to B* is a triple (A, B, r), such that $r \subseteq A \times B$.

(2) A *function from A to B* is a triple (A, B, f) such that $f \subseteq A \times B$, and for all $a \in A$ there is at most one $b \in B$ such that $(a, b) \in f$.

(a) Instead of '(A,B,f)', we write '$f{:}A \to B$', and pronounce this as 'f is a function from A to B'. If A and B are understood, we speak of the function f instead of the function (A,B,f).

(b) Instead of $(a,b) \in f$ we write $f(a)=b$, and in general we write $f(a)$ to denote the unique element b such that $(a,b) \in f$.

If $f(a)=b$, we say that b is the *image* of a under f, and that a is an *inverse* of b under f. Note that an element of A has at most one image under f, whereas an element of B may have zero or more inverses.

The notation $f(a)\downarrow$ means that there exists an element b such that $f(a)=b$. The notation $f(a)\uparrow$ means that a does not have an image under f, i.e. there is no b such that $(a,b) \in f$.

(3) Consider a function $f{:}A \to B$

(a) A is called the *domain* of (A,B,f).

(b) B is called the *codomain* of (A,B,f).

(c) $dom(f) \triangleq \{x \in A \mid f(x)\downarrow\}$; this set $dom(f)$ is also called the domain of f.

(d) $range(f) \triangleq \{y \mid \text{there is an } x \in A \text{ such that } f(x)=y\}$; this set is called the range of f.

(e) The function (A,B,f) is called *total* if $dom(f)=A$.

(f) The function (A,B,f) is called *surjective* if $range(f)=B$.

(g) The function (A,B,f) is called *injective* if for all $b \in B$ there is at most one $a \in A$ such that $f(a)=b$. In other words, if $f(x) \neq f(y)$ whenever $x \neq y$, for all $x, y \in A$.

(h) The function (A,B,f) is called *bijective* if it is injective and surjective.

(i) For $X \in \{\text{in, sur, bi}\}$, Xjective functions are also called Xjections.

A *number-theoretic* function is a function $f{:}N^n \to N^k$ for some $n \geqslant 0$ and $k \geqslant 1$; such a function is called 'n-ary' to indicate the number of arguments the function takes. Note that a total function of the form $f{:}N^0 \to N^k$ is a triple $(\{\varnothing\},\ N^k,\ f \subseteq \{\varnothing\} \times N^k)$. Thus $f=\{(\varnothing,a)\}$ for some $a \in N^k$; the function effectively selects an element of N^k.

The *inverse* of a binary relation (A,B,r) from A to B is a binary relation (B,A,r^{-1}) from B to A, where $r^{-1} \triangleq \{(b,a) \mid (a,b) \in r\}$. The inverse of a function from A to B also is a binary relation from B to A, but this is not necessarily a function. However, if $f{:}A \to B$ is a total bijection, then its inverse is a total bijection too. The inverse is denoted by $f^{-1}{:}B \to A$.

Functions can be compared in many ways. The following two ways of comparison are of importance in this book.

Let $f,g{:}R^+ \to R^+$ be two real-valued functions, where R^+ is the set of all nonnegative real numbers.

(1) f is of *O-order* g if for every $a \in R^+$ there exists a real number $m_a \in R^+$ such that $f(x) \leqslant a.g(x)$ for all $x \geqslant m_a$

The notation '$f = O(g)$' means that f is of O-order g.

(2) f is of *Ω-order* g if for every $a \in R^+$ there exists a real number $m_a \in R^+$ such that $f(x) \geqslant a.g(x)$ for all $x \geqslant m_a$

The notation '$f = \Omega(g)$' means that f is of Ω-order g.

Let f and g be defined by $f(x) = x^2$ and $g(x) = x^4$. Then $f = O(g)$ and $g = \Omega(f)$, but $g \neq O(f)$ and $f \neq \Omega(g)$.

The notation $f = O(g)$ is somewhat misleading, as can be seen from the example above. It is easy to find functions f, g and h such that $f = O(h)$ and $g = O(h)$ but neither $f = g$ nor $f = O(g)$. Similar remarks hold true for Ω.

To specify a function (A, B, f), we must specify the set f, i.e. specify what pairs (a, b) belong to the set f. In cases where the function value $f(x)$ is determined by an expression $E(x)$, it may be advantageous to use the notation $f = \lambda x[E(x)]$, if the domain and codomain of f are understood. For example, we may write $f \triangleq \lambda x[2x + 1]$ to specify the function (N, N, f), where $f = \{(x, y) | y = 2x + 1 \text{ and } x \in N\}$. This is called the *lambda notation.*

The concept 'expression' will be used here as a primitive term. Forming a lambda expression $\lambda x[E(x)]$ from an expression $E(x)$ is called *abstraction*, or more fully *function abstraction*, and we say that the variable x in E is *bound* by λ.

Intuitively, the variable x in the expression represents an arbitrary element of the domain of f, and $E(x)$ specifies the value $f(x)$.

More formally, the function value $f(a)$ obtained as a result of applying the function $\lambda x[E(x)]$ to the argument a, is denoted by $\lambda x[E(x)]a$, and is equal to $E(a)$, the result of evaluating the expression E after substituting a for x in it. There is a fully fledged calculus of lambda expressions; we won't discuss this however, as we only need the idea of abstracting a function from an expression, and of applying an abstracted function to an argument.

The lambda notation can be extended to functions of more than one argument, and vector-valued functions. For example, we write $f \triangleq \lambda xy[(x, x + y, x.y)]$ to abstract the function (N^2, N^3, f), where

$$f = \{((x, y), (p, q, r)) | x, y \in N,\ p = x,\ q = x + y \text{ and } r = x.y\}$$

from the expressions x, $x + y$ and $x.y$. The function application rule is properly extended to handle this more general case.

We also allow expressions to contain the symbol '↑', to specify that the value of the expression is undefined. For example, the function $f: N \rightarrow N$ such that $f(x) = 1$ if x is even, and undefined otherwise, can be defined by function abstraction as follows:

$$f \triangleq \lambda x[\text{if } (x \text{ is even}) \text{ then } 1 \text{ else } \uparrow].$$

1.3 Special functions

In Table 1.2 some number-theoretic functions are defined which are used in the rest of the book.

Table 1.2 Some special functions.

Function	Definition	Name of function
$C_k^n: N^n \to N$	$\lambda x[k]$	constant function
$\pi_i^n: N^n \to N$	$\lambda x[x_i]$	projection function
$\pi_{i \ldots j}^n: N^n \to N^{1+j-i}$	$\lambda x[(x_i, \ldots, x_j)]$	
$I^n: N^n \to N^n$	$\lambda x[x]$	identity function
$z: N \to N$	$\lambda x[0]$	zero function
$+_1: N \to N$	$\lambda x[x+1]$	successor function (sometimes denoted by s)
$\dot{-}_1: N \to N$	$\lambda x[\text{if } (x>0) \text{ then } x-1 \text{ else } 0]$	predecessor function (sometimes denoted by p)
$\dot{-}: N^2 \to N$	$\lambda xy[\text{if}(x>y) \text{ then } x-y \text{ else } 0]$	monus function
$sg: N \to N$	$\lambda x[\text{if } (x>0) \text{ then } 1 \text{ else } 0]$	signum function
$\overline{sg}: N \to N$	$\lambda x[\text{if } (x=0) \text{ then } 1 \text{ else } 0]$	

1.4 Operations on functions

Let (A, B, r) and (C, D, s) be binary relations. The *composition* of these relations is the binary relation (A, D, t) from A to D, where t is defined as follows.

$$t \triangleq \{(a,d) | (\exists c)[(a,c) \in r \text{ and } (c,d) \in s]\}$$

Function composition is a special case of relation composition because functions are relations. We introduce some special cases of function composition.

(1) Let $g: A \to B$ and $f: B \to C$

The composition of f and g is a function from A to C denoted by $f \cdot g$ and defined by $f \cdot g \triangleq \lambda x[f(g(x))]$.

The situation is sketched in Figure 1.1. This diagram is said to *commute*, which means that the two paths from A to C represent the same function. In other words, applying the function $f \cdot g: A \to C$ to an arbitrary element $a \in A$, produces the same result as applying the function $g: A \to B$ to a, producing the value $g(a) \in B$, and then applying $f: B \to C$ to the element $g(a)$; i.e. $f \cdot g(a) = f(g(a))$.

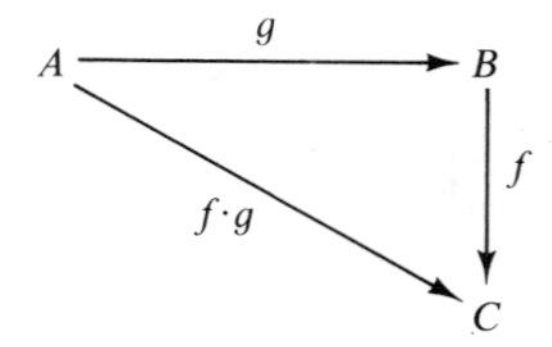

Figure 1.1 The function composition $f \cdot g$.

(2) Let n functions $g_i: A \to B_i$, $1 \leqslant i \leqslant n$, and a function $f: B_1 \times B_2 \times \cdots \times B_n \to C$ be given. The composition of f and the functions g_i is a function from A to C denoted by $f \cdot (g_1, g_2, \ldots, g_n)$ and defined by

$$f \cdot (g_1, g_2, \ldots, g_n) \triangleq \lambda x [f(g_1(x), g_2(x), \ldots, g_n(x))].$$

That is, the composition is defined such that Figure 1.2 commutes.

(3) Let $f: A \to B$ and $g: C \to D$ be functions. The *Cartesian product* of these functions is a function $f \times g: A \times C \to B \times D$ defined by $f \times g \triangleq \lambda xy[(f(x), g(y))]$. Thus, the Cartesian product function is defined such that Figure 1.3 commutes.

In the following definition some lesser-known operations on functions are defined. In particular, the operations of exponentiation and iteration are basic to our discussions.

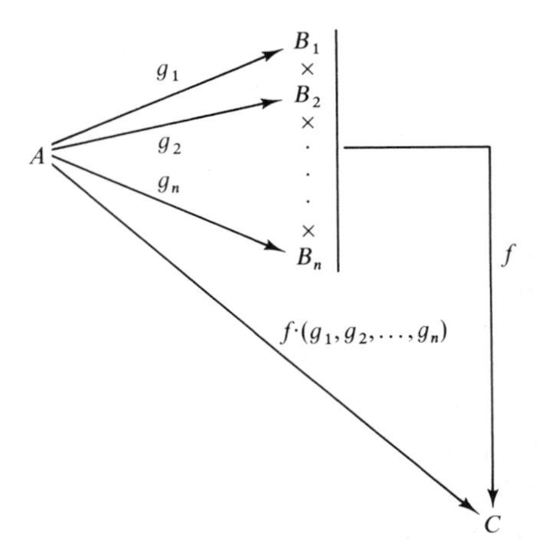

Figure 1.2 The function composition $f \cdot (g_1, g_2, \ldots, g_n)$.

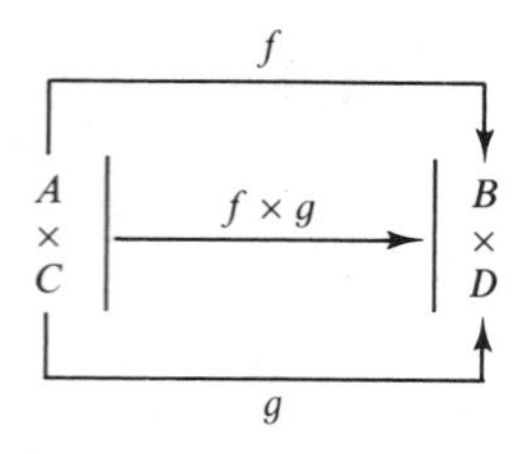

Figure 1.3 The Cartesian product $f \times g$.

Definition 1.1

(1) Let $f: A \to A$. The function $f^m: A \to A$ is defined as follows.

$$f^0 \triangleq \lambda x[x], \text{ i.e. } f^0 \text{ is the identity function on } A$$

$$f^1 \triangleq f$$

$$f^{m+1} \triangleq \lambda x[f(f^m(x))], \text{ for all } m \geqslant 1.$$

(2) Now consider a special case where $A = N^n$, and thus $f: N^n \to N^n$.

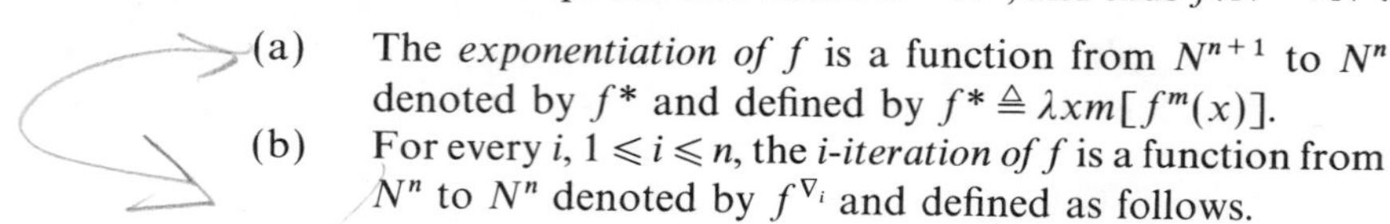

(a) The *exponentiation of f* is a function from N^{n+1} to N^n denoted by f^* and defined by $f^* \triangleq \lambda xm[f^m(x)]$.

(b) For every i, $1 \leqslant i \leqslant n$, the *i-iteration of f* is a function from N^n to N^n denoted by f^{∇_i} and defined as follows.

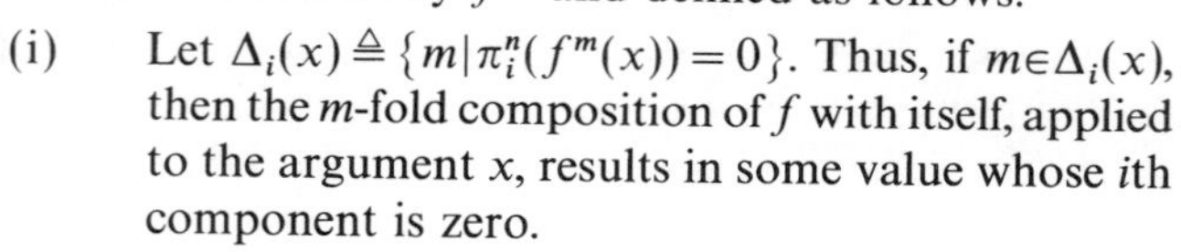

(i) Let $\Delta_i(x) \triangleq \{m \mid \pi_i^n(f^m(x)) = 0\}$. Thus, if $m \in \Delta_i(x)$, then the m-fold composition of f with itself, applied to the argument x, results in some value whose ith component is zero.

(ii) Let $\min \Delta_i(x)$ denote the least element of $\Delta_i(x)$.

$$f^{\nabla_i} \triangleq \lambda x[\text{if } (\Delta_i(x) \neq \varnothing \text{ and } f^m(x){\downarrow} \text{ for all } m \leqslant \min \Delta_i(x)) \text{ then } f^*(x, \min \Delta_i(x)) \text{ else } \uparrow]. \quad \blacksquare$$

Thus, the exponentiation of f corresponds to a given number of applications of f to an argument, whereas the iteration corresponds to repeated application of f until some condition is satisfied, in particular, until the ith component becomes zero.

EXAMPLE 1.1

(1) Let $f \triangleq \lambda xy[(x+1, y)]$ be a number-theoretic function from N^2 to N^2. Then $f^{\nabla_1} = \lambda xy[\uparrow]$.

(2) The function $\lambda xy[x+y]$ is equal to, or may be defined by $\lambda xy[+_1^*(x, y)]$, where $+_1$ is the successor function given in

Table 1.2. Similarly, the monus function can be defined in terms of the predecessor function $\dot{-}_1$. The functions $\dot{-}$ and $\dot{-}_1$ are defined in Table 1.2.

1.5 Relations

A binary relation R *on a set* X is a binary relation (X, X, r). For example, the relation $\leqslant$ is a binary relation on N, formally a triple $(N, N, r_{\leqslant})$, where

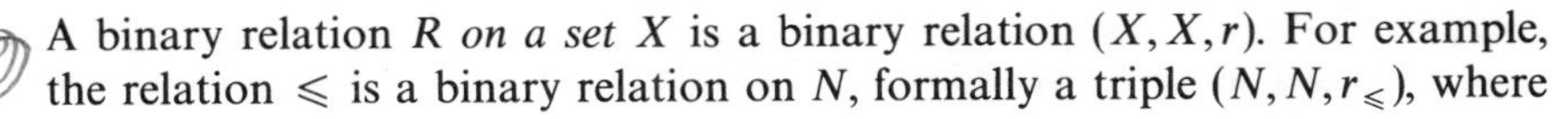

$$r_{\leqslant} = \{(x, y) | (\exists z \in N)[x + z = y]\}.$$

We write xRy instead of $(x, y) \in r$, just as we write $5 \leqslant 7$ instead of $(5, 7) \in r_{\leqslant}$.

The relation R is called

(1) *reflexive*, if xRx for all $x \in X$;
(2) *antireflexive* if xRx for no element $x \in X$;
(3) *symmetric* if xRy implies yRx for all $x, y \in X$;
(4) *antisymmetric* if (xRy and yRx) implies $x = y$, for all $x, y \in X$;
(5) *transitive* if (xRy and yRz) implies xRz, for all $x, y, z \in X$.

Let $R = (X, X, r)$ be a binary relation on a non-empty set X.

(1) R is an *equivalence relation* if R is reflexive, symmetric and transitive.
(2) R is a *partial order relation* if R is reflexive, antisymmetric and transitive.
(3) Elements x and y of X are called *incomparable* if neither xRy nor yRx, and R is a partial order relation.
(4) R is a *total order relation* if R is a partial order relation, and xRy or yRx for every $x, y \in X$.
(5) R is a *quasi-order relation* if R is reflexive and transitive.

Examples of binary relations on N:

(1) Equality between natural numbers is an equivalence relation on N.
(2) $\leqslant$ is a total order relation on N.
(3) Let '$x|y$' mean that x divides y, i.e. there is a $z \in N$ such that $x.z = y$. This relation $|$ is a partial order relation on N^+.
(4) Let '$x \equiv y$' mean that $5|(x - y)$. Then $\equiv$ is an equivalence relation on N.
(5) Let '$x \sim y$' mean that $p|y$ if $p|x$ for every prime number p. Then $\sim$ is a quasi-order relation on N^+.

An equivalence relation R on a set X partitions the set X into *equivalence classes.* The equivalence class containing the element $x \in X$ is denoted by $[x]_R$ and defined by

$$[x]_R \triangleq \{y \in X \mid xRy\}$$

It is easy to prove that either $[x]_R = [y]_R$, in which case xRy, or $[x]_R \cap [y]_R = \varnothing$.

The set of all equivalence classes is called the *factor set of* X *modulo* R, and is denoted by X/R. Thus $X/R = \{[x]_R \mid x \in X\}$.

Consider for example the relation $\equiv$ of the example above, i.e. $x \equiv y$ if and only if $5 \mid (x-y)$. There are 5 mutually different equivalence classes, as specified below.

$$[0]_{\equiv} = \{5x \mid x \in N\}$$

$$[1]_{\equiv} = \{5x+1 \mid x \in N\}$$

$$[2]_{\equiv} = \{5x+2 \mid x \in N\}$$

$$[3]_{\equiv} = \{5x+3 \mid x \in N\}$$

$$[4]_{\equiv} = \{5x+4 \mid x \in N\}$$

Thus $N/\equiv\, = \{[0]_{\equiv}, [1]_{\equiv}, [2]_{\equiv}, [3]_{\equiv}, [4]_{\equiv}\}$. Note that this representation of the factor set is not unique, we might also write for example

$$N/\equiv\, = \{[25]_{\equiv}, [16]_{\equiv}, [37]_{\equiv}, [103]_{\equiv}, [1989]_{\equiv}\}$$

A quasi-order relation R on a set X induces an equivalence relation $\equiv_R$ defined as follows:

$$x \equiv_R y \text{ if and only if } (xRy \text{ and } yRx)$$

The relation R also induces a partial order relation $\leqslant_R$ on the factor set $X/\equiv_R$ defined as follows:

$$[x]_{\equiv_R} \leqslant_R [y]_{\equiv_R} \text{ if and only if } xRy$$

Consider the relation $\sim$ of the example above, i.e. $x \sim y$ if for all prime numbers p, $p \mid y$ if $p \mid x$. The statement $x \equiv_{\sim} y$ means that x and y have the same prime divisors. Thus for each finite set $\{p_1, p_2, \ldots, p_m\}$ of mutually different prime numbers there is an equivalence class

$$[p_1 \cdot p_2 \cdots p_m]_{\equiv_{\sim}} = \{p_1^{x_1} \cdot p_2^{x_2} \cdots p_m^{x_m} \mid 1 \leqslant x_i, \text{ for all } i,\ 1 \leqslant i \leqslant m\}.$$

For example,

$$[5]_{\equiv_{\sim}} = \{5, 25, 125, \ldots\}$$

$$[6]_{\equiv_{\sim}} = \{6, 12, 18, 24, 36, \ldots\}$$

Also, $[5]_{\equiv_{\sim}} \leqslant_{\sim} [15]_{\equiv_{\sim}}$ and $[6]_{\equiv_{\sim}} \leqslant_{\sim} [30]_{\equiv_{\sim}}$, but not $[5]_{\equiv_{\sim}} \leqslant_{\sim} [6]_{\equiv_{\sim}}$, and neither $[6]_{\equiv_{\sim}} \leqslant_{\sim} [5]_{\equiv_{\sim}}$. Thus $[5]_{\equiv_{\sim}}$ and $[6]_{\equiv_{\sim}}$ are incomparable.

The pair $(X, \leqslant)$, where $\leqslant$ is a partial order relation on the set X, is called a *partially ordered set*. Let $a \in X$.

(1) a is a *maximal element* of $(X, \leqslant)$, if there is no $x \in X$ such that $a \leqslant x$.

(2) a is called the *greatest element* of $(X, \leqslant)$ if $x \leqslant a$ for all $x \in X$.

(3) a is called a *minimal element* of $(X, \leqslant)$ if there is no $x \in X$ such that $x \leqslant a$.

(4) a is called the *least element* of $(X, \leqslant)$ if $a \leqslant x$ for all $x \in X$.

Consider the ordered set $(X/\equiv_{\sim}, \leqslant_{\sim})$ discussed above. $[1]_{\equiv_{\sim}} = \{1\}$ is the least element and thus the only minimal element. There are no maximal elements and there is no greatest element.

Let $(X, \leqslant)$ be a partially ordered set and let $V \subseteq X$.

(1) An element $a \in X$ is a *lower bound* of V, if $a \leqslant v$ for all $v \in V$.

(2) An element $a \in X$ is an *upper bound* of V, if $v \leqslant a$ for all $v \in V$.

(3) An element $a \in X$ is the *greatest lower bound* of V, denoted by glb(V), if a is a lower bound of V, and every other lower bound b of V precedes a, i.e. $b \leqslant a$.

(4) Similarly, an element $a \in X$ is the *least upper bound* of V, denoted by lub(V), if a is an upper bound of V that precedes every other upper bound, i.e. if b is an upper bound of V, then $a \leqslant b$.

Clearly, glb(V) and lub(V) are unique if they exist.

Let $(X, \leqslant)$ be a partially ordered set.

(1) $(X, \leqslant)$ is an *upper semi-lattice* if every subset $V \subseteq X$ with two elements has a least upper bound, i.e. lub(V) exists.

(2) $(X, \leqslant)$ is a *complete upper semi-lattice* if lub(V) exists for every non-empty subset $V \subseteq X$.

(3) $(X, \leqslant)$ is a *lower semi-lattice* if glb(V) exists for every subset $V \subseteq X$ with two elements.

(4) $(X, \leqslant)$ is a *complete lower semi-lattice* if glb(V) exists for every non-empty subset $V \subseteq X$.

(5) $(X, \leqslant)$ is a *lattice* if it is both an upper and a lower semi-lattice.

(6) $(X, \leqslant)$ is a *complete lattice* if it is both a complete upper and a complete lower semi-lattice.

Examples of lattices:

(1) Every totally ordered set is a lattice. Consider for example the totally ordered set $(N, \leqslant)$. If $x \leqslant y$ then $\text{glb}(\{x, y\}) = x$, and $\text{lub}(\{x, y\}) = y$.

(2) A standard example of a complete lattice is the partially ordered set $(P(A), \subseteq)$, where A is an arbitrary set, and $P(A)$ is its power set. In this case we have

(a) $\text{lub}(\{X, Y\}) = X \cup Y$
(b) $\text{glb}(\{X, Y\}) = X \cap Y$

Let $B \subseteq P(A)$, then

(a) $\text{lub}(B) = \bigcup_{X \in B} X$, and

(b) $\text{glb}(B) = \bigcap_{X \in B} X$.

Thus $(P(A), \subseteq)$ is a complete lattice.

A lattice can also be defined in the following way. Let X be a set, and $+$ and . two binary operations on X, i.e. functions from X^2 to X. The structure $(X, +, .)$ is a lattice if the following axioms are satisfied.

(1) The operations are *idempotent*: $a + a = a$ and $a.a = a$ for all $a \in X$.

(2) The operations are *commutative*: $a + b = b + a$ and $a.b = b.a$ for all $a, b \in X$.

(3) The operations are *associative*: $(a + b) + c = a + (b + c)$ and $(a.b).c = a.(b.c)$ for all $a, b, c \in X$.

(4) The operations have the *absorption* property: $a + (a.b) = a$ and $a.(a + b) = a$ for all $a, b \in X$.

The corresponding partial order relation can be defined as follows:

$a \leqslant b$ if and only if $a.b = a$

(The same relation is defined by '$a \leqslant b$ if and only if $a + b = b$'.)

The standard example is $(P(A), \cup, \cap)$. It is easily seen that the operations satisfy the above axioms, and that the partial order defined by these operations is the usual set inclusion relation $\subseteq$.

1.6 Finite and infinite sets

Two non-empty sets A and B are *equipotent* if there exists a total bijective function from A to B. The empty set is equipotent with itself, and with no other set.

Obviously, two finite sets are equipotent if they have the same number of elements. Thus equipotence of sets may be considered as a generalization of the concept 'same number of elements' for finite sets.

EXAMPLE 1.2

(1) The sets N and Z are equipotent, because the function

$$\lambda x[\text{if}(x \geqslant 0) \text{ then } 2x \text{ else } 2(-x)-1]$$

is a total bijection from Z to N.

(2) The set R of all real numbers is equipotent with the set of all real numbers between 0 and 1 denoted by $\langle 0,1\rangle = \{x \in R \mid 0 < x < 1\}$. The function $\lambda x[\tan((x-\frac{1}{2})\pi)]$ is a total bijection from $\langle 0,1\rangle$ to R.

(3) The sets N and $N \times N$ are equipotent, because the function $\lambda xy[\frac{1}{2}(x+y)(x+y+1)+y]$ is a total bijection from N^2 to N. It can be shown that N^n is equipotent with N for every $n > 0$.

Definition 1.2

Let $[n]$ denote a subset of N defined as follows.

$$[0] \triangleq \varnothing$$

$$[n] \triangleq \{x \in N \mid 1 \leqslant x \leqslant n\}, \text{ for all } n \geqslant 1$$

Let A be any set.

(1) A is *finite* if it is equipotent with the set $[n]$ for some $n \in N$.

(2) A is *infinite* if it is not finite.

(3) A is *countable* if it is empty or if there is a total surjective function from N to A.

(4) A is *countably infinite* if there is a total bijection from N to A.

(5) A is *uncountable* if it is not countable. ■

Thus the elements of a non-empty, countable set can be indexed by the natural numbers, and can therefore be presented as a sequence $e_0, e_1, e_2, \ldots$.

This sequence necessarily contains duplicates, if the set is finite. Countably infinite sets can always be presented in this way without duplications.

Theorem 1.1

The set $P(N)$ of all subsets of N is uncountable.

Proof

Assume that $P(N)$ is countable. Then there exists a surjective function $f: N \rightarrow P(N)$. Define the set A as follows:

$$A \triangleq \{x \in N \mid x \notin f(x)\}$$

Clearly $A \subseteq N$, hence there exists some n such that $A = f(n)$.

Assume $n \in A$. Then $n \in f(n)$, because by assumption $f(n) = A$. However, if $n \in f(n)$, then $n \notin A$ by the definition of A.

Assume $n \notin A$. Then $n \notin f(n)$, because $A = f(n)$. On the other hand, $n \notin f(n)$ implies $n \in A$.

Therefore, if $P(N)$ is countable, then there is an $n \in N$ such that $n \in A$ and also $n \notin A$, a contradiction.

Consequently, $P(N)$ is uncountable. ■

The above proof is due to Georg Cantor; it is an application of the diagonal method invented by him. The term 'diagonal' refers to the pictorial representation of the definition of A, as sketched in Figure 1.4.

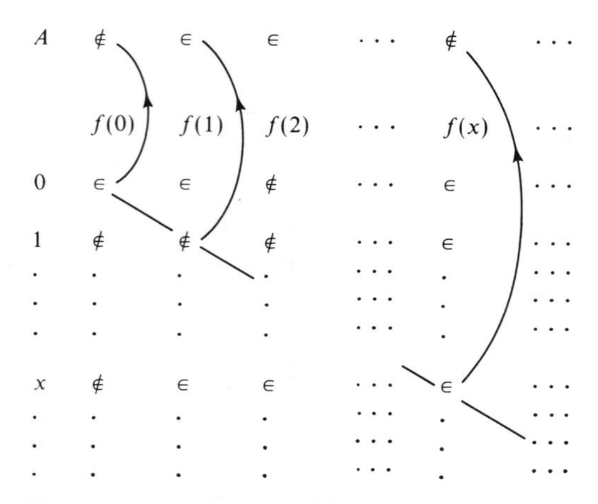

Figure 1.4 Construction of the set A, using diagonal method.

Theorem 1.2

The set R of all real numbers is uncountable.

Proof

Because R and $\langle 0, 1 \rangle$ are equipotent, it suffices to prove that $\langle 0, 1 \rangle$ is uncountable.

Every real number $a \in \langle 0, 1 \rangle$ can be represented as an infinite string $0.a_1 a_2 \ldots$. The representation is not unique, for example $0.100\ldots$ and $0.0999\ldots$ represent the same number. If, however, all digits a_i are neither nines nor zeros, the representation is unique.

Assume that $\langle 0, 1 \rangle$ is countable. Then there is a surjective function $f: N \to \langle 0, 1 \rangle$. Let $0.a_{n,1} a_{n,2} \ldots$ denote the representation of the real number $f(n)$. Define a representation $0.b_1 b_2 \ldots$ as follows.

$$b_i \triangleq \text{if } a_{i,i} = 4 \text{ then } 5 \text{ else } 4 \text{ for all } i, i \geqslant 1$$

Let b denote the real number represented by $0.b_1 b_2 \ldots$. This representation of b is unique.

Because the representation of b differs from the representation of $f(n)$ for all n, and the representation of b is unique, it follows that $b \neq f(n)$, for all $n \in N$. On the other hand, there is an $n \in N$ such that $b = f(n)$, because f is surjective.

We have a contradiction and conclude that $\langle 0, 1 \rangle$, and therefore the set R of all real numbers, is uncountable. ■

1.7 The natural numbers

The natural numbers can be introduced axiomatically using the axiom system of Giuseppe Peano (1858–1932). The primitive terms are the natural number 0 and the successor function s. The axioms are as follows:

Axiom 1 0 is a natural number.

Axiom 2 0 is not the successor of any natural number, i.e. $s(n) \neq 0$ for all n.

Axiom 3 For every natural number n there exists a uniquely determined natural number $s(n)$.

Axiom 4 $s(n) = s(m)$ implies $n = m$, for all natural numbers n and m.

Axiom 5 (induction axiom) For every subset A of the set of all natural numbers we have: if

(1) $0 \in A$, and
(2) $s(n) \in A$ if $n \in A$, for every natural number n

then $A = N$.

It is easy to define the customary arithmetic operations of addition and multiplication in terms of the successor function. Familiar properties, such as commutativity and associativity, may then be proved by induction.

An immediate consequence of the induction axiom is the *principle of finite induction*, which can be formulated as follows.

A statement $P(n)$ is true for all natural numbers n, if

(1) $P(0)$ is true, and
(2) $P(s(n))$ is true if $P(n)$ is true, for all natural numbers n.

An alternative formulation of the principle of finite induction is as follows.

A property P is valid for all natural numbers if

(1) P is valid for 0, and
(2) P is valid for n if P is valid for all natural numbers k, $0 \leqslant k < n$.

The principle of finite induction also justifies the use of recursive definitions. For example, the sum Σ_n of the first n terms of the sequence $a_1, a_2, \ldots$ can be recursively defined as follows:

$$\Sigma_0 \triangleq 0,$$

$$\Sigma_{n+1} \triangleq a_{n+1} + \Sigma_n$$

Assume that we wish to define an object $A(n)$ for all natural numbers n. A recursive definition of $A(n)$ satisfies the following.

(1) $A(0)$ is uniquely defined, independently of the objects $A(m)$, $m > 0$, and
(2) $A(n+1)$ is uniquely defined in terms of the objects $A(0), A(1), \ldots, A(n)$, but independent of the objects $A(m)$, $m > n+1$.

It is a matter of routine to prove that in this case all the objects $A(n)$ are uniquely defined.

A variant of this scheme is:

(1) $A(0)$ to $A(k)$, $k \geqslant 0$, are uniquely defined independent of each other, and
(2) $A(n+1)$ is uniquely defined in terms of the objects $A(0)$ to $A(n)$, for all $n \geqslant k$.

For example, the sequence of so-called Fibonacci numbers, 1, 1, 2, 3, 5, 8, 13, ... can be defined by the above variant of the scheme of recursive definition, as follows:

$$F_0 \triangleq 1,$$

$$F_1 \triangleq 1, \text{ and}$$

$F_{n+2} \triangleq F_n + F_{n+1}$, for all $n \geqslant 0$.

1.8 Graphs

A *graph* $G = (V, E)$ is a structure consisting of a finite non-empty set V and a finite set E. The elements of V are called *vertices* or *nodes* of G; the elements of E are called *edges* of G. In a *directed graph* $E \subseteq V \times V$, thus an edge is an ordered pair (u, v) of vertices; u is called the *start point* and v is called the *end point* of the edge (u, v). In an *undirected graph* an edge is an unordered pair $\{u, v\}$ of vertices of the graph; u and v are called the *end points* of the edge $\{u, v\}$. Unordered pairs of the form $\{u, u\}$ are allowed, so unordered pairs of vertices *are not sets of vertices*. The notation $\{u, v\}$ is thus somewhat misleading.

Graphs can be represented pictorially by drawing points for the vertices, and connecting points if there is an edge between the corresponding vertices, as in Figure 1.5.

In the following, various graph concepts are introduced for undirected graphs. It is left to the reader to rephrase the definitions for the case of directed graphs.

If $\{u, v\}$ is an edge, then this edge is said to be *incident* with the vertices u and v, the vertices are said to be incident with the edge, and the vertices u and v are said to be *adjacent*. The *degree* $d(v)$ of a vertex v is the number of edges incident with v, where an edge $\{v, v\}$ is counted twice. In Figure 1.5 $d(3) = 2$ and $d(4) = 4$. Still more definitions are required.

(1) An *edge-labelled undirected graph* is a structure $((V, E), L, f)$, where (V, E) is an undirected graph, L is a non-empty set, the elements of which are called labels, and $f: E \rightarrow L$ is an assignment of labels to the edges of the graph. Vertex-labelled graphs are defined in a similar way.

(2) The undirected graphs (V, E) and (W, T) are *isomorphic* if there is a total bijection $f: V \rightarrow W$, such that $\{u, v\} \in E$ if and only if $\{f(u), f(v)\} \in T$, for all $u, v \in V$.

(3) The undirected graph (W, T) is a *subgraph* of the undirected graph (V, E) if $W \subseteq V$ and $T \subseteq E$. The subgraph is the *maximal subgraph of* (V, E) *determined by* W, if $T = \{\{u, v\} \in E \mid u, v \in W\}$.

The following concepts concern the way the vertices of an undirected graph are connected by the edges of the graph. Let $G = (V, E)$ be an undirected graph, and $u, v \in V$ two vertices of G.

(1) A *path* from u to v in G is a sequence $u = v_1, v_2, v_3, \ldots, v_{n+1} = v$ of vertices, such that $\{v_i, v_{i+1}\} \in E$ for all i, $1 \leqslant i \leqslant n$. The path is said to *connect* u and v; the *length* of the path is the number of edges used, n in the sequence above.

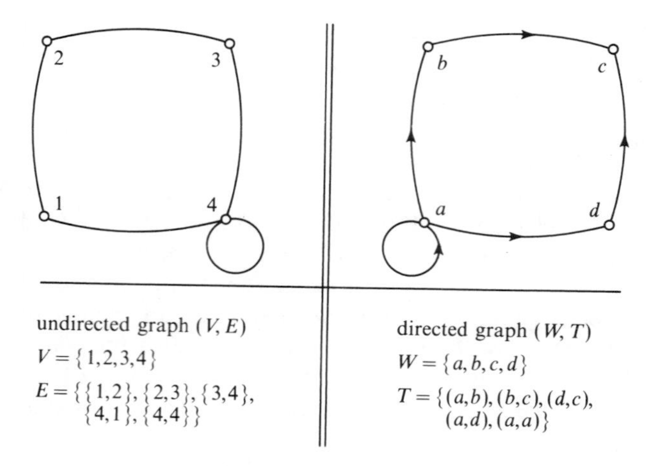

Figure 1.5 A pictorial representation of directed and undirected graphs.

(2) A *circuit* is a path from u to u for some vertex u in G.
(3) An *Euler circuit* is a circuit which includes every edge of G exactly once.
(4) A *Hamiltonian circuit* is a circuit which includes every vertex of the graph exactly once.
(5) A *chain* is a path from u to v, such that all the vertices, except possibly u and v, are mutually different.
(6) A *closed chain* is a chain which is a circuit.
(7) The graph G is *connected* if there is a path from u to v in G for every pair $u, v \in V$ of vertices of G.
(8) The graph is called an *Euler graph* if it contains an Euler circuit.
(9) The graph is called a *Hamiltonian graph* if it contains a Hamiltonian circuit.
(10) A *complete* graph on a set of k vertices, also called a *k-clique*, is a graph (V, E), where V has k elements and $E = \{\{u, v\} \mid u, v \in V, u \neq v\}$.

Theorem 1.3

A connected graph is an Euler graph if and only if the degree of every vertex of the graph is even. ■

In spite of the similarity between the definition of a Hamilton graph and an Euler graph, no simple condition characterizing Hamilton graphs is known. This is one of the major unsolved problems in graph theory. Most theorems stating a sufficient condition for a graph to be a Hamilton graph have the form 'If G has sufficiently many edges, then G is a Hamilton graph'. The following theorem is an example of this type.

Theorem 1.4

If G is an undirected graph with n vertices and $d(v) \geqslant \frac{1}{2}n$ for every vertex v of G, then G is a Hamilton graph. ■

A generalization of the Hamilton circuit problem is the so-called Travelling Salesman Problem. A salesman wishes to visit a given set of cities and return to his starting point, while travelling the least possible distance. The Travelling Salesman Problem can be reformulated in terms of edge-labelled graphs. The graph corresponds to a network of roads, and the label of an edge corresponds to the distance between the two cities represented by the end points of the edge. The graph may be directed or undirected, corresponding to one way or two way roads.

Many combinatorial optimization problems can be phrased as directed or undirected graph problems.

Chapter 2
Introduction to Effective Computability

In this chapter we will discuss the basic notions underlying any discussion of computability; that is to say computation and method of computation, or algorithm. We will touch upon the following questions:

- What is a 'computation'? What are the objects manipulated in a computation?
- What is an 'algorithm'? What is the relation between algorithms and computations?
- What is the purpose of a theory of algorithms? Is its scope necessarily limited to numerical computation? Does a valid theory of algorithms in some sense accurately describe the limitations of the power of computers in general?

Turning to the words themselves, everybody has some idea of what a computation is: something like calculating or doing arithmetic. The word stems from the Latin word 'putare', meaning to think. The current meaning is more related to manipulation than to thought.

The word 'algorithm' is a Latin corruption of 'al-Khuwarizmi', from the name of abu-Jafar Mohammed ibn-Musa al-Khuwarizmi, a ninth-century Persian mathematician. It now stands for any method or procedure of computation, usually involving a sequence of steps (*Collins English Dictionary* 1979).

The word 'theory' is derived from Greek '*θεορειν*', which means to look at, to contemplate. Its current meaning is 'that department of a technical subject which consists in the knowledge or statements of the facts on which it depends, or of its principles and methods as distinguished from the practice of it' (*Oxford English Dictionary*).

2.1 Computations and data types

From a philosophical point of view it can be argued that computers, calculators and the like cannot do arithmetic, the argument being that these electromechanical devices do not possess a notion such as number and therefore certainly cannot calculate. Be this as it may, it is our daily experience that these devices behave in a way which we humans can interpret as being equivalent to doing arithmetic.

We will not pursue the philosophical points but simply identify computing with mechanically producing perceptible input/output behaviour.

Clearly, all objects used as input or output, or used while computing, must be identifiable within a finite time. This means that every input or output of a computation and every object involved in a computation is assembled from a finite number of elementary building blocks. In addition, the number of mutually different elementary building blocks must be finite. Examples of such building blocks are the line elements used for number representation on the display of a pocket calculator, the digits and the dot in the usual decimal representation of numbers, and the characters used by a line printer.

The twin facts that every object is assembled from a finite number of elementary components and that there are only finitely many mutually different such components imply that the objects used in computing can always be represented by finite strings over a finite alphabet, and computing can be identified with manipulating such strings of symbols. The choice of a particular alphabet of symbols is immaterial and can be made to fit the occasion.

Another point of importance is the level of abstraction in the description of a computation. Consider for example the computation of the successor $x + 1$ of the natural number x. Even in this simple case the following levels of abstraction are possible.

(1) Suppose that natural numbers are represented in binary, and that x has the representation $a_1 a_2 \cdots a_n$. The representation of $x + 1$ is computed as follows:

> Search for the rightmost occurrence of the symbol 0.
> If there is none, then the binary representation of $x + 1$ is $10 \cdots 0$, that is, a 1 followed by n 0s.
> If at least one 0 appears in $a_1 a_2 \cdots a_n$, and a_i is the rightmost 0, then the binary representation of $x + 1$ is equal to $a_1 a_2 \cdots a_{i-1} 10 \cdots 0$ ($n - i$ zeros).

(2) We can also describe the computation on some higher level of abstraction. For the above example: 'add one', or 'work in the binary representation of numbers and add one'.

We will almost exclusively use the latter alternative. But if we take the cost of computation into account, it will sometimes be necessary to use the former alternative as well.

In the preceding examples it was assumed that there were base operations, presumably mechanically executable, such as searching for the rightmost symbol or adding one. The computation consisted of a combination of these base operations. Thus computing can be identified on any level of abstraction with the execution of a sequence of base operations. The level of abstraction determines the objects and the base operations. We must keep in mind, however, that an abstraction can be computationally valid only if the base operations are effectively executable and the set of all objects used in computing is finitely representable, i.e. all objects can be written as finite strings over a finite alphabet of symbols.

A more detailed discussion of computation requires the specification of the base operations and of the set of objects used in computing, that is to say the specification of the *underlying algebra.*

Definition 2.1

An *algebra* **A** is a pair $\mathbf{A} = (A, F)$ where A is a non-empty set and F a set of total functions $f: A^{n_f} \to A$, $n_f \geqslant 0$, called the *operators* of the algebra.

The set A is called the domain of the algebra **A**. The elements of the set A are the objects upon which computations are performed and are called data elements of type **A**. The operators of **A**, i.e. the functions in F, specify what actions can be performed on the data elements.

An algebra (A, F) is a *simple data type*, if F is a finite set, i.e. only finitely many operators are provided, and A is a countable set, a necessary and sufficient condition to ensure that the data elements can be represented as finite strings over a fixed finite alphabet. A simple data type is denoted by $(A; f_1, f_2, \ldots, f_k)$ instead of $(A, \{f_1, f_2, \ldots, f_k\})$. ■

EXAMPLE 2.1

(1) $(N; \times, +, 0, 1)$ is a simple data type with domain the set N of all natural numbers, two binary operators and two so-called nullary operators 0 and 1. These signify specific elements of N. Calling them nullary operators instead of specific given elements is a mere technicality.

Assuming that N denotes the natural numbers this specification says that we are considering the set of natural numbers equipped with the binary operations of multiplication and addition, and assume the numbers 0 and 1 to be given.

Note, however, that the specification says nothing about $\times$ apart from its being a binary operator. Properties such as $a \times 1 = a$ for all $a \in N$ must be specified or proved separately.

(2) $(R; \times, +, \pi, 1, 0)$ consisting of the real numbers equipped with multiplication and addition and given specific real numbers π ($= 3.1415\ldots$), one and zero.

Note that the real numbers form an uncountable set and the set R is therefore not finitely representable as a set of finite strings over a finite alphabet. Therefore this algebra is not a simple data type.

Once we have fixed a simple data type, we know what data elements and what operators are at our disposal. Now we can compute by repeatedly applying operators to given or previously computed data elements. We think in general of a computation as any sequence of applications of operators on elements of the given domain. Some of these elements must be given initially, as inputs to the computation, the remaining elements are obtained as (intermediate) results of the computation.

The simplest computations can be described using expressions – *polynomial expressions* to be precise.

Consider for example the algebra $(N; \times, +, 1, 0)$, the natural numbers with multiplication, addition, one and zero. Examples of expressions are:

(1) $(1+1+1)\times(1+1+1+1)$

(2) $(x_1+x_2)\times(x_2+x_3\times x_1)$

(3) $1+(x_1+x_2)\times(x_3+1)+0\times 1.$

These expressions can be evaluated in the obvious way. Expression (2) from the above list, for example, can be evaluated as follows:

$a := x_1 + x_2;$

$b := x_3 \times x_1;$

$c := x_2 + b;$

$d := a \times c.$

The input consists of the values of the variables x_1, x_2 and x_3 and the output of the value of d. The description of the evaluation method is itself an example of a so-called *straight-line program*, a program without repeat- and branch-statements. Thus, the sequence of assignments executed is fixed and independent of the input values.

Polynomial expressions can be defined for any algebra, and the expressions can always be evaluated using straight-line programs as in the above example. Straight-line programs do not get you very far, however, because the same predetermined sequence of operations is executed for any given input. The function $\lambda xy[x^y]$ for instance cannot be computed by a straight-line program over the data type $(N; \times, +, 1, 0)$. Thus, a more powerful programming language is needed. In particular, statements to specify repetition, such as while statements, are required.

To allow Boolean expressions needed for the while statements, we equip the underlying data type with one or more predicates. Note that predicates, such as $\lambda xy[x=y]$, return truth values, TRUE or FALSE. These values generally do not belong to the domain of the data type, hence, they certainly cannot be manipulated. To cope with this problem, many-sorted algebras, i.e. algebras having more than one domain, will be introduced.

EXAMPLE 2.2

$(N, B; i, +, 0, 1, Z, \wedge, \vee, \neg)$ is a many-sorted algebra having two sorts: the natural numbers N and the Booleans $B = \{\text{TRUE}, \text{FALSE}\}$. The operators are the identity operator $i = \lambda x[x]$ and the binary operator of addition on N; also, there are given two specific elements 0 and 1 from N; furthermore on the set B we have $\wedge$ (*and*), $\vee$ (*or*), and $\neg$ (*not*), which are the customary operators on truth values, and finally there is a predicate $Z: N \to B$, testing for zero: $Z \triangleq \lambda x[x=0]$. We now have data elements of two sorts:

- the elements that belong to the set N, which are data elements of type N (= natural number)
- the elements that belong to the set B, which are data elements of type B (= Boolean)

What we have here is a *compound data type*, a concept of fundamental importance in programming. Usually in a compound data type, one sort is of special interest, and is denoted by TOI (= Type Of Interest). This is the main type, the others being considered auxiliary. In the preceding example the TOI is N, the domain which contains the objects used in computing, whereas the Booleans form an auxiliary domain, whose elements are used in controlling the computation.

To define properly a (compound) data type, it is not sufficient to state its sorts of values and its operators; it is also necessary to give an unambiguous description of the properties of the operators.

We now give the formal definition of many-sorted algebras and compound data types.

Definition 2.2

A many-sorted algebra is a tuple (S, F), where S is a collection of non-empty sets, the sorts or domains of the algebra, and where F is a collection of total functions $f: D_1 \times D_2 \times \cdots \times D_{n_f} \to D_{n_f+1}$, D_1 to D_{n_f+1} being sorts of the algebra.

A many-sorted algebra (S, F) is a *compound data type* if

(1) S is finite, i.e. there are only finitely many sorts, including the set $B = \{\text{TRUE}, \text{FALSE}\}$. Furthermore all sorts are countable.

(2) F is finite, i.e. there are only finitely many operators. Furthermore, F contains the identity function $i_D: D \to D$ for every domain D in S.

(3) One sort is identified as the TOI.

Compound data types are denoted by $(D_1, \ldots, D_n; o_1, \ldots, o_m)$ instead of $(\{D_1, \ldots, D_n\}, \{o_1, \ldots, o_m\})$ where D_1 is the TOI. ■

A compound data type (S, F) determines the following.

(1) The values used in computations. These are the elements of the sorts $D \in S$ occurring in the data type.

(2) The actions which can be performed on given values. These are determined by the operators $f \in F$.

(3) The properties of values that can be determined. These correspond to the operators having as range the set of truth values $\{\text{TRUE}, \text{FALSE}\}$, and can therefore be identified with predicates. These predicates provide the means to specify computations in which the sequence of operations performed depends on the given values.

(4) One of the sorts of the data type is identified as the TOI, thus distinguishing the main type from the auxiliary types.

We will not need compound data types in their full generality, but only a few simple cases, namely:

(1) two sorts, namely the natural numbers and the Booleans with the natural numbers as the TOI

(2) three sorts, an alphabet V, the set V^* of all strings over this alphabet and the Booleans with the set V^* as the TOI.

2.2 Algorithms

An algorithm is a finite description specifying a computation over a certain data type for any given input. An input to a computation is a set of values from the TOI, to be used as initial values. Algorithms are represented as programs written in an extremely simple programming language called SAL(**S**), (Simple Algorithmic Language over the compound data type **S**). The language is defined relative to a given compound data type $\mathbf{S}=(D_1, D_2, \ldots, D_s; o_1, o_2, \ldots, o_t)$ and has the following syntactical concepts:

(1) variables $x, y, \ldots$

(2) assignment statements:

$$x := o(y_1, \ldots, y_{n_o})$$

for every operator o of the data type. Note that this implies that the given specific elements of the sorts (i.e. the nullary operators), are available as constants in the programming language, and also that we have assignment statements of the form $x := y$, because the data type has the identity function as an operator.

(3) control statements to specify repetition:

$$\textbf{while } p(x_1, \ldots, x_{n_p}) \textbf{ do } S \textbf{ od}$$

where S is a sequence of statements, and p an operator in the data type whose range is the set {TRUE, FALSE}, and can thus be identified with a predicate.

The assignment statement $x := y$ (forced by the requirement that one of the operators o_i is equal to the identity function) is incorporated for two reasons:

- to facilitate data transport in the course of some computation
- these simple statements avoid the otherwise necessary requirement that the algebra has operators to select a value from some tuple of values.

EXAMPLE 2.3

Consider for example the compound data type $\mathbf{N}=(N, \{T, F\}; i, p, s, 0, \neq_0)$. That is, the TOI is the set N of all natural numbers equipped with three unary operators, the identity operator i, the predecessor operator p, the successor operator s, the nullary operator 0 and one predicate $\neq_0(x)$, which is true if x is nonzero, together with the programming language over this compound data type denoted by

SAL(**N**). The following is a SAL(**N**) program which computes the function $\lambda x[2x]$.

```
input: x; {x∈N}
output: y; {y = 2x}
method:
   y:= x;
   while ≠₀(x) do
      y:= s(y);
      x:= p(x)
   od
```

This example shows the purpose of specifying the TOI, namely to specify the character, that is the domain and codomain, of computable functions. Input and output values are elements of the TOI, and the SAL(**S**) program computes a function $\text{TOI}_S^n \to \text{TOI}_S^k$, for some $n \geq 0$ and $k > 0$. In the current example the TOI is the set N, and the program computes a function from N to N.

It is clear that the elementary arithmetical functions of addition, multiplication and exponentiation are also SAL(**N**) computable, i.e. can be computed by some SAL(**N**) program.

An important question will be: what functions are not SAL(**N**) computable and are we satisfied with that? It requires some elaboration, but it can be shown beyond any reasonable doubt that a function $N \to N$ which is not SAL(**N**) computable, is not effectively computable in any reasonable sense of the term.

The set of all functions $N^n \to N^k$ which can be computed by SAL(**N**) programs can be defined precisely. This set is denoted by F(SAL(**N**)). The intention is that F(SAL(**N**)) is precisely the set of all number-theoretic functions which are effectively computable in any reasonable intuitive sense. It can be argued that this is indeed the case.

Thus the language SAL(**N**) together with all the necessary definitions can be considered to be a model of computation; not because it would model human computation activities, but because it determines a subset of the set of all number-theoretic functions, and calls functions belonging to this subset 'computable'. The model is reasonable because its formal notion of computability coincides with intuition.

It makes sense to talk about the time complexity of a SAL(**N**) program. We can count the number of operations which are performed in the computation on the input x. Under the assumption that all the operations of the given (compound) data type take unit time, the number of operations performed indeed gives the time required for the computation.

The time complexity of the program of the example above on input x is equal to $1 + 3x + 1$.

- The assignment $y := x$ takes unit time.
- The while statement takes 3 units for the execution of the predicate $\neq_0$ and the operations p and s. Thus the whole while loop takes $3x$ units of time.
- Then follows the final evaluation of $\neq_0(x)$, which returns FALSE.

This brings the total to $3x + 2$ operations.

It does not make sense to talk about the space complexity of the program. Within this domain of computation there is no representation of numbers and therefore no concept such as the space required to store some number. Thus the only space complexity measure we have is the number of variables used in the program, which is certainly not a reasonable measure of space complexity.

EXAMPLE 2.4

Let us now consider a compound data type of rather low level of abstraction, namely the set of binary representations of natural numbers $BIN \triangleq \{0, 1, 10, 11, 100, 101, 110, 111, \ldots\}$ supplemented with the operators:

- the identity function $i: BIN \to BIN$,
- the function $rightmost: BIN \to BIN$ defined by

$$rightmost(x) \triangleq a, \text{ such that } x = ua \text{ for some } u \in \{0, 1\}^* \text{ and } a \in \{0, 1\}$$

- the function $remainder: BIN \to BIN$ defined by

$$remainder(x) \triangleq [\text{if } (x = u0 \text{ or } x = u1 \text{ and } u \text{ not empty}) \text{ then } u \text{ else } x]$$

- the functions $put_0, put_1: BIN \to BIN$ defined by

$$put_a(x) \triangleq [\text{if}(x = 0) \text{ then } a \text{ else } xa]$$

- the nullary functions 0 and 1, giving the particular elements 0 and 1 from N.
- the predicates $\neq_0$ and $\neq_1$ defined by

$$\neq_a(x) \triangleq [\text{if } x \neq a \text{ then TRUE else FALSE}], \; a = 0, 1$$

$put_0(0) = 0$; $put_0(1) = 10$; $put_1(0) = 1$; $put_1(1) = 11$
appends 1 or 0 to x

In other words, we consider computations over the compound data type $\mathbf{S} \triangleq (BIN, \{\text{TRUE}, \text{FALSE}\}; i, rightmost, remainder, put_0, put_1, 0, 1, \neq_0, \neq_1)$.

The following is an example of a SAL(**S**) program.

```
input: x;
output: y;
method:
   y:= x;
   while ≠0(x) do
      y:= put0(x);
      x:= 0
   od
```

The program computes the function $\lambda x[\text{if}(x=0) \text{ then } 0 \text{ else } x0]$. Note that the number of operations performed in a computation on input x equals $1+1+2+1=5$ if x is nonzero, and $1+1=2$ otherwise. Thus the time complexity of the program is $\lambda x[\text{if}(x=0)$ then 2 else 5].

SAL(**S**) is also a model of computation, related to, but clearly not the same as, SAL(**N**).

Interpreting the strings in BIN as natural numbers written in binary, the above program doubles its input. From an intuitive point of view, this program computes the same function as the preceding SAL(**N**) program. Formally these functions cannot be equal, one is a function $N \to N$, the other a function $BIN \to BIN$. The relation between these functions is expressed more precisely using the concept of a coding.

Definition 2.3

Let W be some non-empty set. A *coding* from N into W is an injection $c: N \to W$. The code can be extended to $\bar{c}: N^n \to W^n$ by defining

$$\bar{c}(x_1, x_2, \ldots, x_n) = (c(x_1), c(x_2), \ldots, c(x_n)).$$

The function $f: N^n \to N^k$ is *incorporated* in the function $g: W^n \to W^k$ relative to the coding c if and only if

$$\{(\bar{c}(x), \bar{c}(y)) | y = f(x)\} \subseteq \{(u, v) | v = g(u)\}$$

Equivalently, f is incorporated in g relative to the coding c, if $g(c(x)) = c(f(x))$ for all $x \in N^n$. That is, if the diagram in Figure 2.1 commutes. ■

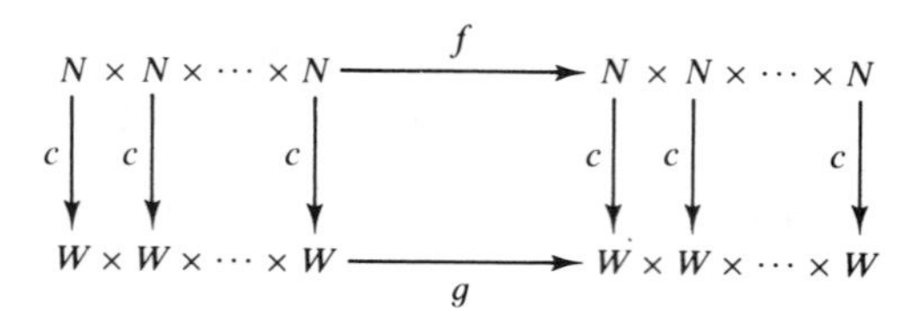

Figure 2.1 The function f is incorporated in the function g relative to the coding c.

Computability relative to a coding is the basic concept in comparing the power of computation models. Let $F(\mathrm{SAL}(\mathbf{X}_i))$ denote the set of all functions which can be computed by $\mathrm{SAL}(\mathbf{X}_i)$ programs, and let $\mathbf{X}_1$ and $\mathbf{X}_2$ be compound data types.

(1) $F(\mathrm{SAL}(\mathbf{X}_1))$ is *incorporated* in $F(\mathrm{SAL}(\mathbf{X}_2))$ relative to the coding c, if for every function f in $F(\mathrm{SAL}(\mathbf{X}_1))$ there is a function g in $F(\mathrm{SAL}(\mathbf{X}_2))$ which incorporates f relative to c.

(2) The computation models $\mathrm{SAL}(\mathbf{X}_1)$ and $\mathrm{SAL}(\mathbf{X}_2)$ are *equivalent* if there exist codings c_1 and c_2 such that

  (a) $F(\mathrm{SAL}(\mathbf{X}_1))$ incorporates $F(\mathrm{SAL}(\mathbf{X}_2))$ via coding c_1
  (b) $F(\mathrm{SAL}(\mathbf{X}_2))$ incorporates $F(\mathrm{SAL}(\mathbf{X}_1))$ via coding c_2.

Consider again Example 2.4 above. As is easily seen $F(\mathrm{SAL}(\mathbf{S}))$ incorporates $F(\mathrm{SAL}(\mathbf{N}))$ relative to the coding $c: N \rightarrow BIN$ which maps zero onto 0, and any number greater than zero onto its binary representation – leading zeros removed. It can be shown that these computation models are equivalent.

In the above, the term 'computation model' is used in a broad sense. It denotes a whole complex of definitions – data types and programming languages, or maybe a class of machines, or any other – finally resulting in the specification of a particular set of functions, namely those functions which are computable according to the model in question. Beauty, naturalness, practical validity and computational power are all aspects of such a model of computations. The computational power of the model is represented by the extension of the set of all functions computable according to the model. Thus, we can compare the power of computation models using the concept 'incorporation relative to some suitable coding'.

2.3 Theory of algorithms: its purpose and scope

The purpose of a theory is to organize the knowledge in some field of endeavour. The purpose of this organizing is efficiency: a large body of knowledge can easily be absorbed if organized well. Systematizing our knowledge about algorithms and algorithmic methods increases our knowledge and leads to significant practical advances. For example the theory can provide yardsticks to compare algorithms and problems; it can show that the search for a sorting algorithm using less than $O(n.\log(n))$ comparisons is futile; it can show that there can be no time-optimal algorithm for matrix multiplication; and so forth.

A theory of algorithms must begin with appropriate definitions of its subjects, viz. computations and algorithms. It must concern itself with the study of particular algorithms and problems, but also with more abstract questions such as:

- do noncomputable functions exist?
- are all noncomputable functions equally noncomputable?
- what are valid measures of the cost of computation? Can you always get better results for more money?

When developing a theory, you are free to define any concept in any way you wish. If you want your theory of algorithms to be of practical significance however, you must argue that your definitions, especially those of a computation and of an algorithm are practically significant and fully grasp the intuitive notions.

We have given some definitions of computation and algorithm in the preceding sections, and argued their practical validity. It remains to show that they are fully equivalent to the intuitive notions. This cannot, of course, be proved rigorously, because intuitive notions are imprecise by definition. These statements can only be validated by experiment as is the case in physics. We assume that our formal notion of computability fully captures the intuitive notion and then try to find some example disproving our assumption. Searching is done haphazardly by trying any function, and more systematically by trying to capture the intuitive notion in some other way, e.g.:

- try and define the class of computable functions mathematically by using effectiveness-preserving operations
- try and define a class of machines instead of a class of programs as your model of 'mechanically executable'
- try and define a class of string-rewriting systems, in view of the fact that computation is essentially symbol manipulation

and so forth.

Many alternatives have indeed been tried:

(1) The recursive functions of Gödel and Herbrand are among the oldest proposals (1931). This set is defined to be the smallest set of number-theoretic functions which contains a certain set of base functions, such as the identity function and the constant functions,

and which is closed with respect to a given set of functional operators. These base functions are so simple that they certainly may be considered to be effectively computable and the functional operators are evidently constructive, hence one may assume that they preserve effective computability. Intuitively, all recursive functions are effectively computable. In fact one can prove that every recursive function is SAL(**N**) computable and vice versa.

(2) Another possibility is to use idealized mechanical computing devices. According to this definition a function is computable if it can be computed by a particular machine belonging to the class of machines considered. Thus instead of a program you take a machine. Examples of this are the Turing machine (1936) and the register machine of Shepherdson and Sturgis (1963).

(3) Of the many other definitions we finally mention the normal algorithms of Markov (1962) and the canonical systems of Post (1943). These computation models are based on the idea that computing is in fact string manipulation, so the data type consists of the set of all strings over some finite alphabet and has simple rewrite rules as operations.

All these alternatives have been shown to be mutually equivalent, in the sense that they define the same set of computable functions relative to some codings. This adds credibility to the belief that every *intuitively* computable function is also computable according to a definition by one of the computation models. This belief is known as the generalized thesis of Church (Alonzo Church, an American logician). Church stated that a function is intuitively computable if and only if it is λ-definable, that is, computable according to some particular model. This statement has subsequently been generalized, in view of the fact that all models which have been studied have been shown to be mutually equivalent. We will refer to this generalized statement as 'Church's thesis'.

It is evident that Church's thesis cannot be proved, because intuitive notions are of necessity mathematically imprecise. Church's thesis is therefore not a mathematical statement and hence evades methods of mathematical proof. Nevertheless at the moment almost all mathematicians, logicians and computer scientists believe Church's thesis to be true.

From the preceding discussion, especially the equivalence of all sensible definitions of effectiveness, it follows that the theory of algorithms is not limited to numerical computation, even if presented in terms of number-theoretic functions. The theory accurately describes the limitations of the power of computing devices in the broadest possible sense.

2.4 Theory of algorithms: its historical development

Algorithms have been used from ancient times. The theory of algorithms also existed long before the advent of general-purpose electronic processors as we know them. Think only of the calculation of tables of logarithms, tables of primes, the calculation of planet positions and all the calculation necessary for ship navigation. A strong need within mathematics to understand what could and what could not be done using effective methods gave impetus to the development of the theory of algorithms.

At the turn of the century, Gottlob Frege (1848–1925), a German mathematician, believed that mathematics could and should be deduced from principles of logic. He developed concise and precise notation, and started to develop mathematics in this precise language, insisting on full and precise proofs. Soon it was found that the system contained a contradiction. Obviously that makes any theory utterly worthless.

Faced with such serious problems. David Hilbert (1862–1943), one of the greatest mathematicians of our age, started the study of these formalized theories themselves. He insisted not only on an explicit and unambiguous notation, but also required that it always be possible to check in a finite number of 'mechanical' steps whether a sequence of formulae is a proof. Until the 1930s, most mathematicians, Hilbert the most prominent among them, firmly believed that all of mathematics could be developed within these restrictions. In his famous 1900 address entitled *Mathematical Problems* and in an address *On the Infinite* in 1925, Hilbert firmly stated this belief. Hilbert and his school spent a lot of effort searching for an effective procedure for checking within a finite number of steps the truth or falsity of mathematical propositions.

However, Hilbert's intuition about the provability of mathematical statements was wrong. In 1931 Kurt Gödel proved as a consequence of his famous incompleteness theorem that elementary arithmetic is algorithmically undecidable. This means that there does not exist an algorithm for proving the truth or falsity of any proposition of elementary arithmetic.

Since then the existence of many algorithmically undecidable theories and algorithmically unsolvable problems has been proved. An example of such a problem is the so-called 10th problem of a list of 23 problems Hilbert stated in his famous 1900 lecture (some of these problems remain unsolved to this day). The 10th problem is the question whether an arbitrary polynomial equation $P(x_1, \ldots, x_n) = 0$ with integer coefficients has an integer solution. Such an equation is called a Diophantine equation. Matijacevic proved in 1970 that there does not exist an effective procedure deciding whether an arbitrary Diophantine equation has a solution or not.

It is evident that in order to prove negative results like these, it is necessary to know precisely what is meant by a computation, a computable function and an algorithm or an effective procedure.

In daily practice a definition is hardly ever necessary, since generally there is little or no difference of opinion whether a proposed procedure may be considered effectively executable or not. This even applies if the procedure is described in imprecise terms.

To achieve general agreement about the non-existence of an effective procedure, however, it is necessary to make this notion mathematically precise. The starting point is:

(1) An algorithm is a finite description of a computation process. It is a finite text, written in an algorithmic language, in which every sentence has an unambiguous meaning.

(2) The operations mentioned in the algorithm must be executable in a mechanical way by a computing agent within a finite amount of time and belong to the repertory of base operations of the computing agent.

(3) The data elements used in a computation must be recognizable within a finite time by the computing agent (otherwise it would be impossible to compute with these data). Thus every data element must be representable by a finite string over a finite alphabet.

All of these requirements put together and phrased in modern terminology lead to the definition of data types and programming languages as has been done in the preceding sections.

Before the advent of the powerful computing machines we are used to, the theory of algorithms mainly concerned itself with the characterization of what could be done *in principle*. The cost to be assigned to the computations was of no consequence whatever. This picture has changed dramatically. In practice one must pay in hard cash for every second the computation requires; memory must also be paid for.

Study of the time and space requirements of computation is the realm of complexity theory, a branch of the theory of algorithms. It concerns specific algorithms for specific problems. Specific programs are analysed with respect to their need for computing resources such as time or space. More difficult and more interesting is the study of specific problems. How many comparisons are necessary and sufficient to determine both the greatest and the least element of a set? These can be very difficult questions, especially as regards lower bounds.

Complexity theory also concerns itself with more abstract questions such as:

- do there exist arbitrarily complex problems?
- suppose you have two functions f and g such that for all $n, f(n) < g(n)$. Is there a problem requiring more than $f(n)$ but at most $g(n)$ time?

Problems which can in principle be solved by algorithmic methods, but only at insurmountable cost, are not very well solved from a practical point of view. The computing community generally agrees to calling a problem solution feasible if the time required for it is at most a polynomial function of the size of the input. Many practical problems of combinatorial nature, for instance scheduling problems, appear to have no polynomial solution, although it is possible to check in polynomial time whether a proposed answer actually is a solution. The question of whether the class of problems solvable in polynomial time is equal to the class of problems verifiable in polynomial time, the $\mathbf{P} \stackrel{?}{=} \mathbf{NP}$ question, is one of the most prominent unsolved problems of complexity theory.

Complexity theory is more sensitive to the model of computation than the theory of algorithms in general. To obtain meaningful results in complexity theory it is necessary to take the set of all strings over some fixed finite alphabet as the TOI of the data type, in order to take representation into account.

2.5 Finite representability

It has been argued in the preceding sections that the string representation of objects is essential to computing. Using **N** (Example 2.3) as the domain of computation is thus a level of abstraction beyond actual manipulation.

In reality, manipulations are performed on representations of numbers, i.e. strings of some sort, which can be viewed as denoting numbers written in the unary, binary, decimal or even Roman number system. This is also the case in **S** where the carrier is *BIN*, a set of strings. But in **N** you are concerned with abstract entities which you cannot manipulate in reality.

It can be advantageous to assume particular functions, such as predecessor and successor, to be base operators, available to the computing agent, without having to worry about the representation of numbers. When taking the cost of computation into account you pay for the convenience by misleading results. This is due to the fact that there is an implicit representation.

In the domain **N** you have a finite number of elements of N at your disposal. These are the constant 0 and the input values. Any other number used in the computation can only be constructed using the operators: e.g. the number 5 is $s(s(s(s(s(0)))))$. A shorter way of arriving at the number 5 is not available. Any number actually occurring in a computation is constructed using such a polynomial expression in the input values and the given constants.

With respect to complexity, the domain behaves as if the numbers were represented as polynomial expressions. This accounts for the fact that adding 117 to x requires 117 steps, doubling x requires x steps, etc. Also, in terms of the domain of computation, the length of the representing polynomial expression is the only reasonable measure of space.

Thus in complexity theory it is important to choose a domain which charges the cost of computing in accordance with current practice. Consequently the domain is a domain of strings, in which numbers are represented in the binary, the decimal, or any other convenient radix system, so that for instance doubling requires $\log(x)$ steps instead of x steps, and the space required to store x is also $\log(x)$ instead of x.

2.6 References

Historically the first class of computable number-theoretic functions ever defined was the class of primitive recursive functions. This class was obtained from all possible constant functions and the successor function by means of substitution, identification of arguments and the primitive recursion schema.

Ackermann (1928) proved in 1928 that there exists an intuitively effectively computable function (now called the *Ackermann function*) which is not primitive recursive. In the early 1930s several definitions of the concept of 'computable functions' were proposed and still others appeared later.

It is remarkable that all these definitions or computation models proved to be equivalent up to a coding.

In 1931 Gödel (1965) defined the concept 'general recursive function', which he partly credited to Herbrand. Church (1936) identified the *intuitive* concept 'computable function' with the *formal* concept 'recursive function', as defined by Kleene. A detailed discussion of Church's thesis can be found in Church (1936).

Later Turing (1936), Markov (1962) and Post (1936) formulated analogous theses for the Turing machine model, Markov algorithms and Post systems. It is remarkable that Turing, Markov and Post formulated their ideas of effective computations and effectively computable functions in terms of strings of symbols. In fact all these computation models are symbol manipulation systems devised from different points of view.

In 1968 Engeler (1968) formulated the idea of a computation over a data type, and represented algorithms as programs in a programming language using this data type. The language SAL(**N**) used in this book is a variant of his language and the language developed by Meyer and Ritchie (1967a). Analogous ideas were developed by Shepherdson and Sturgis, who used an abstract machine called the unlimited register machine and a programming language for this machine.

In 1970 Eilenberg and Elgot (1970) developed the theory of recursive functions over a data type with TOI a set of strings. Brainerd and Landweber (1974) discussed the theory of computation using a string domain and a Pascal-like programming language PL.

Data types, simple and compound data types are in fact universal algebras. These are discussed very thoroughly by Grätzer (1968). A discussion of data types from the programming point of view can be found in many books on the design of algorithms, for instance in Liskov and Guttag (1986).

Concerning the history of the theory of recursive functions and other computation models the reader is referred to Mostowski's excellent book *Thirty Years of Foundational Studies* (1966), to *From Frege to Gödel* by Heyenoort (1981) and to *The Undecidable* by Davis (1965).

Chapter 3
The Programming Language SAL(N)

The major aim of this chapter is to develop the precise and rigorous definitions of the concepts introduced in Chapter 2. First of all we define a particular compound data type **N** and a particular programming language SAL(**N**) over this data type. This programming language will be used in the chapters to come as our model of computation.

Both the syntax of SAL(**N**), specifying which computational procedures are legal SAL(**N**) programs, and the semantics of SAL(**N**), specifying the computations induced by a given SAL(**N**) program, will be defined rigorously.

3.1 The compound data type N

SAL(**N**) consists of the programming language SAL – to be defined in the next sections – and the following compound data type.

Definition 3.1

The compound data type **N** is the many-sorted algebra $\mathbf{N} \triangleq (N, B; 0, +_1, \dot{-}_1, \neq_0)$ consisting of:

(1) Two sorts:

(a) $N = \{0, 1, 2, 3, \ldots\}$, the set of all natural numbers, and
(b) $B = \{\text{TRUE}, \text{FALSE}\}$, the set of truth values.

(2) and four operators:

(a) a nullary operator 0, i.e. a given particular element of N, namely 0.

(b) $+_1 : N \to N$, the successor function. It assigns to the natural number x its successor $x + 1$. (We will write $x + 1$ instead of $+_1(x)$ as soon as possible.)

(c) $\dot{-}_1 : N \to N$, the predecessor function. This function assigns to the positive natural number $x + 1$ its predecessor x, and to the natural number 0, which does not have a proper predecessor, it assigns 0.

(d) $\neq_0 : N \to B$, a test for zero, $\neq_0(x)$ is FALSE if x equals zero and TRUE otherwise. ■

The above definition presupposes knowledge of the properties of the natural numbers, such as $\dot{-}_1(+_1(x)) = x$ for all $x \in N$, and what natural numbers are in the first place. This could be avoided by adding axioms, stating the required properties. Since we want to develop the theory of computation and not the theory of natural numbers, we will refrain from doing so, and instead tacitly assume knowledge about natural numbers.

The above algebra has only the simplest of operations. We cannot add, multiply, divide or whatever. All these will have to be implemented by programs using only successor and predecessor as primitive operations. This would be an intolerable inconvenience were our purpose the development of a practical programming language. As regards computability there are but two points concerning the primitive operations:

(1) The operators must be intuitively effective, and this must be indisputably evident.

(2) The operators must be chosen so as to simplify proving properties of programs. This requirement however, is a matter of convenience, not of principle.

To avoid the presentation of lengthy but trivial programs, we will introduce in Section 3.5 a shorthand, equivalent to procedures, in this way at least partially meeting the objection raised above.

3.2 Syntax of SAL(N)

In this section we will define the set of all legal SAL(**N**) programs. In Chapter 2 the language has already been described informally. This is not sufficient and a precise definition is required, because SAL(**N**) programs are and must be mathematical objects, intended to capture intuitive notions, but not themselves intuitive notions.

The syntax of SAL is defined using a variant of the so-called

Backus–Naur form (BNF). This device can be used to define sets of words over a finite alphabet. Such sets, called syntactical categories, are denoted by their names enclosed in the brackets ⟨ and ⟩. For instance the expression

$$\langle \mathit{sequence} \rangle ::= a \,|\, ab \,|\, abb$$

pronounced as 'the category *sequence* is defined to consist of *a*, *ab* or *abb*', defines the set, or syntactical category $\langle \mathit{sequence} \rangle \triangleq \{a, ab, abb\}$.

Names of sets may also be used in the definition of other sets. For example the expression

$$\langle \mathit{word} \rangle ::= \langle \mathit{sequence} \rangle \,|\, \langle \mathit{sequence} \rangle b$$

states that the syntactical category ⟨*word*⟩ consists of all strings which are ⟨*sequence*⟩s or are formed by concatenating a string from the set ⟨*sequence*⟩ and the character *b*. Thus $\langle \mathit{word} \rangle \triangleq \{a, ab, abb, abbb\}$.

BNF definitions may also be recursive, i.e. the name of a set may be used in the definition of the set itself. The expression

$$\langle \mathit{b\text{-}word} \rangle ::= a \,|\, aa \,|\, \langle \mathit{b\text{-}word} \rangle b$$

is an example of such a recursive BNF definition. The set ⟨*b-word*⟩ is equal to the set $\{ab^n, aab^n \,|\, n = 0, 1, \ldots\}$.

In the following definition of the syntax of SAL the strings **od**, **do** and **while**, printed in bold face, should be considered single symbols.

The syntax of SAL(N)

$$\langle \mathit{identifier} \rangle ::= \{x_i \,|\, i = 1, 2, 3, \ldots\}.$$

$$\begin{aligned}
\langle \mathit{assignment} \rangle ::= \; & \langle \mathit{identifier} \rangle := 0 \,| \\
& \langle \mathit{identifier} \rangle := \langle \mathit{identifier} \rangle \,| \\
& \langle \mathit{identifier} \rangle := \langle \mathit{identifier} \rangle + 1 \,| \\
& \langle \mathit{identifier} \rangle := \langle \mathit{identifier} \rangle \dot{-} 1.
\end{aligned}$$

$$\begin{aligned}
\langle \mathit{statement} \rangle ::= \; & \langle \mathit{assignment} \rangle \,| \\
& \mathbf{while} \; \langle \mathit{identifier} \rangle \neq 0 \; \mathbf{do} \; \langle \mathit{sequence} \rangle \; \mathbf{od}.
\end{aligned}$$

$$\begin{aligned}
\langle \mathit{sequence} \rangle ::= \; & \langle \mathit{statement} \rangle \,| \\
& \langle \mathit{sequence} \rangle ; \langle \mathit{statement} \rangle.
\end{aligned}$$

$$\langle \mathit{program} \rangle ::= \{(n, k, p, \langle \mathit{sequence} \rangle) \,|\, n \geqslant 0, k \geqslant 1 \text{ and } n, k \leqslant p\}.$$

In the syntax of SAL x_i is called an identifier. More precisely, x_i is a variable whose denotation is left unspecified. The customary denotation using a decimal subscript is tacitly assumed. If arbitrary identifiers are allowed, technical difficulties in defining the meaning of a program result. For instance, let there be two identifiers *JOHN* and *MARY*, denoting input variables. The program computes a function $N^2 \rightarrow N^k$, for some k, but which one is the first coordinate, *JOHN* or *MARY*? The labour of rigorously defining the correspondence between variables and identifiers denoting variables can be saved by using numbered variables.

Note that in the language SAL the operations $+_1(x)$, $\dot{-}_1(x)$, and $\neq_0(x)$ of the underlying compound data type are denoted by $x+1$, $x \dot{-} 1$, and $x \neq 0$.

In the above, a SAL program is defined as a quadruple (n, k, p, P) consisting of three natural numbers and a sequence P of SAL statements. The variables occurring in the sequence P are called the program variables. The numbers n, k specify that $x_1, \ldots, x_n$ and $x_1, \ldots, x_k$ are the input and output variables and the number p specifies that all these variables and all the program variables, i.e. the variables occurring in the sequence P, belong to the set $\{x_1, \ldots, x_p\}$.

The requirements $n \geqslant 0$ and $k \geqslant 1$ express that every SAL program may have an input, but must have an output. The general idea is that, given an input value, a program determines a computation on this input value, which terminates after some finite time producing a supposedly useful result. With this view, nonterminating computations are a waste of time, which indeed they often are. But there also are systems or programs which run forever, do not have terminating computations, but are useful while running, e.g. computer operating systems, real-time clocks, control systems, etc. These programs are outside the scope of our presentation. They do not, however, offer new computational possibilities, because their behaviour can be computed, in our sense of the word, by a program which has inputs s and t giving the start condition at time zero and a time t; the program then computes and outputs the condition of the system at time t.

The above definition defines the mathematical object 'SAL(**N**) program'. SAL(**N**) programs generally are not easy to read. This is partly due to the sparseness of the set of operations, but also to the poor facilities for controlling the computation. We have two possibilities: sequential execution and repetition as defined by the while statement; but no if-statement and no for-statement. Remember however, that SAL is not designed for ease of programming, but to be sufficiently powerful and as simple as possible. Since for-statements and if-statements can be implemented using only while-statements (see Section 3.5 on the use of macros), they are omitted, facilitating (especially inductive) proofs about programs.

3.3 Intuitive semantics of SAL(N) and example programs

The meaning of a program, and in particular of a SAL(**N**) program, is the set of all computations described by the program. The set of all pairs (initial value, final value) determined by these computations is the input/output relation computed by the program. This is always a functional relationship, since the computation proceeds deterministically. (Computations which proceed non-deterministically will be discussed in Section 9.4. These computations are especially important in the theory of complexity, discussed in Chapter 11 onwards.)

The computations associated with a SAL(**N**) program (n, k, p, P) consist of the repeated application of operators of the underlying compound data type **N**, on a state vector $x \in N^p$ representing the values of all the p program variables. Which operator is to be applied will be formally defined in the next section. For the time being, we assume that a pointer, e.g. your index finger, points at the relevant position in the program text.

(1) Let $a \in N^n$ be an input value. Then the state vector x is initialized to $(a_1, a_2, \ldots, a_n, 0, 0, \ldots, 0)$. That is, input variables get their value as determined by the input, all other program variables are set to zero. This initializing of all program variables which are not input variables is not always implemented in real-life programming systems. Such initializing however is required to ensure that the result obtained by running the program depends solely on the input values. We insist on this property in order to get a manageable definition of the meaning of a SAL(**N**) program.

The actual computation starts on this initialized state vector with the first statement of the program, i.e. initializing is not considered part of the computation. The computation proceeds sequentially in accordance with the program text and the meaning of the individual statements.

(2) There are two types of statements, assignment statements and while statements.

(a) An assignment is of one of the forms:

$$\begin{aligned} x_i &:= 0, \\ &:= x_j, \\ &:= x_j + 1 \text{ or} \\ &:= x_j \dot{-} 1 \end{aligned}$$

The execution of these statements transforms the current value of the state vector by replacing respectively the ith component value by 0, the value of the jth component, $+_1$(value of the jth component), or $\dot{-}_1$(value of the jth component). This transformation is considered a single step in the computation.

(b) The sequence S in a while statement **while** $x_i \neq 0$ **do** S **od** is to be executed as long as the while condition $x_i \neq 0$ remains true. If a while statement must be executed, the predicate $\neq_0(x_i)$ is evaluated and the pointer in the program text is advanced either to the sequence S or past the closing **od**. Performing an **od** also amounts to evaluating $\neq_0(x_i)$ and appropriately advancing the pointer in the program text. Evaluating the predicate, and appropriately advancing the pointer, is a single step in the computation.

(3) If the pointer in the program text has advanced beyond the last statement of the program text, the computation does not proceed any further, it terminates. The output value is the value in the final state vector of the output variables. Thus, the output value is obtained by projection from the value of the state vector. Executing this projection is not considered part of the computation.

The input/output behaviour of the program is the function which assigns to the input value $a \in N^n$ the result of the computation starting on $(a, 0, \ldots, 0) \in N^p$.

A SAL(**N**) program which has n input variables and k output variables ranging over N computes the partial function $f: N^n \rightarrow N^k$, if for all $x \in N^n$:

(1) if $f(x)$ is defined, then the computation on input x determined by the program terminates, and on termination the value of the output variables is $f(x)$

(2) if $f(x)$ is undefined, then the computation on input x determined by the program does not terminate.

This function is called the function computed by the program. The latter is an abuse of terminology; a program does not 'compute'.

A program is a static object, a string consisting of letters, digits and punctuation marks. This object does not live, love or compute; *we* associate with this object other objects, namely computations – that is certain strings consisting of digits and punctuation marks – and functions. Indiscriminate use of terminology appropriate for humans or human activity, to describe aspects of inanimate objects such as programs, easily leads to would-be questions, such as 'does the program understand what it is doing?' or 'does the program think?'

Since there exist programs and inputs such that the computation induced by the program does not terminate, it does not suffice to consider total functions; partial functions are necessary to capture the meaning of programs.

There does not exist a programming language over **N**, or any other sufficiently powerful compound data type, which avoids this problem. That is, if for every program all computations defined by the program terminate, then the set of functions computable by programs written in this language is unequal to the set of SAL(**N**) computable total functions, and also unequal to the set of all total functions computable in the intuitive sense (see Section 5.4).

EXAMPLE 3.1

(1) The addition function $\lambda x_1 x_2[x_1 + x_2]$ is computed by the following program $(2, 1, 2, P_1)$.

```
P1: while x2 ≠ 0 do
        x1 := x1 + 1;
        x2 := x2 ∸ 1
    od
```

(2) The multiplication function $\lambda x_1 x_2[x_1 . x_2]$ is computed by the program $(2, 1, 4, P_2)$.

```
P2: while x2 ≠ 0 do
        x3 := x1;
        while x3 ≠ 0 do
           x4 := x4 + 1;
           x3 := x3 ∸ 1
        od;
        x2 := x2 ∸ 1
    od;
    x1 := x4
```

(3) The program $(1, 1, 1, P_3)$ determines a function $f: N \to N$ which is undefined for every value of its argument, so dom(f) is empty.

```
P3: x1 := x1 + 1;
    while x1 ≠ 0 do x1 := x1 + 1 od
```

Thus, this everywhere undefined function f is SAL(**N**) computable. There are many other non-total functions which are SAL(**N**) computable, e.g. the functions f_k defined by $f_k \triangleq \lambda x[\text{if}\,(k \text{ divides } x) \text{ then } 1 \text{ else } \uparrow]$, where $\uparrow$ means undefined, are SAL(**N**) computable for every $k \geqslant 0$.

As stated earlier, partial functions are unavoidable. The possibility of nonterminating computations generates many questions, foremost among them the *Halting Problem for* SAL(**N**) *programs*. This is the following:

Given a SAL(**N**) program and a legal input value for this program, does the computation starting with this input value as determined by the given program terminate?

Although one wouldn't guess so from the preceding examples, this is a very difficult question. We will show later that the answer to this question cannot be determined by algorithmic methods. The Halting Problem is *algorithmically unsolvable.*

(4) The function $sg: N \rightarrow N$, defined by $sg \triangleq \lambda x_1[\text{if}(x_1 \text{ is equal to } 0) \text{ then } 0 \text{ else } 1]$, is computed by the program $(1, 1, 2, P_4)$.

```
P4: while x1 ≠ 0 do
        x1 := 0;
        x2 := x2 + 1
    od;
    x1 := x2
```

Certainly, SAL(**N**) is not a very convenient programming language, because it lacks many useful features common in high-level programming languages. These features include more general assignments, various data types such as sets, Booleans, etc., data-structuring facilities such as arrays and lists, and program-structuring facilities such as procedures.

Remember, however, that SAL(**N**) is designed to formalize the intuitive concept *effective procedure*, thereby making it possible to prove theorems about algorithms. In that respect its simplicity is an advantage. In the next section we will formally define the semantics of SAL(**N**). This definition, in particular, would become intolerably complicated were we to incorporate the above-mentioned features.

3.4 Semantics of SAL(N)

The semantics as described in the preceding section is vague only in its specification of which statement is to be executed. To remedy this, the statements of a program will be labelled.

Write a given SAL(**N**) program (n, k, p, P) on successive lines, one entity per line; i.e. a line contains:

- an assignment statement
- a while-head **while** $x_i \neq 0$ **do**, or
- a while-tail **od**.

Number the lines from 1 upwards, place behind each while-head the line

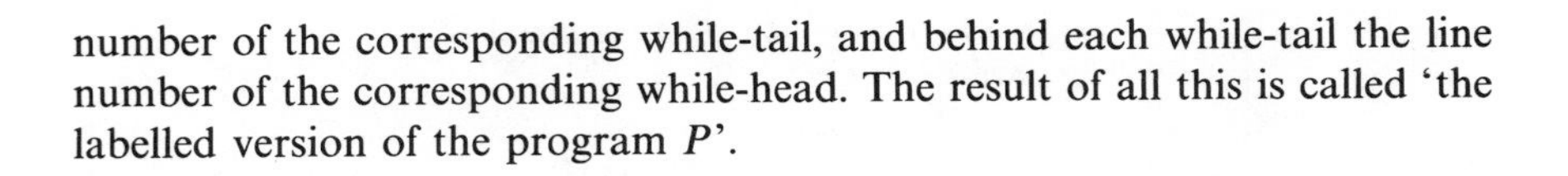

number of the corresponding while-tail, and behind each while-tail the line number of the corresponding while-head. The result of all this is called 'the labelled version of the program P'.

EXAMPLE 3.2

1:	**while** $x_2 \neq 0$ **do**	:4
2:	$x_1 := x_1 + 1;$	
3:	$x_2 := x_2 \dot{-} 1$	
4:	**od**	:1

Definition 3.2

Let $Q = (n, k, p, P)$ be a SAL(**N**) program, with its lines numbered from 1 to f. A *state vector* of Q is a tuple (s, x), $s \in \{1, 2, 3, \ldots, f, f+1\}$, $x \in N^p$. The state vector is called *initial* if $s = 1$, and $x_{n+1} = x_{n+2} = \cdots = x_p = 0$. It is called *final* if $s = f + 1$. ■

In vector (s, x) the first component s determines which statement must be executed next. This component acts as a pointer in the program text. The second component gives the current value of the program variables.

Definition 3.3

Let $Q = (n, k, p, P)$ be a SAL(**N**) program, having its lines numbered from 1 to f. Let V_Q denote the set of all state vectors of Q. A state vector $v_1 = (s, x) \in V_Q$ is transformed into the state vector $v_2 = (t, y) \in V_Q$ in a single computation step, notation $v_1 \vdash_Q v_2$, if and only if:

(1) s: $\langle$*assignment*$\rangle$ occurs in the labelled version of Q, $t = s + 1$, $y_r = x_r$ for all r such that $1 \leqslant r \leqslant p$ and $r \neq i$, the value of y_i depends on the assignment statement as specified in Table 3.1.

(2) Otherwise, the line with number s contains a while-head or a while-tail. In that case $y = x$, and the value of t is determined as follows.

(a) if s: **while** $x_i \neq 0$ **do** :s' occurs in the labelled version of Q, then $t = s + 1$ if $\neq_0(x_i)$ evaluates to TRUE and $t = s' + 1$ otherwise

(b) if s: **od** :s' and s': **while** $x_i \neq 0$ **do** :s occur in the labelled version of Q, then $t = s' + 1$ if $\neq_0(x_i)$ evaluates to TRUE and $t = s + 1$ otherwise ■

This definition specifies the single steps in the computation determined

Table 3.1 Semantics of assignment statements.

⟨*assignment*⟩	Value of y_i
$x_i := 0$	$y_i = 0$
$x_i := x_j$	$y_i = x_j$
$x_i := x_j + 1$	$y_i = +_1(x_j)$
$x_i := x_j \dot{-} 1$	$y_i = \dot{-}_1(x_j)$

by the program. A computation then, is a sequence of single steps $u \vdash_Q u_1 \vdash_Q u_2 \vdash_Q \cdots \vdash_Q v$, where u is an initial state vector. The computation *terminates* with state vector v if it can proceed no further, i.e. if there is no state vector w such that $v \vdash_Q w$. This is the case if and only if v is a final state vector.

The notation $u \vdash_Q^+ v$ means that there is a computation which transforms u into v and consists of $k \geqslant 1$ steps; the notation $u \vdash_Q^* v$ means that u can be transformed into v by a sequence of zero or more computation steps, i.e. if $u \vdash_Q^* v$ then $u \vdash_Q^+ v$ or $u = v$.

Note that the relation $\vdash_Q$ is single-valued but non-total; for every given state vector v there is at most one state vector w such that $v \vdash_Q w$. Therefore $\vdash_Q$ is a partial function from V_Q to V_Q. Since it is easier to work with total functions, we introduce the function $F_Q: V_Q \to V_Q$, which associates with any state vector x the state vector $F_Q(x)$ following x in a computation of Q.

Definition 3.4

Let $Q = (n, k, p, P)$ be a SAL(**N**) program and V_Q the set of all state vectors of Q. The 'follow function' $F_Q: N^{p+1} \to N^{p+1}$ is determined by:

$$F_Q(x) = \begin{cases} y \text{ if } x \vdash_Q y \\ x \text{ if } x \notin V_Q \text{ or there is no } y \text{ such that } x \vdash_Q y \end{cases}$$

In the above, an argument $(s, x_1, \ldots, x_p)$ must be interpreted as a state vector $(s, (x_1, \ldots, x_p))$. ■

Thus we can identify a computation induced by a program $Q = (n, k, p, P)$ with a sequence $v_0, v_1, \ldots$ of state vectors, such that v_0 is an initial state vector and $F_Q(v_i) = v_{i+1}$, for all $i \geqslant 0$. The computation terminates with state vector v_t if $t = \min\{j \mid F_Q(v_j) = v_j\}$. The length of this computation is t. The length of the computation is also called its time complexity, and is denoted by $T_Q(x)$.

Definition 3.5

Let $Q = (n, k, p, P)$ be a SAL(**N**) program having its lines numbered from 1 to f. The meaning of the program is the function $M_Q: N^p \to N^p$

determined as follows:
Let $a \in N^p$ and $(1,a) \vdash_Q s_1 \vdash_Q s_2 \vdash_Q \cdots \vdash_Q (t,b)$ be a computation of Q.

(1) If the computation terminates with (t,b), i.e. (t,b) is a final state vector and thus $t = f + 1$ and $F_Q((t,b)) = (t,b)$, then $M_Q(a) = b$.

(2) If no terminating computation starting with state vector $(1,a)$ exists, then $M_Q(a) = \uparrow$, i.e. undefined.

The function $f_Q: N^n \to N^k$ computed by the program is determined as follows:

$$f_Q \triangleq \lambda x[\pi^p_{1\ldots k}(M_Q(x_1, x_2, \ldots, x_n, 0, 0, \ldots, 0))]$$

where $\pi^p_{1\ldots k}$ is the function $\lambda x_1 x_2 \cdots x_p[(x_1, x_2, \ldots, x_k)]$, projecting a p-dimensional vector on its first k components (see Table 1.2).

Finally, we define the set $F(\text{SAL}(\mathbf{N}))$ of all SAL(**N**) computable functions:

(1) $F^n_k(\text{SAL}(\mathbf{N}))$ is the set of all functions $f: N^n \to N^k$ such that there exists a SAL(**N**) program (n,k,p,P) computing f.

(2) $F(\text{SAL}(\mathbf{N}))$ is the union of all $F^n_k(\text{SAL}(\mathbf{N}))$, $n \geqslant 0$, $k > 0$. ■

In the following we drop the subscript Q from $\vdash_Q$ and F_Q when it is clear which program is involved.

EXAMPLE 3.3

The following program $(2,1,2,P)$ computes the addition function $\lambda xy[x+y]$.

```
P: 1: while x2 ≠ 0 do      :4
   2:       x1 := x1 + 1;
   3:       x2 := x2 ∸ 1
   4: od                   :1
```

On input $(2,3)$ the computation runs as follows:

$$\begin{aligned}(1,(2,3)) &\vdash (2,(2,3)) \vdash (3,(3,3)) \vdash (4,(3,2))\\ &\vdash (2,(3,2)) \vdash (3,(4,2)) \vdash (4,(4,1))\\ &\vdash (2,(4,1)) \vdash (3,(5,1)) \vdash (4,(5,0))\\ &\vdash (5,(5,0))\end{aligned}$$

$$F((1,(2,3))) = (2,(2,3)), \qquad F((2,(2,3))) = (3,(3,3))$$

and so on, until

$$F((5,(5,0))) = (5,(5,0)),$$

$(5,(5,0))$ being a final state vector. The computation halts with this vector. The output is $\pi_1^2((5,0)) = 5$.

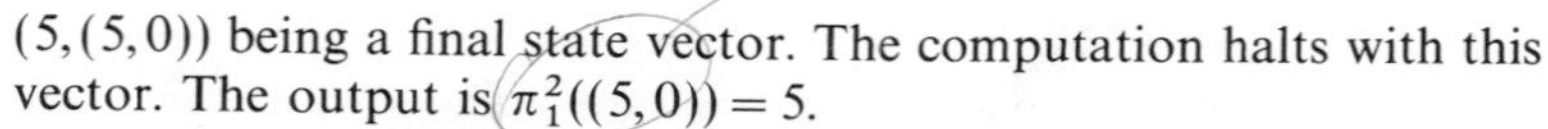

On an arbitrary input $(a,b) \in N^2$, the computation runs as follows:

$$\begin{aligned}
(1,(a,b)) &\vdash (2,(a,b)) \vdash (3,(a+1,b)) \vdash (4,(a+1,b \dot{-} 1)) \\
&\vdash (2,(a+1,b \dot{-} 1)) \vdash (3,(a+2,b \dot{-} 1)) \vdash (4,(a+2,b \dot{-} 2)) \\
&\;\vdots \\
&\vdash (2,(a+i,b \dot{-} i)) \vdash (3,(a+i+1,b \dot{-} i)) \\
&\qquad\qquad \vdash (4,(a+i+1,b \dot{-} i \dot{-} 1)) \\
&\;\vdots \\
&\vdash (5,(a+b,0))
\end{aligned}$$

The computation halts with $(5,(a+b,0))$, since this is a final state vector. The output is $\pi_1^2((a+b,0)) = a+b$.

It is easy to see that $M_{(2,1,2,P)} = \lambda xy[(x+y,0)]$ and that $f_{(2,1,2,P)} = \lambda xy[x+y]$. Formally proving this is another matter. This requires:

(1) showing that for all $(a,b) \in N^2$ there exists a terminating computation, and

(2) showing that in any terminating computation which starts with the state vector $(1,(a,b))$ and terminates with state vector $(5,(p,q))$, we have $p = a+b$ and $q = 0$.

Point (1) can be shown by observing that x_2 can only be decreased, and that an actual decrease occurs at least once every three computation steps. Point (2) can be shown by observing that in any computation $(1,(a,b)) \vdash \cdots \vdash (4,(p,q))$, we have $p+q = a+b$, which can be shown by induction on the length of the computation. Furthermore, the final step in any terminating computation is $(4,(p,q)) \vdash (5,(p,q))$, which is possible if and only if $q = 0$.

Both the meaning of, and the function computed by, the program of the example above, can be expressed in terms of the follow function F using the operators on functions defined in Section 1.4. For this purpose we consider the state vectors to be elements of N^3, i.e. we write e.g. $(2,3,4)$ instead of $(2,(3,4))$. We then have:

(1) the time complexity $T_{(2,1,2,P)}(x,y)=\min\{k|\pi_1^3(F^*(1,x,y,k))=5\}$,

(5 is one plus the number of the highest numbered line in $(2,1,2,P)$),

(2) $M_{(2,1,2,P)}(x,y)=F^*(x,y,T_{(2,1,2,P)}(x,y))$

(3) $f_{(2,1,2,P)}(x,y)=\pi_1^2(M_{(2,1,2,P)}(x,y))$

It is possible to rephrase Definition 3.5 and define $M_{(n,k,p,P)}$ and $f_{(n,k,p,P)}$ directly in terms of the follow function $F_{(n,k,p,P)}$, using the functional operators $*$ and ∇_i, as in the example above. When doing so, the concepts 'computation step' and 'termination' are less naturally represented. Hence the detour. In Section 4.2, the follow function will be discussed further.

3.5 Macro statements and examples

Writing programs in the programming language SAL(**N**) as defined in the preceding sections is a tedious and laborious task. Furthermore, when the programs become larger they tend to become less intelligible. In this respect SAL(**N**) programs resemble programs written in some machine code.

In order to be able to write programs which are easily comprehensible we introduce a shorthand. A SAL(**N**) program written using this shorthand is called a macro program.

First we consider the naming of the program variables and the input and output conventions. Sometimes it may be convenient to denote these variables by arbitrary names (strings of characters and digits beginning with a character) instead of x_1 etc. If arbitrary names are used, the names of the input and output variables have to be declared in a program header, as in the following example.

```
input: x, y;
output: sum; {sum = x + y}
method:

P: count:= y; sum:= x;

    while count ≠ 0 do
      {sum + count = x + y}
      sum:= sum + 1;
      count:= count ∸ 1
    od
    {sum + count = x + y and count = 0}
```

Statements included in braces are comments. These are mainly used to give statements which remain true during the execution of the program.

To convert such a macro program into a SAL(**N**) program, number the n input variables from 1 to n according to the ordering in the declaration and number the remaining program variables from $n+1$ upwards in alphabetical order. Remove the declarations and the comments and then append to the resulting sequence P the k assignments

$$x_1 := x_{i_1}; \ldots; x_k := x_{i_k};$$

where $i_1, \ldots, i_k$ are the numbers assigned to the output variables in the above numbering.

The resulting SAL(**N**) program is called the expansion of the macro program. Obviously it computes the function intended to be computed by the macro program. Expanding the above macro program results in:

(1) numbering the variables:

$$x = x_1,\ y = x_2,\ count = x_3,\ sum = x_4.$$

(2) the SAL(**N**) sequence:

```
x3:= x2; x4:= x1;

while x3 ≠ 0 do
   x4:= x4 + 1;
   x3:= x3 ∸ 1
od;

x1:= x4
```

Another useful shorthand is the macro statement:

$$(z_1, \ldots, z_k) := f(x_1, \ldots, x_n),$$

where $f: N^n \to N^k$ is a SAL(**N**) computable function. Such statements are called macro statements. The above macro statement stands for a sequence S computing the function value $f(x_1, \ldots, x_n) = (y_1, \ldots, y_k)$ and assigning the values of the variables y_i to the variables z_i, $1 \leqslant i \leqslant k$. The values of all other variables occurring in the program containing the macro statement are to remain unchanged. The following example illustrates the conversion of a macro statement into its corresponding SAL(**N**) sequence.

EXAMPLE 3.4

Let $h = \lambda x_1 x_2 [x_1 . x_2]$ be the product function and $f = \lambda x_1 x_2 [x_1 + x_2]$ the sum function. The function f is computed by the program $(2, 1, 2, P_f)$,

P_f: **while** $x_2 \neq 0$ **do**
$x_1 := x_1 + 1;$
$x_2 := x_2 \dot{-} 1$
od

The following is a macro program computing the function h.

input: x, y;
output: z; $\{z = x.y\}$
method:

$\{z = 0$ by assumption$\}$

while $y \neq 0$ **do**
$z := f(z, x);$
$y := y \dot{-} 1$
od

This macro program can be transformed into a SAL(**N**) program P_h by substituting the SAL(**N**) sequence P_f for the macro $z := f(z, x)$, and suitably renaming the variables.

(1) In the macro program substitute x_1 for x, x_2 for y, and x_3 for z.
(2) In the sequence P_f substitute x_4 for x_1 and x_5 for x_2.
(3) This results in the following sequence P_h:

while $x_2 \neq 0$ **do**

$x_4 := x_3;\ x_5 := x_1;$ **while** $x_5 \neq 0$ **do** $x_4 := x_4 + 1;$ $x_5 := x_5 \dot{-} 1$ **od**; $x_3 := x_4;$

$x_2 := x_2 \dot{-} 1$

od

The sequence of statements enclosed in a box is the expansion of the macro statement $z := f(z, x)$.

The expansion of an arbitrary macro statement

$(x_{i_1}, \ldots, x_{i_k}) := f(x_{j_1}, \ldots, x_{j_n})$

runs as follows. Let (n, k, p, P) be a SAL(**N**) program computing the function f and Q a macro program with q program variables in which the above macro statement occurs. The expansion P_Q of the macro statement in Q reads:

P_Q: $x_{q+1} := x_{j_1}; \ldots; x_{q+n} := x_{j_n};$

$x_{q+n+1} := 0; \ldots; x_{q+p} := 0;$

'the sequence obtained from P by substituting x_{q+i} for x_i, $1 \leqslant i \leqslant p$';

$x_{i_1} := x_{q+1}; \ldots; x_{i_k} := x_{q+k}$

The meaning of the SAL(**N**) sequence P_Q is a function from $N^{q+p} \to N^{q+p}$, mapping the vector

$$(x_1, \ldots, x_q, a_{q+1}, a_{q+2}, \ldots, a_{q+p}) \in N^{q+p}$$

onto the vector

$$(y_1, \ldots, y_q, b_{q+1}, b_{q+2}, \ldots, b_{q+p}) \in N^{q+p}$$

for which $(y_{i_1}, \ldots, y_{i_k}) = f(x_{j_1}, \ldots, y_{j_n})$, and $y_i = x_i$ for the remaining indices i, $1 \leqslant i \leqslant q$.

In a macro program P_m containing more than one macro statement, these statements are transformed one after another. In this way the macro program P_m is transformed into a SAL(**N**) program P with the same input and output variables computing the same function as the macro program P_m.

Note that after each replacement of a macro statement by a SAL(**N**) sequence, the number of program variables is enlarged by a number of (auxiliary) variables. In actual computing practice this would lead to a waste of memory space.

EXAMPLE 3.5

(1) The addition function is computed by

while $x_2 \neq 0$ **do** $x_1 := x_1 + 1; x_2 := x_2 \dot{-} 1$ **od**.

(2) The subtraction function is computed by

while $x_2 \neq 0$ **do** $x_1 := x_1 \dot{-} 1; x_2 := x_2 \dot{-} 1$ **od**.

(3) The maximum of x and y is computed by

$$z := (x \dot{-} y) + y.$$

(4) The absolute difference $|x - y|$ of x and y is computed by

$$z := (x \dot{-} y) + (y \dot{-} x)$$

Note that the preceding two examples show a generalization in that expressions instead of assignments are used.

(5) The function $sg = \lambda x[\text{if } x > 0 \text{ then } 1 \text{ else } 0]$ is computed by

$$y := x \dot{-} 1; x := x \dot{-} y.$$

(6) The function $\overline{sg} = \lambda x[\text{if } x = 0 \text{ then } 1 \text{ else } 0]$ is computed by

$$1 \dot{-} sg(x).$$

(7) The equality function $equal = \lambda xy[\text{if } x = y \text{ then } 1 \text{ else } 0]$ is computed by

$$\overline{sg}(|x - y|).$$

(8) The product of x and y is computed by the macro program

while $y \neq 0$ **do** $z := z + x$; $y := y \dot{-} 1$ **od**.

Regarding the flow of control, the following shorthands will be used:

(1) **if** $x \neq 0$ **then** S_1 **else** S_2 **fi**

This macro stands for:

$r_0 := 1 \dot{-} x$; $r_1 := 1 \dot{-} r_0$;

$\{r_0$ and r_1 are new variables$\}$

$\{$if $x \neq 0$ then ($r_0 = 0$ and $r_1 = 1$) else ($r_0 = 1$ and $r_1 = 0$)$\}$

while $r_1 \neq 0$ **do** S_1; $r_1 := 0$ **od**;

while $r_0 \neq 0$ **do** S_2; $r_0 := 0$ **od**

(2) **do** x **times** S **od**

This expression stands for:

$r := x$; $\{r$ is a new variable$\}$

while $r \neq 0$ **do** S; $r := r \dot{-} 1$ **od**

This macro statement is called a *times statement*. The variable x is the loop variable, the sequence S the scope of the times statement, and **do** the head and **od** the tail of the times statement.

The if statement is very intuitive, and will be used frequently.

The times statement also can be more intuitive than a while statement. Its main use, however, is in the definition of a class of programs (so-called loop programs), which compute total functions. For this class of programs there is no Halting Problem. Note that any while statement **while** P **do** S **od** which is executed at most z times, can be replaced by a times statement:

$t := t + 1$;

{t is a variable not occurring in **while** P **do** S **od**}

do z **times**

  if $P \wedge t = 1$ **then** S **else** $t := 0$ **fi**

od

Nevertheless, the times statement is less powerful than the while statement (see Section 4.5).

Clearly, many other useful language features might be introduced. For example the macro statements

do $f(x)$ **times** S **od**,

while $f(x) \neq 0$ **do** S **od**,

if $f(x) \neq g(x)$ **then** S_1 **else** S_2 **fi**,

if $f(x) \neq 0$ **then** S_1 **fi**.

Here f and g are SAL(**N**) computable functions and S, S_1, and S_2 sequences. These statements can be expanded for instance in the following way:

$w := f(x)$; **do** w **times** S **od**,

$w := f(x)$; **while** $w \neq 0$ **do** S; $w := f(x)$ **od**,

$w := |f(x) - g(x)|$; **if** $w \neq 0$ **then** S_1 **else** S_2 **fi**,

$w := sg(f(x))$; **do** w **times** S **od**.

EXAMPLE 3.6

(1) The function $divides = \lambda xy[\text{if}(x, y \geqslant 1 \text{ and } x \text{ divides } y) \text{ then } 1 \text{ else } 0]$ is computed by the following macro program:

```
input: x, y;
output: z;
method:

  if (x ≠ 0 ∧ y ≠ 0) then

     t:= x;
     while t < y do t:= t + x od;
     if t = y then z:= 1 fi
  fi
```

(2) The function *prime* is defined by

$$prime = \lambda x[\text{if}(x \text{ is a prime number}) \text{ then } 1 \text{ else } 0]$$

The function *prime* is computed by the program

```
input: x
output: y
method:

  if x ⩽ 1 then
     y:= 0
  else
     count:= x − 2;
     y:= 1; d:= 2;

     do count times

        if divides(d, x) = 1 then y:= 0 fi;
        d:= d + 1
     od
  fi
```

With the macro statements introduced in this section programming in SAL(**N**) is comparable with programming a higher-level language. In fact SAL(**N**) extended with macro statements is essentially Pascal without declarations and data types.

It is clear that macros help considerably in proving that relatively complex functions are SAL(**N**) computable.

3.6 Exercises

Definitions of various functions used in the following exercises can be found in Table 1.2.

3.1 Prove that the functions sg, $\overline{sg}$, π_i^n, $\pi_{i\ldots j}^n$, I^n and C_r^n are SAL(**N**) computable.

3.2 Prove that the following functions are SAL(**N**) computable.

(a) $\lambda xy[x \dot{-} y]$
(b) $\lambda xy[x+y]$
(c) $\lambda xy[x.y]$
(d) $\lambda xy[\text{if}(x=0) \text{ then } 0 \text{ else } x^y]$
(e) $\lambda x[\text{the least } k \text{ such that } x<2^k]$
(f) $\lambda xy[x \bmod y]$

3.3 *POL* is the set of all functions which are SAL(**N**) computable in polynomial time. Prove that *POL* is closed with respect to function composition and Cartesian product of functions. Why is *POL* not closed with respect to iteration and exponentiation? Give examples such that $f \in POL$ and f^* and f^{∇_i} are not in *POL*.

3.4 A prime number is a number without proper divisors. Arranged in increasing order, the prime numbers form an infinite sequence 2, 3, 5, 7, 11, 13,.... The function $p \triangleq \lambda x[\text{the } x\text{th prime number in the above sequence}]$ is SAL(**N**) computable. Prove this.

3.5 Prove that the inverse f^{-1} of a SAL(**N**) computable bijective function $f: N \rightarrow N$ is SAL(**N**) computable. (The inverse of a function is defined in Section 1.2.)

3.6 The language SAL(**N**) can be extended with a goto statement, resulting in a language denoted by GOTOSAL(**N**). A goto statement is of the form '*goto AA*', where *AA* is a natural number. The execution of the statement '*goto AA*' causes the statement with label *AA* to be executed next.

For example

S_1;

AA: S_2;

S_3;

goto AA;

S_4

This program fragment with goto statement can be simulated in SAL(**N**) as follows:

$c := 1;$

while $c \leqslant 5$ **do**

  if $c = 1$ **then** $S_1;\ c := c + 1$ **fi**;

  if $c = 2$ **then** $S_2;\ c := c + 1$ **fi**;

  if $c = 3$ **then** $S_3;\ c := c + 1$ **fi**;

  if $c = 4$ **then** $c := 2$ **fi**;

  if $c = 5$ **then** $S_4;\ c := c + 1$ **fi**

od

Prove that every GOTOSAL(**N**) computable function is also SAL(**N**) computable. Therefore, goto statements do not increase the 'computational power' of SAL(**N**).

3.7 Prove that the functions $\overline{sg}$, sg, $\lambda xy[x + y]$ and $\lambda xy[x.y]$ are computable in polynomial time.

3.8 The programming language RATAL(**Q**) is defined as follows.

⟨*constant*⟩::= ⟨*rational number*⟩

⟨*operator*⟩::= + | − | . | /

⟨*assignment*⟩::= ⟨*identifier*⟩:= ⟨*constant*⟩ |
  ⟨*identifier*⟩:= ⟨*identifier*⟩⟨*operator*⟩⟨*identifier*⟩

⟨*program*⟩::= ⟨*assignment*⟩ |
  ⟨*program*⟩; ⟨*assignment*⟩

Define the semantics of the language RATAL(**Q**).

3.9 Show that every computation with rational numbers can be programmed in SAL(**N**). You can use the fact that every rational number can be represented as a triple of natural numbers (x, a, b), where $x = 0$ if the rational number is not negative and $x = 1$ otherwise. The magnitude of the rational number is represented by a/b.

Show that all RATAL(**Q**) computable functions are SAL(**N**) computable in polynomial time relative to this coding.

3.10 Define a compound data type **R** of real numbers; take the usual arithmetical operations as operations in the data type. Define an algorithmic language over this data type **R** and define the semantics of this language.

3.11 Define a compound data type **G** and a programming language SAL(**G**) in such a way that computations with the Gaussian integers = $\{a+bi \mid a, b \in Z \text{ and } i^2 = -1\}$ can be performed.

3.12 Prove that there exists at least one function $f: N \rightarrow N$ which is not SAL(**N**) computable.

Extend the computation model SAL(**N**) in such way that this function is computable in the extended computation model. Is the extended model of computation any better than the original?

3.13 Extend SAL(**N**) with the times statement '**do** x **times** S **od**' and define the semantics of this extended language.

Then prove that the programs P_1 and P_2 given below define the same function, i.e. that P_1 and P_2 are equivalent.

P_1: *input*: x, y, z
output: x, y
method:

do x **times** S **od**

P: *input*: x, y, z
output: x, y
method:

$u := x;$

while $u \neq 0$ **do**
$u := u \dot{-} 1;$
S
od

3.14 g_i, $i = 1, 2, 3, \ldots$ is a system of functions recursively defined as follows:

$$g_1 \triangleq \lambda x[x + x + (1 \dot{-} x)]$$

$$g_2 \triangleq \lambda x[2^x]$$

$$g_{n+1} \triangleq \lambda x[\text{if}(x=0) \text{ then } 1 \text{ else } g_n(g_n(x))], \quad \text{for all } n \geqslant 2.$$

This system is called a system of Ackermann functions.

Prove the following properties:

(a) $g_{n+1}(x) = g_n(g_n \cdots (g_n((1))\cdots) = g_n^x(1)$, for all $n \geqslant 1$ and $x \geqslant 0$.
(b) The function g_i is SAL(**N**) computable for all $i \geqslant 1$.

(c) The function g_i is a monotonically increasing function for all $i \geqslant 1$.

(d) The function $\lambda xy[g_x(y)]$ is SAL(N) computable.

3.15 Let P be a program with n input variables which consists solely of assignment statements and 'times' macro statements, but such that the times macro statements are not nested. Prove that there are natural numbers A and B such that $T_P(x) < A|x| + B$, for all $x \geqslant 0$, where $|x|$ denotes the maximum of $\{x_1, x_2, \ldots, x_n\}$.

3.7 References

The Backus–Naur notation was used to define the syntax of the programming language ALGOL 60, see Naur (1963). The programming language SAL is similar to the language introduced by Meyer and Ritchie (1967a). The idea of a computation over a certain data type was introduced by Engeler (1968).

Although Turing machines (Turing, 1936) are introduced as single-purpose machines capable of computing string functions, the formalism can also be interpreted as a programming language. Programmable Turing machines were introduced by Hao Wang (1957); for this concept see, for example, Hermes (1961). More details about machine-oriented computation models can be found in Chapter 10.

Chapter 4
The Recursive Functions

4.1 Definitions and elementary properties

In the preceding chapters effectiveness was characterized as programmability using the language SAL(**N**). A SAL(**N**) program, the formal equivalent of an algorithm, induces a computation for every input x.

This computation model has the advantage that it closely follows the intuition of most computer science students about effective procedure, computation and effectively computable function.

Quite another approach, perhaps more appealing to mathematicians than to computer scientists, is the definition of the set of recursive functions. The general idea is to select some functions which are considered effectively computable, and then to select some operators on functions which preserve effectiveness. By repeatedly applying these operators, a set of computable functions is obtained. Such a definition may be considered to constitute a model of computation in the broad sense of the word 'model' as used in the discussion on the equivalence of models of computation (Section 2.2).

Assume given a non-empty set B of functions and a non-empty set F of *functional operators*; these are operators which can be applied to functions and deliver functions as results. The set $C(B,F)$, the *closure of B with respect to F*, is defined as follows:

(1) $C(B,F)$ contains all the functions in B

(2) $C(B,F)$ contains all functions constructed from these by finitely often applying operators belong to F

(3) $C(B,F)$ contains no other functions.

Equivalently, $C(B,F)$ is the smallest set containing B and *closed* with respect to all the functional operators in F. (A set X is said to be *closed with respect to an operator* if the result of applying the operator to elements of the set X is an element which also belongs to the set X.) The functions in B are called base functions, because all other functions in $C(B,F)$ are constructed from these functions, using the given operators.

Of natural interest is the case where all base functions are (intuitively) effectively computable and the functional operators intuitively preserve effective computability. When defining the set of recursive functions, the sets B and F are chosen to satisfy these requirements. For example, let

- B consist of

  the identity functions $I^n: N^n \to N^n$, $I^n = \lambda x[x]$,
  the projection functions $\pi_i^n: N^n \to N$, $\pi_i^n = \lambda x_1 \cdots x_n[x_i]$,
  and the constant functions $C_r^n: N^n \to N$, $C_r^n = \lambda x[r]$, for all $n, r \geqslant 1$ and i, $1 \leqslant i \leqslant n$;

- F consist of

  function composition,
  and Cartesian product of functions. (These operators are defined in Section 1.4.)

Then all functions in $C(B,F)$ certainly are SAL(**N**) computable. However, $C(B,F)$ does not contain simple arithmetical functions such as addition; it is not a very powerful model of computation. To obtain a more powerful model, we must include more base functions or more operators or both.

Note that a particular function can be obtained in more than one way. For example, $C_2^3 = \pi_1^2 \cdot (C_2^3, \pi_2^3)$. We have a comparable situation regarding SAL(**N**) computable functions. For any such function there is a countably infinite number of SAL(**N**) programs computing this particular function, because you can add the statement $x_1 := x_1$ to any program any number of times without altering the function computed.

The similarity between expressions denoting recursive functions, and programs denoting SAL(**N**) computable functions, is more than superficial. The expressions denoting recursive functions are in fact polynomial expressions over an algebra, even a simple data type. That is, these expressions stand for straight-line programs over this particular data type. Consider the algebra (A, O), where

- A is the set of all functions $N^n \to N^m$, $n, m \geqslant 1$.

- $O \triangleq B \cup F$, i.e. the operations in F supplemented by a nullary operation for each element of B.

Then the set $C(B, F)$ is precisely the set of functions determined by polynomial expressions over this algebra having zero variables. Each such expression determines exactly one element of A, which is constructed from elements of B by applying operations from F finitely often.

The algebra (A, O) together with the programming language consisting of all straight-line programs constitutes a computation model in which we have:

- complicated objects, namely functions, as objects used in the computation. To have practical validity, these functions, which are infinite objects, must be effectively specified in some finite way.
- complicated operations transforming the functions given as arguments into the functions to be delivered as results. To have practical validity the operations must be effective, in that given finite effective descriptions of the argument functions a finite effective specification of the result function must be produced.
- extremely simple program structure. You only need straight-line programs having zero input values and one output value. This single output value is a function from A. The program is a particular finite specification of this function. It is a practically valid specification if both the objects and the operations are practically valid.

The domain A is uncountable, therefore (A, O) is not a data type. Taking the set $C(B, F)$ instead of the set of all functions, makes $(C(B, F), O)$ into a data type if the sets B and F are well chosen.

Below, we give a precise definition of the closure concept. It is easily seen that this definition is equivalent to the more informal definition given above.

Definition 4.1

Let B be any countable set of functions and F any countable set of operators on functions.

$$A_0 \triangleq B.$$

$$A_{i+1} \triangleq \{o(f_1, \ldots, f_{n_o}) \mid o \in F, f_1, \ldots, f_{n_o} \in \bigcup_{0 \leqslant j \leqslant i} A_j\}, \text{ for all } i \geqslant 0.$$

$$C(B, F) \triangleq \bigcup_{i \geqslant 0} A_i.$$

$C(B, F)$ is called the *closure* of B with respect to F. ■

To show that all functions in $C(B, F)$ have some property P, we proceed by *induction on the construction* of $C(B, F)$, as suggested by the above definition. That is:

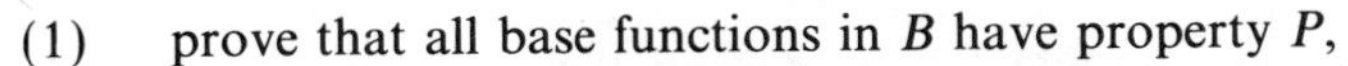

(1) prove that all base functions in B have property P,

(2) prove that all functions in the set A_{n+1} have property P whenever all functions in all sets A_i, $i \leqslant n$, have the property P.

It then follows by finite induction on n, that all functions in $C(B, F)$ have the particular property P.

Definition 4.2

The set *REC* of all *partial recursive* functions is the closure $C(B, F)$ where:

(1) the set B of base functions consists of

- (a) the successor function $+_1: N \rightarrow N$,
- (b) the predecessor function $\dot{-}_1: N \rightarrow N$,
- (c) the identity functions $I^n: N^n \rightarrow N^n$, for all $n \geqslant 0$,
- (d) the projection functions $\pi_i^n: N^n \rightarrow N$, for all $n \geqslant 1$ and i, $1 \leqslant i \leqslant n$,
- (e) the zero function $z: N \rightarrow N$, $z = \lambda x[0]$,

(2) the set F of operators consists of

- (a) composition of functions,
- (b) Cartesian product of functions,
- (c) exponentiation of functions,
- (d) iteration of functions.

The above functions and operators are defined in Sections 1.3 and 1.4 on mathematical preliminaries. In particular, the operations of exponentiation and iteration are defined in Definition 1.1 (see page 8).

Finally, *TREC* denotes the set of all *total recursive* functions in *REC* and REC_k^n ($TREC_k^n$) will denote the set of all partial (total) recursive functions, $f: N^n \rightarrow N^k$. ■

From the definition of the iteration operator it is clear that *REC* contains partial functions. The term 'recursive' is explained by the fact that in earlier definitions other functional operators were used, one of which is 'primitive recursion'. This operator is defined as follows:

The function $f: N^{n+1} \rightarrow N$ is defined by primitive recursion from $g: N^n \rightarrow N$ and $h: N^{n+2} \rightarrow N$ if

$$f(x,0)=g(x), \text{ and}$$
$$f(x,y)=h(x,y,f(x,y\dot{-}1)) \qquad \text{for all } y>0.$$

The second clause of the above definition states that to compute the value of $f(x,y)$, you should compute $f(x,y\dot{-}1)$ and then use this value where necessary in the further calculation needed to find the value of $f(x,y)$. Hence the term 'recursive'.

EXAMPLE 4.1

The following are examples of recursive functions:

(1) Addition, $\lambda xy[x+y]=\lambda xy[+_1^*(x,y)]$.

(2) Subtraction, $\lambda xy[x\dot{-}y]=\lambda xy[\dot{-}_1^*(x,y)]$.

(3) Multiplication, $\lambda xy[x\cdot y]=\lambda xy[\pi_1^2(\oplus^*(0,x,y))]$, where $\oplus$ is defined as follows:

$$\lambda xy[x\oplus y]\triangleq\lambda xy[(+_1^*(x,y),y)]$$

(4) All the constant functions $C_i^n\triangleq\lambda x[i]$, $i\geqslant 0$.

$$C_0^n=z\cdot\pi_1^n$$
$$C_{i+1}^n=+_1\cdot C_i^n,\ i\geqslant 0.$$

(5) The functions $sg\triangleq\lambda x[\text{if}(x>0) \text{ then } 1 \text{ else } 0]$, and $\overline{sg}\triangleq\lambda x[\text{if}(x=0) \text{ then } 1 \text{ else } 0]$.

$$\overline{sg}=\dot{-}\cdot(C_1^1,I^1)$$
$$sg=\dot{-}\cdot(C_1^1,\overline{sg})$$

(6) The absolute difference function $\lambda xy[|x-y|]$. This function is defined by the expression

$$+\cdot(\dot{-}\cdot(\pi_1^2,\pi_2^2),\dot{-}\cdot(\pi_2^2,\pi_1^2)).$$

(7) Let $f\triangleq\lambda xyz[(x,y+1,x-y^2)]$. Then

$$f=I^3\cdot(\pi_1^3,+_1\cdot\pi_2^3,\dot{-}\cdot(\pi_1^3,\times\cdot(\pi_2^3,\pi_2^3))).$$

Furthermore, $f^{\nabla_3}(x,0,1)=(x,y+1,0)$, where y is the least integer such that $x\dot{-}y^2=0$. Thus the following function *csqrt* computes the ceiling of the square root of x:

$$csqrt\triangleq\dot{-}_1\cdot\pi_2^3\cdot f^{\nabla_3}\cdot(I^1,z,C_1^1)$$

This is a recursive total function. Had we defined $f(x, y, z) = (x, y+1, |x - y^2|)$, then the resulting function $csqrt(x)$ would produce $\sqrt{(x)}$ for x a perfect square, and would be undefined for all other values of x.

Finally, let us consider the *minimization operator* μ, also called the μ-operator. It may only be used with a total function as an argument and then produces a partial function as a result.

Let $f: N^{n+1} \to N$ be any total function. The function $\mu f: N^{n+1} \to N$ is defined as follows:

$$\mu f \triangleq \lambda xy[\text{if}(f(x,z) \neq y \text{ for all } z \geqslant 0) \text{ then } \uparrow \text{ else the smallest } z \text{ such that } f(x,z) = y]$$

If f is a total recursive function, then μf is a partial recursive function. To prove this, consider the auxiliary function $H: N^{n+3} \to N^{n+3}$ defined as follows:

$$H(x, y, z, w) \triangleq (x, y, z+1, |y - f(x,z)|), \text{ for all } x \in N^n,\ y, z, w \in N.$$

Clearly, $H^{\nabla_{n+3}}(x, y, 0, 1) = (x, y, \mu f(x, y) + 1, 0)$, supposing at least that there is some z such that $f(x, z) = y$. If such a z does not exist, then $H^{\nabla_{n+3}}(x, y, 0, 1) = \uparrow$. Hence,

$$\mu f = \dot{-}_1 \cdot \pi^{n+3}_{n+2} \cdot H^{\nabla_{n+3}} \cdot (I^{n+1}, C_0^{n+1}, C_1^{n+1}).$$

Thus, if f is a total recursive function, then so is H. Consequently, μf is a partial recursive function.

If we are not satisfied with the fact that the minimization operator may only be applied to total functions, we may proceed as follows:

- let $f: N^{n+1} \to N$ be any partial function
- let $V(x, y) \triangleq \{z \mid f(x, z) = y\}$
- define the function $(\bar{\mu} f)$ as follows:

$$(\bar{\mu} f)(x, y) \triangleq \text{if } V(x, y) \neq \varnothing \text{ then } \min V(x, y) \text{ else } \uparrow.$$

The problem with this definition is that there are partial recursive functions f such that $\bar{\mu} f$ is a partial non-recursive function. One such function will be examined in Section 6.1. The problem is caused by the test '$V(x, y) \neq \varnothing$', which cannot be performed algorithmically.

We can extend the applicability of the μ-operator to non-total recursive functions, such that the resulting function is partial recursive, in the following way:

- let $f:N^{n+1} \rightarrow N$ be any partial function
- let $V(x, y) \triangleq \{z \mid f(x, z) = y\}$
- define the function $\mu^p f$ as follows:

$$\mu^p \triangleq \lambda xy[\text{if}(V(x, y) \neq \varnothing \text{ and } (\forall z \leqslant \min V(x, y))[f(x, z)\downarrow]) \\ \text{then } \min V(x, y) \text{ else } \uparrow].$$

The preceding proof that $\mu f \in REC$ if $f \in TREC$, with minor modifications, shows that $\mu^p f \in REC$ whenever $f \in REC$.

4.2 The equivalence theorem

In this section we will prove that $REC = F(\text{SAL})$. This theorem will be referred to as the 'equivalence theorem'. The computation models are equivalent relative to the identity as a coding. In later chapters we will meet other models of computation, and other equivalence theorems, which do require nontrivial codings.

Proving the equivalence theorem is somewhat laborious. For the sake of clarity, the proof will be presented using a sequence of lemmas.

Lemma 4.1

All recursive functions are SAL(**N**) computable, i.e. $REC \subseteq F(\text{SAL}(\mathbf{N}))$.

Proof

This is proved by showing that for every $f \in REC$ there is a SAL(**N**) program P_f, which computes f. The validity of this assertion is proved by induction on the construction of REC. We have to consider the following cases:

- It is clear that the base functions $+_1$, $\dot{-}_1$, π_i^n and z are SAL(**N**) computable.
- If the functions $f:N^n \rightarrow N^k$ and $g:N^m \rightarrow N^r$ are recursive and SAL(**N**) computable then $f \times g:N^{n+m} \rightarrow N^{k+r}$ is also SAL(**N**) computable, since the following macro program computes $f \times g$:

input: $x_1, \ldots, x_n, x_{n+1}, \ldots, x_{n+m}$;
output: $z_1, \ldots, z_k, z_{k+1}, \ldots, z_{k+r}$;
method:

$(z_1, \ldots, z_k) := f(x_1, \ldots, x_n)$;

$$(z_{k+1},\ldots,z_{k+r}):=g(x_{n+1},\ldots,x_{n+m})$$

- Let $f{:}N^n \rightarrow N^k$ and $g_i{:}N^r \rightarrow N^{n_i}$ $1 \leqslant i \leqslant m$, $n_i > 0$ for all i and $n_1+n_2+\cdots+n_r=n$, be recursive functions which are SAL(**N**) computable. The composition of these functions $f\cdot(g_1,\ldots,g_m)$ is recursive and SAL(**N**) computable, because it is computed by the following macro program

  input: $x_1,\ldots,x_r$;
  output: $z_1,\ldots,z_k$;
  method:

  $u_1:=g_1(x);\ \ldots;\ u_m:=g_m(x);$

  $z:=f(u)$

- Assume that $f{:}N^n \rightarrow N^n$ is a recursive SAL(**N**) computable function. Then the functions f^* and f^{∇_i} $(1 \leqslant i \leqslant n)$ are also SAL(**N**) computable, since these functions are computed by the macro programs:

f^*: *input*: $x_1,\ldots,x_{n+1}$;
output: $x_1,\ldots,x_n$;
method:

do x_{n+1} **times**

$(x_1,\ldots,x_n):=f(x_1,\ldots,x_n)$

od

f^{∇_i}: *input*: $x_1,\ldots,x_n$;
output: $x_1,\ldots,x_n$;
method:

while $x_i \neq 0$ **do**

$(x_1,\ldots,x_n):=f(x_1,\ldots,x_n)$

od ∎

Proving the converse of the above lemma is the really laborious task. First consider the follow function associated with a SAL(**N**) program $Q=(n,k,p,P)$. Let the program have its statements numbered from 1 to f as described in Section 3.4. The follow function $F_Q{:}N^{p+1} \rightarrow N^{p+1}$ is defined by cases (see Definitions 3.3 and 3.4 on pages 45 and 46):

$F_Q(s,x_1,x_2,\ldots,x_p)=(t,y_1,y_2,\ldots,y_p)$, such that

(1) if $s=0$ or $s>f$ then $t=s$ and $y_r=x_r$ for all r, $1 \leqslant r \leqslant p$.

(2) if s: <*assignment*> occurs in the labelled version of P, then $t=s+1$

Table 4.1 Semantics of assignment statements.

⟨*assignment*⟩	Value of y_i
$x_i := 0$	$y_i = 0$
$x_i := x_j$	$y_i = x_j$
$x_i := x_j + 1$	$y_i = +_1(x_j)$
$x_i := x_j \dot{-} 1$	$y_i = \dot{-}_1(x_j)$

and $y_r = x_r$ for all r such that $1 \leqslant r \leqslant p$ and $r \neq i$, and the value of y_i depends on the assignment statement as specified in Table 4.1.

(3) Otherwise, the line with number s contains a keyword. In that case $y_r = x_r$ for all r, $1 \leqslant r \leqslant p$ and the value of t is determined as follows:

(a) if s: **while** $x_i \neq 0$ **do** :s' occurs in the labelled version of P, then $t = s + 1$ if $\neq_0(x_i)$ evaluates to TRUE and $t = s' + 1$ otherwise;
(b) if s: **od** :s' and s': **while** $x_i \neq 0$ **do** :s occur in the labelled version of P, then $t = s' + 1$ if $\neq_0(x_i)$ evaluates to TRUE and $t = s + 1$ otherwise.

In a particular program, each label corresponds to a fixed statement, while-head, or while-tail. The function F_Q can be written:

$$F_Q(s, x_1, \ldots, x_p) = \begin{cases} (s, x_1, \ldots, x_p) & \text{if } s = 0 \\ F_1(s, x_1, \ldots, x_p) & \text{if } s = 1 \\ F_2(s, x_1, \ldots, x_p) & \text{if } s = 2 \\ \vdots & \\ F_f(s, x_1, \ldots, x_p) & \text{if } s = f \\ (s, x_1, \ldots, x_p) & \text{if } s > f \end{cases}$$

where F_j is the transformation corresponding to line j in the labelled version of the program (n, k, p, P). (See also Example 4.5(4) on page 81.) Therefore we have the following:

Lemma 4.2

The follow function $F_Q: N^{p+1} \to N^{p+1}$ is a total recursive function for any SAL(**N**) program $Q = (n, k, p, P)$. ■

Lemma 4.3

The time complexity function T_Q is a partial recursive function for any SAL(**N**) program $Q = (n, k, p, P)$.

Proof

If the computation determined by Q that starts with state vector $(1, x)$ terminates, then $T_Q(x)$ gives the length of this computation. $T_Q(x)$ is undefined otherwise.

Assume that the program has its lines numbered from 1 to f. The computation terminates once a final state vector, i.e. a state vector of the form $(f + 1, y)$, is reached.

Consider the auxiliary function $H: N^{p+3} \rightarrow N^{p+3}$

$$H(d, c, s, x) = ((f + 1) \dot{-} t, c + 1, t, y), \text{ where } (t, y) = F_Q(s, x).$$

Then

$$T_Q(x) = \pi_2^{p+3} \cdot H^{\nabla_1}(1, 0, 1, x).$$

and therefore T_Q is a partial recursive function. ∎

Lemma 4.4

All SAL(**N**) computable functions are partial recursive functions.

Proof

Let $Q = (n, k, p, P)$ be any SAL(**N**) program. Then

$$M_Q(x) = \pi_{2 \ldots p+1}^{p+1}(F_Q^*(1, x, T_Q(x)))$$

where $\pi_{2 \ldots p+1}^{p+1}(x_1, \ldots, x_{p+1}) = (x_2, \ldots, x_{p+1})$.

Thus M_Q is partial recursive, and therefore f_Q is also partial recursive. ∎

Combining Lemmas 4.1 and 4.4, we arrive at the *equivalence theorem*.

Theorem 4.1

A number-theoretic partial function $f: N^n \rightarrow N^k$ is SAL(**N**) computable if and only if it is a partial recursive function. ∎

The importance of the equivalence theorem is that it establishes that two fairly different ways of characterizing the intuitive concept of effectiveness capture the same intuitive concept. This is strong evidence that both capture the intended concept of effectiveness.

4.3 Examples of recursive functions and recursive operators

In this section a number of recursive functions and recursive functional operators will be discussed. These will be used in later sections.

4.3.1 Arithmetical order relations

In Example 4.1 (page 64) we have seen that addition, subtraction, absolute difference and $sg = \lambda x[\text{if } x > 0 \text{ then } 1 \text{ else } 0]$, are recursive functions. Thus the order relations as defined below are recursive as well.

$$less \triangleq \lambda xy[\text{if } x < y \text{ then } 1 \text{ else } 0] = \lambda xy[sg(y \dot{-} x)]$$

$$equal \triangleq \lambda xy[\text{if } x = y \text{ then } 1 \text{ else } 0] = \lambda xy[\overline{sg}(|x - y|)]$$

$$greater \triangleq \lambda xy[\text{if } x > y \text{ then } 1 \text{ else } 0] = \lambda xy[sg(x \dot{-} y)]$$

4.3.2 Division and logarithm

In Example 3.6 (page 54) we have seen that the function

$$divides \triangleq \lambda xy[\text{if } (x \neq 0 \text{ and } (\exists q)[q.x = y]) \text{ then } 1 \text{ else } 0]$$

is SAL(**N**) computable. By the equivalence theorem the function *divides* is recursive.

Turning to the division operation itself:

$$x/y \triangleq \text{if } y \neq 0 \text{ and } divides(y, x) \text{ then } q \text{ such that } q.y = x \text{ else } \uparrow$$

To see that this is a partial recursive function, consider the function $f: N^5 \to N^5$ defined by:

$$f(x, y, z, q, d) \triangleq (x, y, z + y, q + 1, |x - z| + \overline{sg}(y))$$

Clearly $f \in TREC^5_5$. Furthermore,

$$/ = \dot{-}_1 \cdot \pi^5_4 \cdot f^{\nabla_5} \cdot (\pi^2_1, \pi^2_2, C^2_0, C^2_0, C^2_1).$$

To obtain the integer approximation (*floor*) of division.

$$x \% y \triangleq \text{if } y \neq 0 \text{ then } q \text{ such that } q.y \leqslant x < (q + 1).y, \text{ else } 0$$

replace the above function f by the function f_a:

$$f_a(x, y, z, q, d) \triangleq (x, y, z + y, q + 1, (x \dot{-} z).sg(y))$$

As to the function $\log(x)$, we only consider the integer approximation (*floor*) of the base two logarithm, which is denoted by $\lfloor \log_2(x) \rfloor$. The following program computes $\lfloor \log_2(x) \rfloor$.

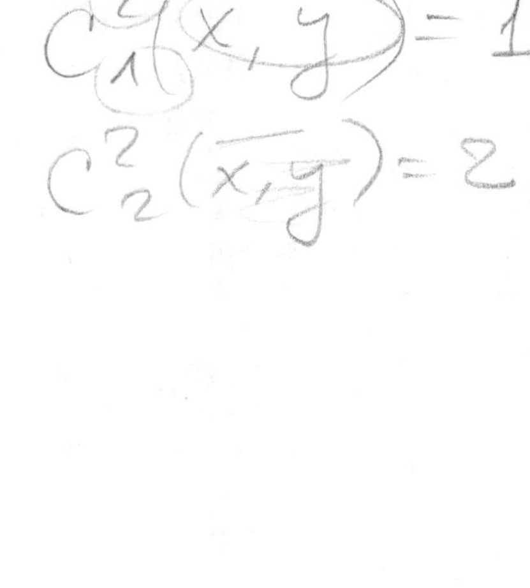

input: x
output: $k\{2^k \leqslant x < 2^{k+1}\}$
method:

$\{k = 0\}$

$g := sg(x)$;

while $x \neq 0 \wedge g \leqslant x$ **do**

$\{x \neq 0, g = 2^k, 2^k \leqslant x\}$

$g := g + g$;

$k := k + 1$

$\{x \neq 0, g = 2^k, 2^{k-1} \leqslant x\}$

od;

$\{$if $x = 0$ then $k = 0$ else $2^{k-1} \leqslant x < 2^k\}$

if $x \neq 0$ **then** $k := k \dot{-} 1$ **fi**

The above program computes the function $\lambda x[\text{if } x = 0 \text{ then } 0 \text{ else } \lfloor \log_2(x) \rfloor]$. We will write $\log(x)$ to denote this function. Since it is SAL(**N**) computable, the function is also partial recursive by the equivalence theorem.

4.3.3 Generating primes

In Example 3.6 (page 54) we have seen that

$$prime \triangleq \lambda x[\text{if } (x \text{ is a prime number}) \text{ then } 1 \text{ else } 0]$$

is SAL(**N**) computable, and thus by the equivalence theorem recursive. Let $p_0, p_1, p_2, \ldots$ denote the sequence $2, 3, 5, \ldots$ of all prime numbers, and $p \triangleq \lambda x[p_x]$ the function generating the set of all primes in ascending order. Consider the program:

input: x
output: $y\{y = p_x\}$
method:

$\{z = 0\}$

$y := 2$;

while $z < x$ **do**

$\{p_z \leqslant y$ and $z < x\}$

```
        y:= y + 1;
        if prime(y) then z:= z + 1 fi
    od
```

The above macro program clearly computes the function p, which is therefore recursive. Note that the sequence in the scope of the while statement in the above program is executed $p_x - 2$ times. Since an upper bound for p_x is known, namely $p_x < 2^{2^x}$, and this upper bound can be computed by a times program, the above program computing the function p can be converted into a times program using the technique described in Section 4.5. (A proof of the upper bound can be found in many textbooks on number theory, for example in Hardy, G. H. and Wright, E. M., *An Introduction to the Theory of Numbers*, Oxford: Clarendon Press.)

4.3.4 Bounded minimization

In Section 4.1 the minimization operator has been introduced. This operator takes a function $f \in TREC$ as an argument, and produces the function $\mu f \in REC$ as an answer. Recall from Section 4.1 the definition:

> Let $f: N^{n+1} \to N$ be any total function. The function $\mu f: N^{n+1} \to N$ is defined as follows:
>
> $$\mu f \triangleq \lambda xy[\text{if } (f(x,z) \neq y \text{ for all } z \geq 0) \text{ then } \uparrow \text{ else the smallest } z \text{ such that } f(x,z) = y]$$

Limiting the search space (the set of values of z considered) to some finite subset of N, i.e. the set $\{x \in N \mid x \leq b\}$ for some bound b, gives us the bounded minimization operator μ_b, which is defined as follows:

> Let $f: N^{n+1} \to N$ be any total function. The function $\mu_b f: N^{n+1} \to N$ is defined as follows:
>
> $$\mu_b f \triangleq \lambda xy[\text{if } f(x,z) \neq y \text{ for all } z, 0 \leq z \leq b \text{ then } 0 \text{ else the smallest } z \text{ such that } f(x,z) = y]$$

Let $f \in TREC$ and $b \in N$. The following macro program computes the function $\mu_b f$.

```
input: x, y
output: z
method:
   {z = 0, u = 0}
```

while $f(x,u) \neq y \wedge u < b$ **do** $u := u + 1$ **od**;

if $f(x,u) = y$ **then** $z := u$ **fi**

Therefore, by the equivalence theorem, if $f \in TREC$, then for any bound $b \in N$ the function $\mu_b f$ belongs to REC.

We have an infinite number of bounded minimization operators, one for each bound $b \in N$. Alternatively, we might consider a single bounded minimization operator $\mu_B: F \times N \to F$, $\mu_B(f,b) \triangleq \mu_b f$, where $F = \{f: N^n \to N^k \mid n, k \geqslant 1\}$ is the set of all total number-theoretic functions. This is a recursive operator in that given any bound b and any SAL(**N**) program P, we can effectively determine a SAL(**N**) program P_b, which computes $(\mu_b f_P)$, if at least f_P is a total function, i.e. P terminates for all its legal inputs. Because a SAL(**N**) program is not a natural number, it is not permissible to say that an operator which takes a program as an argument is recursive or SAL(**N**) computable. To remedy this, we need an effective correspondence between programs and natural numbers. The correspondence will be established in Definition 5.2 (page 101), and discussed further in Chapter 5. The fact that μ_B may be considered a recursive operator relative to a correspondence between natural numbers and SAL(**N**) programs will be shown in Example 5.5 (page 112).

The fact that such a correspondence exists makes it possible that programs manipulate programs, even themselves. On the one hand this brings about simplifications. For instance, one can show the existence of a universal program: a program which can simulate the behaviour of any other program. Nowadays this is quite familiar: any general-purpose computer is an implementation of such a universal program. On the other hand, the power of programs to manipulate programs brings about many problems which can be shown to be unsolvable by algorithmic methods. An example is the Halting Problem for SAL(**N**) programs we have already mentioned in item 3 of Example 3.1 (page 43).

Since any solution method can be described in some programming language, there are only countably many such solution methods. Because there are uncountably many problems, it is obvious that problems exist which cannot be solved by algorithmic methods. It is less obvious how to construct examples of such unsolvable problems. The power of programs to manipulate other programs brings about many of these unsolvable problems.

4.4 Predicates and sets

The objects of computability theory, as presented up to now, are number-theoretic functions. Sometimes it is more convenient or more natural to describe a problem in terms of sets or predicates. For example, it is more natural to talk of the predicate *divides*, instead of the function *divides*: $N^2 \to \{0, 1\}$.

A predicate with n variables ranging over domains D_1 to D_n determines in a natural way a total function $D_1 \times D_2 \times \cdots \times D_n \to \{\text{TRUE, FALSE}\}$. Note that we do not have 'partial predicates'; any n-tuple $(d_1, d_2, \ldots, d_n) \in D_1 \times D_2 \times \cdots \times D_n$ either does or does not satisfy the predicate; there is no 'undefined' or 'unknown' in between.

From a theoretical point of view, there is nothing against the introduction of an entity 'partial predicate'. However, the established meaning of the word 'predicate', both in logic and in mathematics, encompasses predicates to be totally defined objects. To avoid deviation from established use, we will require predicates to be total: for any n-tuple of legal parameter values the predicate evaluates either to TRUE or to FALSE.

By establishing a correspondence between functions, predicates and sets, we can make properties of functions carry over to properties of predicates and sets.

Definition 4.3

(1) Let $f: N^n \to N$ be any partial function. The predicate P_f corresponding to f is defined as follows:

$$P_f \triangleq \lambda x[sg(f(x)) = 1]$$

(2) Let P be any number-theoretic predicate having n arguments. The set $V_P \subseteq N^n$ corresponding to P is defined as follows:

$$V_P \triangleq \{x \in N^n \mid P(x) = \text{TRUE}\}$$

(3) Let V be any subset of N^n. The function $\chi_V: N^n \to N$ corresponding to V is defined as follows:

$$\chi_V \triangleq \lambda x[\text{if } x \in V \text{ then } 1 \text{ else } 0]$$

χ_V is called the *characteristic function* of V. ■

Note that if $f \in TREC$, then it is possible to effectively determine whether or not $P_f(x) = \text{TRUE}$. Similarly the value $\chi_{P_f}(x)$ can be effectively computed; this function is also recursive. (The characteristic function is total by definition.)

Note carefully, that predicates and sets belong to quite different categories and must not be confused, let alone identified. On the other hand, the above definition gives a one-to-one correspondence between them. We can use the one or the other, whichever is the more convenient in the situation at hand.

The correspondence between functions and sets (or predicates) is less tight. Every partial function defines a predicate, but different functions may determine the same predicate. Thus a function cannot be recovered from a predicate.

EXAMPLE 4.2

Consider the following function $f: N^2 \to N$

$$f \triangleq \lambda xy[\text{if } (y \text{ is even}) \text{ then } x + y \text{ else } \uparrow]$$

Then

$$f(2,7) = \uparrow, \quad \text{thus } P_f(2,7) = \text{FALSE}$$
$$f(3,8) = 11, \quad \text{thus } P_f(3,8) = \text{TRUE}$$
$$f(0,0) = 0, \quad \text{thus } P_f(0,0) = \text{FALSE}$$
$$f(0,2) = 2, \quad \text{thus } P_f(0,2) = \text{TRUE}$$

The characteristic function of the predicate P_f is

$$\chi = \lambda xy[\text{if } (y \text{ is even and } x + y \text{ greater than } 0) \text{ then } 1 \text{ else } 0].$$

Definition 4.4

Let K be any class of functions $N^n \to N$. The class $PRED(K)$ of number-theoretic predicates is defined as follows:

$$PRED(K) \triangleq \{P_f \mid f \in K\}$$

(1) A predicate P is called *recursively decidable* if and only if $P \in PRED(TREC)$.

(2) A predicate is called *recursively enumerable*, or *recursively semi-decidable*, if and only if $P \in PRED(REC)$. ■

The term 'decidable' is self-descriptive: it can effectively be decided whether or not a given tuple of parameter values satisfies the predicate. If a predicate P is recursively decidable, it is possible to list all x such that $P(x) = \text{TRUE}$, and it is also possible to list all values for which $P(x) = \text{FALSE}$, using *recursive enumeration functions.* That is, there are functions $f, g: N \to N^n$ such that:

$$\{x \mid P(x) = \text{TRUE}\} = \{f(u) \mid u \in N\}$$

and

$$\{x \mid P(x) = \text{FALSE}\} = \{g(u) \mid u \in N\}$$

If P is recursively enumerable, i.e. $P \in PRED(REC)$, the set $\{x \mid P(x) = \text{TRUE}\}$ can be enumerated by a total recursive function. The set $\{x \mid P(x) = \text{FALSE}\}$, however, can in general not be enumerated by a total recursive function. In Chapter 6 the relation between recursive decidability and recursive enumerability will be discussed further.

For the time being, let us look at some examples of recursively decidable predicates.

EXAMPLE 4.3

(1) Consider the following functions.

$$greater \triangleq \lambda xy[sg(x \dot{-} y)]$$
$$less \triangleq \lambda xy[sg(y \dot{-} x)]$$
$$equal \triangleq \lambda xy[sg((x \dot{-} y) + (y \dot{-} x))]$$

These are total recursive functions, as we have seen in Section 4.3. Therefore the predicates $P_{greater}$, P_{less} and P_{equal}, commonly denoted by '>', '<' and '=', are recursively decidable.

(2) The predicate $ODD \triangleq \lambda x[(\exists k \in N)[x = 2k + 1]]$ is a recursively decidable predicate. Its characteristic function χ_{ODD} is computed by the following program:

```
input: x
output: y
method:
  while x ≠ 0 do
    x:= x ∸ 1;
    if P_equal(x, 0) then y:= y + 1 fi;
    x:= x ∸ 1
  od
```

(3) Consider /, the partial operation of division, and its integer approximation %. We have seen in Section 4.3 that / is a partial recursive function, and % is a total recursive function. (By definition we have $x \% 0 = 0$.) Then

$$P_{/}(x, y) = \text{TRUE iff } y \neq 0 \text{ and } y \text{ is a divisor of } x$$

Since the function $/$ is in REC, it follows that $P_/$ is a recursively enumerable predicate. But because $P_/$ is logically equivalent to $P_{\%}$, and $\% \in TREC$, we have that $P_/ \in PRED(TREC)$ and therefore $P_/$ is a recursively decidable predicate.

Operations on predicates will now be considered.

4.4.1 Substitution

Let Q be an n-ary predicate, and assume given n m-ary functions $g_i: N^m \to N$. Let P be the predicate obtained from Q by substituting $g_1(z_1, z_2, \ldots, z_m)$ for x_1, $g_2(z_1, z_2, \ldots, z_m)$ for x_2, ..., and $g_n(z_1, z_2, \ldots, z_m)$ for x_n. We say in this case that P is obtained from Q by *substitution.*

EXAMPLE 4.4

Let the predicate $Q = \lambda xy[x = y]$ and the functions $g_1 = \lambda xyz[x + y]$ and $g_2 = \lambda xyz[x.y.z]$ be given. The result of substituting $g_1(x, y, z)$ for x and $g_2(x, y, z)$ for y is the predicate $P = \lambda xyz[x + y = x.y.z]$.

Theorem 4.2

Let K be any class of number-theoretic functions. If K is closed with respect to function composition then $PRED(K)$ is closed with respect to substitution.

Proof

Let $Q = P_f \in PRED(K)$ have n arguments x_1 up to x_n. The predicate obtained by substituting $g_i(z_1, \ldots, z_m)$ for x_i, $g_i \in K$, $1 \leqslant i \leqslant n$, is the predicate P_h, where $h = f \cdot (g_1, \ldots, g_n)$. Since K is closed with respect to composition, it follows that $P_h \in PRED(K)$, and $PRED(K)$ is therefore closed with respect to substitution. ■

Corollary 4.1

Both the recursively decidable and the recursively enumerable predicates are closed with respect to substitution. ■

4.4.2 Propositional connectives

These are the connectives 'and' ($\wedge$), 'or' ($\vee$), and 'not' ($\neg$) used to construct new propositions from given ones. Other logical connectives, such as the

Table 4.2 Truth functional definition of propositional connectives.

x	y	$\neg x$	$x \wedge y$	$x \vee y$
FALSE	FALSE	TRUE	FALSE	FALSE
FALSE	TRUE	TRUE	FALSE	TRUE
TRUE	FALSE	FALSE	FALSE	TRUE
TRUE	TRUE	FALSE	TRUE	TRUE

'if–then' construct, can be defined in terms of these three. We use the propositional connectives in their usual meaning as given in Table 4.2.

The closure properties of the class of recursively decidable predicates with respect to the propositional connectives are different from those of the class of recursively enumerable predicates (see Table 4.3).

Only the closure properties of recursively decidable predicates will be proved here with those of the recursively enumerable predicates being approached in Chapter 6, as the latter proofs require some tools not yet discussed.

Let $R = P_f$ and $S = P_g$ be recursively decidable predicates, i.e. $f, g \in TREC$.

(1) Let $h = \lambda x[1 \dot{-} f(x)]$. Then $h \in TREC$ and $\neg R = P_h$ and therefore $\neg R$ is recursively decidable.

(2) Let $h = \lambda x[f(x) + g(x)]$. Then $h \in TREC$ and $R \vee S = P_h$ and therefore $R \vee S$ is recursively decidable.

(3) Let $h = \lambda x[f(x).g(x)]$. Then $h \in TREC$ and $R \wedge S = P_h$ and therefore $R \wedge S$ is recursively decidable.

Table 4.3 Closure properties of recursively decidable and recursively enumerable predicates with respect to propositional connectives.

Connective	Recursively decidable predicates	Recursively enumerable predicates
$\neg$	closed	not closed
$\wedge$	closed	closed
$\vee$	closed	closed

4.4.3 Quantification

Another way to construct new predicates from given ones is the use of quantifiers. We consider four quantifiers:

(1) $(\forall x)P(x)$, 'for all x $P(x)$', the *unbounded universal quantifier*.

(2) $(\forall x \leqslant z)P(x)$, 'for all x less than or equal to z $P(x)$', the *bounded universal quantifier*.

(3) $(\exists x)P(x)$, 'there exists an x such that $P(x)$', the *unbounded existential quantifier*.

(4) $(\exists x \leqslant z)P(x)$, 'there is an x less than or equal to z such that $P(x)$', the *bounded existential quantifier*.

The closure properties of the recursively decidable and the recursively enumerable predicates with respect to these quantifiers are given in Table 4.4.

Proving the closure properties given in Table 4.4 is largely beyond our current capabilities and will be deferred to Chapter 6. But it is easy to see that the recursively decidable predicates are closed with respect to bounded quantification.

Let $Q = P_f$ be a recursively decidable predicate having $n+1$ arguments. Consider the following program:

input: $x \in N^n, z \in N$
output: $y \in N$
method:

while $t \leqslant z$ **do**

$y := y + f(x, t)$;

$t := t + 1$

od

Table 4.4 Closure properties of recursively decidable and recursively enumerable predicates with respect to bounded and unbounded quantification.

Quantifier	Recursively decidable predicates	Recursively enumerable predicates
$\forall$	not closed	not closed
bounded $\forall$	closed	closed
$\exists$	not closed	closed
bounded $\exists$	closed	closed

Let h be the function computed by this program. Because $f \in TREC$, we have that $h \in TREC$ and $P_h = (\exists t \leqslant z)Q(x,t)$. Thus the class of recursively decidable predicates is closed with respect to bounded existential quantification. A similar program can be used to show that the recursively decidable predicates are also closed with respect to bounded universal quantification.

We conclude this section with some examples showing the use of predicates.

EXAMPLE 4.5

(1) The class of recursively decidable predicates is closed with respect to $\rightarrow$, logical implication. The truth-functional definition of the logical implication is shown in Table 4.5. As can be seen from this definition $x \rightarrow y$ is logically equivalent to $\neg x \vee y$. Because the class of recursively decidable predicates is closed with respect to $\neg$ and $\vee$, it is closed with respect to logical implication as well.

Table 4.5 Truth table for $x \rightarrow y$.

x	y	$x \rightarrow y$
FALSE	FALSE	TRUE
FALSE	TRUE	TRUE
TRUE	FALSE	FALSE
TRUE	TRUE	TRUE

(2) The predicate $PRIME \triangleq \lambda x[x \text{ is a prime number}]$ is a decidable predicate, because

$$PRIME = \lambda x[x > 1 \wedge (\forall z \leqslant x)[divides(z,x) \rightarrow (z = 1 \vee z = x)]],$$

the predicates $<$, $=$, $\rightarrow$, and *divides* are recursively decidable, and the class of recursively decidable predicates is closed with respect to $\wedge$, $\vee$, $\rightarrow$, and bounded universal quantification.

(3) The bounded minimization operator applied to predicates. In Sections 4.1 and 4.3 we have seen the bounded and unbounded minimization operators, which take functions as arguments and produce functions as results. We now introduce a minimization operator that operates on predicates and returns natural numbers. The intention is to use this operator if we want to define a function value to be the least z having some particular property.

Let $Q = P_f$ be a recursively decidable predicate having $n+1$ arguments, and let the function h be defined as follows:

$$h \triangleq \lambda xy[\text{if } (Q(x,z) = \text{FALSE for all } z \leqslant y) \text{ then } z+1 \text{ else the least } z \text{ such that } Q(x,z) = \text{TRUE}]$$

Instead of the above, we will write: $h \triangleq \lambda xy[(\mu z \leqslant y)Q(x,z)]$. The following program computes the function h.

input: $x \in N^n$, $y \in N$
output: $z \in N$
method:

while $z \leqslant y \wedge f(x,z) = 0$ **do** $z := z+1$ **od**

From the fact that $f \in TREC$, it follows that $h \triangleq \lambda xy[(\mu z \leqslant y) Q(x,z)]$ also is total recursive. When applied to recursively enumerable predicates, h is not necessarily recursive. The same problems we discussed in relation to the μ-operator in Section 4.1 occur here.

(4) Definition of a function by cases. Let $f_i: N^n \to N^k$ be a sequence of total functions and P_i a sequence of recursively decidable predicates, $1 \leqslant i \leqslant m$. Assume further that for every $x \in N^n$ there is precisely one i such that $P_i(x) = \text{TRUE}$.

The following is a definition by cases of the function $f: N^n \to N^k$ from the functions f_i and predicates P_i, $1 \leqslant i \leqslant m$.

$$f(x) \triangleq \begin{cases} f_1(x) & \text{if } P_1(x) = \text{TRUE} \\ f_2(x) & \text{if } P_2(x) = \text{TRUE} \\ & \vdots \\ f_m(x) & \text{if } P_m(x) = \text{TRUE} \end{cases}$$

Let χ_i be the characteristic function of the predictate P_i, $1 \leqslant i \leqslant m$. Then we can write

$$f(x) = f_1(x).\chi_1(x) + f_2(x).\chi_2(x) + \cdots + f_m(x).\chi_m(x)$$

Therefore, f is a total recursive function.

Without using the name, we have made use of definition by cases in Section 4.2 to define the follow function.

(5) Every natural number x can be uniquely factored into a product of primes:

$$x = p_0^{x_0}.p_1^{x_1} \cdots$$

where $p_0 = 2$, $p_1 = 3$, $p_2 = 5, \ldots$ is the sequence of all primes in ascending order. As we have seen in Section 4.3, there is a total recursive function p generating the primes in this order.

Below we define the function $q: N^2 \rightarrow N$, such that $q(x, k)$ gives the exponent of p_k in the decomposition into primes of x. The function q is defined as follows.

$$q(x,k) \triangleq \begin{cases} (\mu z \leqslant x)[\neg\, divides(p_k^{z+1}, x)] & \text{if } x \neq 0, \\ 0 & \text{otherwise} \end{cases}$$

The function q is defined by cases using total recursive functions and recursively decidable predicates (two mutually exclusive predicates), and therefore q is a total recursive function.

4.5 Subrecursive hierarchies

Up to now, we have divided the set of all number-theoretic functions into two classes:

- the SAL(**N**) computable functions
- those which are not SAL(**N**) computable.

We will have to study each of these two subclasses.

Recursion theory largely concerns itself with this borderline, and with structuring the class of noncomputable functions. An important question is, for instance, whether all noncomputable functions are 'equally noncomputable'. That is, if one simply assumes that the Halting Problem of SAL(**N**) programs is solvable, does it follow that all previously noncomputable functions have become computable? If not, what functions do remain noncomputable?

The study of the class of SAL(**N**) computable functions is the realm of complexity theory. The class of all computable functions is divided into complexity classes, consisting of functions, equally difficult to compute. Complexity theory studies the structure imposed on the class of all computable functions by the amount of time or space required to compute these functions. This leads to theorems such as: 'For every rational number k, and every real number $\varepsilon > 0$, there is a function which can be computed by a SAL(**N**) program P with time complexity $T_P(n) \leqslant n^{k+\varepsilon}$, $n \geqslant 0$, but there is no SAL(**N**) program Q with time complexity T_Q such that $T_Q(n) \leqslant n^k$ for all $n \geqslant 0$ computing this function'.

Within recursion theory there also are hierarchies of computable functions. These depend on program structure, at least in our 'programming over a data type' model of computation.

- One class, strictly included in *TREC*, is obtained by considering loop programs, that is, programs without while statements, which may however have arbitrarily nested times statements.
- An infinite subhierarchy thereof is obtained by limiting the nesting depths of the times statements.

We will summarize the state of this subrecursive world. Proofs will only be sketched. Full proofs can be assembled following the sketches, from facts and methods which have already been presented in preceding sections.

Definition 4.5

A *loop program* is a SAL(**N**) macro program, which has no while statements, but which may have an arbitrary number of times statements. A *0-loop program* is a loop program without times statements; 0-loop programs are sequences of assignment statements. A *k-loop program* is a loop program in which the nesting depth of times statements is at most k.

More formally,

$$\langle \textit{0-loop sequence} \rangle ::= \langle \textit{assignment} \rangle \mid \langle \textit{0-loop sequence} \rangle ; \langle \textit{0-loop sequence} \rangle$$

$$\begin{aligned} \langle i+1\textit{-loop sequence} \rangle ::= {} & \langle \textit{i-loop sequence} \rangle \mid \\ & \mathbf{do}\ \langle \textit{identifier} \rangle\ \mathbf{times}\ \langle \textit{i-loop sequence} \rangle\ \mathbf{od} \mid \\ & \langle (i+1)\textit{-loop sequence} \rangle ; \langle (i+1)\textit{-loop sequence} \rangle \end{aligned}$$

$$\langle \textit{k-loop program} \rangle ::= \{ (n, m, p, \langle \textit{k-loop sequence} \rangle) \mid n \geqslant 0, m \geqslant 1, n, k \leqslant p \}$$

Finally,

- L_i is the set of all number-theoretic functions which can be computed by i-loop programs, and
- $L \triangleq \bigcup_{i \geqslant 0} L_i$, is the set of all number-theoretic functions which can be computed by loop programs. ■

The following hierarchy can be shown to exist:

$$L_2 \subset L_3 \subset \cdots \subset L_i \subset L_{i+1} \subset \cdots \subset L \subset TREC$$

In the above all the inclusions are strict.

The above classes can also be defined as the closures $C(B, F)$ of well-chosen sets B of base functions, and F of functional operators, as in Definition 4.1.

Definition 4.6

The set $PRIM$ is the closure $C(B, F)$, where:

(1) The set B of base functions consists of

(a) the successor function $+_1: N \to N$,
(b) the predecessor function $\dot{-}_1: N \to N$,
(c) the identity functions $I^n: N^n \to N^n$, for all $n \geqslant 1$,
(d) the projection functions $\pi_i^n: N^n \to N$, for all $n \geqslant 1$ and i, $1 \leqslant i \leqslant n$,
(e) the zero function $z: N \to N$, $z = \lambda x[0]$.

(2) The set F of operators consists of

(a) composition of functions,
(b) Cartesian product of functions,
(c) exponentiation of functions.

The functions in $PRIM$ are called *primitive recursive functions.* ■

The above definition is almost the same as Definition 4.2. The only difference is that the functional operator 'iteration' is permitted in the construction of partial recursive functions, (see Definition 4.2), but it may not be used to construct primitive recursive functions. Thus, all primitive recursive functions are total.

Theorem 4.3

A function is primitive recursive if and only if it can be computed by a loop program, in other words, $L = PRIM$. ■

The proof of the above theorem closely follows the proof of the equivalence theorem $REC = F(\mathrm{SAL}(\mathbf{N}))$ (Section 4.2).

To characterize the class L_i of functions computable by i-loop programs, a bounded version of function exponentiation is used. Below we describe a characterization of the class L_2. The method can be generalized so as to obtain characterizations of the classes L_i. Only a sketch of how to do that will be given.

Definition 4.7

(1) $exp \triangleq \lambda x[2^x]$

(2) $MAJ \triangleq \{\lambda x[exp^*(x, k)] \mid k \geqslant 0\}$

(3) The function $h: N^n \to N^k$ is obtained by *MAJ-bounded exponentiation* from the function f, if

(a) $h = f^*$, and

(b) the function h is *bounded by* a function from MAJ, that is, there is a function $b \in MAJ$ such that $|h(x)| \leqslant b(|x|)$ for all $x \in N^n$, where $|x|$ denotes the largest component of the vector x. ■

We now define a class E of so-called *elementary* functions. The definition is almost the same as the definition of $PRIM$, but we use MAJ-bounded exponentiation instead of arbitrary exponentiation.

Definition 4.8

The set E is the closure $C(B, F)$ where:

(1) The set B of base functions consists of

- (a) the successor function $+_1: N \to N$,
- (b) the predecessor function $\dot{-}_1: N \to N$,
- (c) the identity functions $I^n: N^n \to N^n$, for all $n \geqslant 1$,
- (d) the projection functions $\pi_i^n: N^n \to N$, for all $n \geqslant 1$ and i, $1 \leqslant i \leqslant n$,
- (e) the zero function $z: N \to N$, $z = \lambda x[0]$,
- (f) the function *exp*.

(2) The set F of operators consists of

- (a) composition of functions,
- (b) Cartesian product of functions.
- (c) MAJ-bounded exponentiation of functions.

The functions in the class E are called *elementary functions.* ■

It can be shown that $E = L_2$, a function is elementary if and only if it can be computed by a 2-loop program. In proving this, the problem is that every application of MAJ-bounded exponentiation gives rise to a times statement (as in the proof of $PRIM = L$). It is not immediately obvious how to reduce the nesting depth.

Let P be any arbitrary SAL(**N**) program. We construct another SAL(**N**) program Q which is equivalent to P, i.e. it computes the same function, $f_P = f_Q$. The program Q consists of two parts:

- The first part computes T_P, the time complexity function of P. The structure of this part, as regards while and times statements, is unrestricted.
- The second part performs the actual computation, $y = f_P(x)$. This part is a 2-loop program.

In the example below it is shown how to obtain the program Q from a given program P.

EXAMPLE 4.6

A sample program P and its equivalent program Q are given in Figure 4.1.

```
P: S₁;
  while u ≠ 0 do
    S₂;
    while v ≠ 0 do
      S₃;
      while w ≠ 0 do
        S₄
      od;
      S₅
    od
  od;
  while z ≠ 0 do
    S₆
  od;
  S₇
```

This program P is transformed into:

```
Q: t:= T_P(x); l₀:= 1;
  {l₁ up to l₄ are zero}
  do t times
    if l₀ = 1 then S₁ fi;
    if l₀ = 1 ∧ u ≠ 0 then l₁:= l₀; l₀:= 0 fi;
      if l₁ = 1 then S₂ fi;
      if l₁ = 1 ∧ v ≠ 0 then l₂:= l₁; l₁:= 0 fi;
        if l₂ = 1 then S₃ fi;
        if l₂ = 1 ∧ w ≠ 0 then l₃:= l₂; l₂:= 0 fi;
          if l₃ = 1 then S₄ fi;
        if l₃ = 1 ∧ w = 0 then l₂:= l₃; l₃:= 0 fi;
        if l₂ = 1 then S₅ fi;
      if l₂ = 1 ∧ v = 0 then l₁:= l₂; l₂:= 0 fi;
    if l₁ = 1 ∧ u = 0 then l₀:= l₁; l₁:= 0 fi;
    if l₀ = 1 ∧ z ≠ 0 then l₄:= l₀; l₀:= 0 fi;
      if l₄ = 1 then S₆ fi;
    if l₄ = 1 ∧ z = 0 then l₀:= l₄; l₄:= 0 fi;
    if l₀ = 1 then S₇ fi;
    if l₀ = 1 then l₀:= 0 fi
  od
```

Figure 4.1 Reducing the nesting depth of program P results in program Q.

Table 4.6 Assignments of program P as executed by program Q.

Loop indicator					Statement	Executed
l_0	l_1	l_2	l_3	l_4		
1	0	0	0	0	S_1	yes
0	1	0	0	0	S_2	yes
0	0	1	0	0	S_3	yes
0	0	0	1	0	S_4	yes
0	0	0	1	0	S_5	no
0	0	0	1	0	S_6	no
0	0	0	1	0	S_7	no
0	0	0	1	0	S_1	no
0	0	0	1	0	S_2	no
0	0	0	1	0	S_3	no
0	0	0	1	0	S_4	yes
0	0	0	1	0	S_5	no
.	.	.	.	.	.	.
.	.	.	.	.	.	.
.	.	.	.	.	.	.
1	0	0	0	0	S_7	yes
0	0	0	0	0	no more	—

The nesting depth of program P is 3. The statements S_i, $1 \leqslant i \leqslant 7$, are simple assignment statements. There are altogether 4 while loops.

Program Q first computes $T_P(x)$. No statement of P can be executed more than $T_P(x)$ times. The second part of Q consists of a single times statement. Within the scope of this times statement occur all the assignment statements of P, guarded by a loop indicator l_i. An assignment is executed only if its loop indicator is on. The while-heads and -tails of P are transformed into (guarded) assignments, which pass an activity token, the value 1, from one loop indicator to the next.

Execution proceeds as indicated in Table 4.6.

The technique sketched in the example above can be used to transform any arbitrary program. Because the if statement can be implemented as a 1-loop program, we have that part 2 of the transformed program, i.e. the large times loop, has nesting depth 2. Thus we have:

Theorem 4.4

For all $i \geqslant 2$ and all $f: N^n \rightarrow N^k$, f can be computed by an i-loop program if and only if f can be computed by a SAL(**N**) program P, such that the time complexity function of P is bounded by some function in L_i. In other words, for some $h \in L_i$, $T_P(x) \leqslant h(x)$, for all $x \in N^n$. ■

This theorem can be proved using the technique explained in the example above.

We return now to the case $i = 2$. By induction on the structure of E it can be shown that every elementary function is bounded by some function in MAJ. By induction on the structure of 2-loop programs, it can be shown that the time complexity function of a 2-loop program is bounded by some function in MAJ.

Theorem 4.5

A function is elementary if and only if it can be computed by a 2-loop program. In other words, $E = L_2$.

Proof

Given an elementary function f, a program computing it is constructed in the same way as was done for a recursive function in Section 4.2.

Simultaneously it is shown that the time complexity function of this program is bounded by some function in MAJ. Therefore the program can be transformed into a 2-loop program.

If a function is computed by a loop program, it is primitive recursive by Theorem 4.3. If the function belongs to L_2, the time complexity and thus the value of the output are bounded by some function in MAJ. Thus any exponentiations are bounded. ■

Essential in the definition of the class of elementary functions are the function exp and the class MAJ. These can be generalized.

Definition 4.9

(1) $exp_3 \triangleq \lambda x[exp_2^*(x, x)]$, where exp_2 is the function exp defined in Definition 4.7.

(2) $MAJ_3 \triangleq \{\lambda x[exp_3^*(x, k)] \mid k \geqslant 0\}$

(3) The function $h: N^n \rightarrow N^k$ is obtained by MAJ_3-bounded exponentiation from the function f, if

(a) $h = f^*$, and

(b) the function h is bounded by some function from MAJ_3, that is, there is a function $b \in MAJ_3$ such that $|h(x)| \leqslant b(|x|)$ for all $x \in N^n$, where $|x|$ denotes the largest component of the vector x. ■

The function exp_3 cannot be elementary, because it is not bounded by any function in MAJ. It is easy, however, to write a 3-loop program which computes exp_3. This provides us with an example of a function in L_3 but not in L_2. Thus the inclusion $L_2 \subset L_3$ is indeed strict.

By using exp_3, MAJ_3 and MAJ_3-bounded exponentiation, we can define a class E_3 and show that $E_3 = L_3$, and that a function is in E_3 if and only if it can be computed by a SAL(**N**) program whose time complexity function is bounded by some function in E_3.

In general, we can use the function exp_i to define the function exp_{i+1}, and using exp_{i+1} define the class MAJ_{i+1} and MAJ_{i+1}-bounded exponentiation, and the class E_{i+1}. It can be shown that $exp_{i+1} \in L_{i+1}$, and $exp_{i+1} \notin L_i$. Thus, for all $i \geqslant 2$, we have that L_i is strictly included in L_{i+1}.

Finally, to see that $PRIM$ is a proper subset of $TREC$, consider the function

$$LARGE \triangleq \lambda x[exp_{x+2}(x+2)].$$

Clearly, $LARGE \notin L_i$ for all $i \geqslant 2$, because $LARGE$ strictly majorizes all functions in MAJ_i, for all $i \geqslant 2$. Since by Church's thesis $LARGE \in TREC$, it follows that $PRIM$ is strictly included in $TREC$. The hierarchy,

$$E = L_2 \subset L_3 \subset \cdots \subset L_i \subset L_{i+1} \subset \cdots \subset L = PRIM$$

is not very discriminative. The class E contains functions computable in linear time, but also functions computable only in exponential time. The order of magnitude of the difference between the functions exp_i and exp_{i+1} is enormous. In Chapter 11 on complexity theory, far more discriminative hierarchies will be established.

In practice, solution methods which require exponential time are not considered feasible. Thus problems which can be solved in reality are a proper subclass of the elementary problems. Complexity issues about these more practical problems are discussed in Chapter 12.

4.6 Exercises

4.1 Prove that the following functions are primitive recursive.

(a) λxy[the maximum of x and y]

(b) λxy[the minimum of x and y]

(c) λxyz[if ($z = 0$ or $z = 1$) then 0 else $x + y(\bmod z)$]

(d) λx[the number of 1s in the binary representation of x]

(e) The characteristic function χ_A of A, where the set A is recursively defined as follows:

(i) $0 \in A$,

(ii) if $n \in A$ then $2.3^n \in A$,

(iii) A is the smallest set satisfying the above conditions.

4.2 Prove that the function

$$h \triangleq \lambda xn[\text{the } (n+1)\text{th significant digit in the decimal representation of } x^2]$$

is elementary.

4.3 Consider the following three-place predicate Q.

$$Q(x, y, z) \triangleq (\exists u)[x \leqslant u \leqslant y \text{ and } u^2 = z]$$

(a) Prove that Q is a primitive recursive predicate.

(b) Prove by using the appropriate substitutions that the predicates R_1, R_2, R_3 and R_4 are primitive recursive.

$$R_1(x, y) \triangleq [x = y^2]$$

$$R_2(x, y) \triangleq [x = 7]$$

$$R_3(x, y) \triangleq [x \text{ is a perfect square}]$$

$$R_4(x, y) \triangleq [x \leqslant y]$$

4.4 Prove that each of the following predicates is (primitive) recursive.

$$P(x, y, z) \triangleq [z \neq 0,\ z \neq 1 \text{ and } x = y (\operatorname{mod} z)].$$

$$Q(x, y) \triangleq [x, y \neq 0 \text{ and } x \text{ and } y \text{ are relatively prime}].$$

$$R(x, y, z) \triangleq [z \text{ is the least common multiple of } x \text{ and } y].$$

$$T(x, y, z) \triangleq [z \text{ is the greatest common divisor of } x \text{ and } y].$$

4.5 Let $g: N \to N$ be a primitive recursive function. Consider the function f defined as follows.

$$f(x) \triangleq \min\{y \mid 0 \leqslant y \leqslant x \text{ and } g(y) = \min\{g(i) \mid 0 \leqslant i \leqslant x\}\}$$

Is f a partial recursive function? Is f a total function? Is f a primitive recursive function?

4.6 There exists a partial recursive function $f: N \to \{0, 1\}$ such that the time complexity of SAL(**N**) programs computing f cannot be bounded by any total recursive function. In other words, if P is a SAL(**N**) program and $f_P = f$, then there does not exist a total recursive function g such that $T_P(x) \leqslant g(x)$ for all $x \in N$. Prove this.

4.7 Prove that for every elementary function f there exists a SAL(**N**) program (n, k, p, P) and a natural number k such that $T_P(x) \leqslant g^*(|x|, k)$ for all $x \geqslant 0$, where $g = \lambda x[2^x]$.

4.8 Let P be an n-place primitive recursive predicate and f_P its characteristic function. Express each of the following predicates and functions in terms of P and f_P.

(a) The predicate $Q \triangleq \lambda x_1 \cdots x_n[(\forall y \leqslant x_n)P(x_1, \ldots, x_{n-1}, y)]$, and its characteristic function.

(b) The predicate $R \triangleq \lambda x_1 \cdots x_n[(\exists y \leqslant x_n)P(x_1, \ldots, x_{n-1}, y)]$, and its characteristic function.

4.7 References

The primitive recursive functions as defined in this section, were called recursive in 1931 by Gödel (1965) in the proof of his famous incompleteness theorem for arithmetic. This theorem states that for any, in a technical sense 'reasonable', set of axioms of arithmetic, there exists a true proposition about arithmetic which cannot be proved in this axiom system.

The primitive recursive functions have been studied by Dedekind (1930), Hilbert (1934), Ackermann (1928), and many others. Ackermann exhibited an intuitively computable function which he proved to be recursive, but not primitive recursive. This function is now known as the Ackermann function.

In 1934 Gödel used a system of equations to define a class of functions called general recursive functions; this class contains the set of all primitive recursive functions as a subclass. Kleene (1936) proved that the class of recursive functions could be defined as the smallest set which contains the successor function, the projection functions $\pi_i^n (1 \leqslant i \leqslant n)$ and the zero function $z = \lambda x[0]$, and which is closed with respect to the primitive recursion scheme, function composition and the μ-operator. Nowadays the recursive functions are commonly defined in this way, see for example Hermes (1961) and Hennie (1976).

Asser (1960) and Eilenberg and Elgot (1970) defined the recursive functions over string domains. Eilenberg and Elgot also replaced the primitive recursion scheme and the μ-operator by iteration and exponentiation; they also introduced recursive vector functions $f: N^n \rightarrow N^k$. In this book we follow this line, since iteration and exponentiation are the 'natural' functions assigned by the formal semantics of SAL(**N**) to the times and while statements.

The equivalence between the primitive recursive functions and the set of functions computable by loop programs was proved by Meyer and Ritchie (1967b).

Early work on subrecursive hierarchies can be found in Peter (1936). The classes L_i were first defined by Kalmar (1943); the equivalence of the various definitions has been proved by Grzegorczyk (1953).

The loop hierarchy and its relation to Grzegorczyk's hierarchy were studied by Meyer and Ritchie (1967b).

Recent interest in these subrecursive hierarchies originates from the realization that these classes may be considered as complexity classes. This point of view was studied by Ritchie (1963) and Cobham (1964), and gave rise to Ritchie's hierarchy (based on memory requirements of the computation) and Cleave's hierarchy (based on the number of jumps executed during the computation on a so-called universal register machine). Results about the relations between the various hierarchies can be found in Meyer and Ritchie (1967a), Constable and Borodin (1972) and Tsichritzis (1970).

Chapter 5
Basic Tools of Recursion Theory

Up to now, statements, programs, and data elements have been totally distinct. These are indeed entities of quite different categories.

Nevertheless, it is possible to define a one-to-one correspondence between programs and natural numbers. This can be done in an effective way. That is, given a number, we can compute the program coded by this number, and given a program we can determine the number corresponding to this program. We can even simulate the computations of this program. We will describe a *universal program*, capable of simulating the computations determined by other programs.

Because terminating computations are finite sequences of vectors of natural numbers and SAL(**N**) programs can only manipulate natural numbers, we must represent these sequences as natural numbers.

We will start with the problem of encoding fixed-length tuples of natural numbers as natural numbers.

5.1 Cantor numbering

A Cantor numbering of k-tuples of natural numbers is a bijection between N^k and N. Such a Cantor numbering can be used to systematically enumerate all possible k-tuples of natural numbers. In the above, k is a particular number.

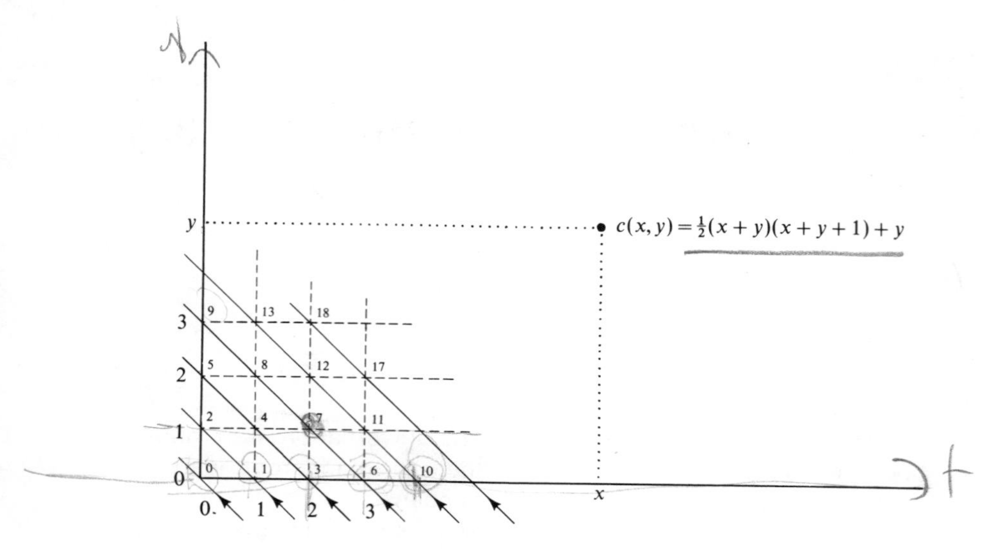

Figure 5.1 Cantor numbering of N^2.

Thus a Cantor numbering cannot be used to enumerate all possible finite sequences of natural numbers. That is the purpose of Gödel numberings, see Section 5.2.

Consider the problem of representing ordered pairs of natural numbers (x, y) by single natural numbers, i.e. finding a simple and effectively computable bijection $c: N^2 \to N$. This is equivalent to finding a route on an $N \times N$ grid in such a way that every point (x, y) of the grid is visited exactly once. There are many ways to do this; a well-known method is shown in Figure 5.1.

The pair (x, y) is represented by the grid point with Cartesian coordinates x and y. The points along the diagonal lines are sequentially numbered 'from the lower right to the upper left'. The origin $(0, 0)$ has number 0, $(1, 0)$ has number 1, $(0, 1)$ has number 2, and so forth.

The code number assigned in this way to the ordered pair (x, y) is denoted by $c(x, y)$. Thus

$$c(0,0) = 0,$$

$$c(1,0) = 1,\ c(0,1) = 2,$$

$$c(2,0) = 3,\ c(1,1) = 4,\ c(0,2) = 5,$$

and so forth. The idea of this numbering scheme is due to Georg Cantor, who used it to prove that the set of rational numbers is countable.

Clearly, $c(x, 0) = 0 + 1 + 2 + 3 + 4 + \cdots + x$. To find the number of (x, y), move down along the diagonal to $(x + y, 0)$, compute the code number of $(x+y,0)$ and move up again to (x,y). Thus

$$c(x,y) = c(x+y,0) + y = \tfrac{1}{2}(x+y).(x+y+1) + y$$

The inverse $c^{-1}: N \to N^2$ can be found as follows. Let the function d be defined by

$$d \triangleq \lambda z[(\mu t \leqslant z)[c(t,0) > z] \dot{-} 1].$$

Thus d finds the diagonal on which the point (x,y) with code number z is situated. Now we move up the diagonal to find it. The y-coordinate is computed using the function u defined below.

$$u \triangleq \lambda z[z \dot{-} c(d(z),0)]$$

Thus the inverse c^{-1} can be defined as follows.

$$c^{-1} \triangleq \lambda z[(d(z) \dot{-} u(z), u(z))]$$

The pair (c, c^{-1}) is called a Cantor numbering scheme. The scheme described above will be assumed throughout, but a more suggestive notation will be used:

σ_1^2: $N^2 \to N$ for coding,

σ_2^1: $N \to N^2$ for the inverse,

$\sigma_{2,i}^1$: $N \to N$ for the ith component of the inverse.

Obviously all these functions are (easily) computable. Furthermore they satisfy the following equalities:

$$\sigma_1^2(\sigma_{2,1}^1(z), \sigma_{2,2}^1(z)) = z \qquad \text{for all } z \geqslant 0$$

and

$$\sigma_{2,1}^1(\sigma_1^2(x,y)) = x \quad \text{and} \quad \sigma_{2,2}^1(\sigma_1^2(x,y)) = y \qquad \text{for all } x, y \geqslant 0.$$

Using the above functions, Cantor schemes can be devised for ordered triples, quadruples, quintuples etc.

First consider the ordered triples of natural numbers (x,y,z). Every triple (x,y,z) can be represented by an ordered pair $(x,(y,z))$. A Cantor scheme (σ_1^3, σ_3^1) between N and N^3 can thus be defined as follows:

$$\sigma_1^3(x,y,z) \triangleq \sigma_1^2(x, \sigma_1^2(y,z)) \qquad \text{for all } x,y,z \in N.$$

$$\sigma_3^1(w) \triangleq (\sigma_{2,1}^1(w), \sigma_{2,1}^1(\sigma_{2,2}^1(w)), \sigma_{2,2}^1(\sigma_{2,2}^1(w))) \qquad \text{for all } w \in N.$$

$$\sigma_{3,i}^1(w) \triangleq \pi_1^3(\sigma_3^1(w)) \qquad \text{for } i = 1,2,3 \text{ and for all } w \in N.$$

Obviously this technique can be used repeatedly. In that way one can construct a Cantor coding scheme between N and N^{k+1}, given a Cantor scheme between N and N^k.

- $\sigma_1^{k+1}(x_1,\ldots,x_{k+1}) = \sigma_1^2(x_1, \sigma_1^k(x_2,\ldots,x_{k+1}))$, for all $x_1,\ldots,x_{k+1} \in N$.
- $\sigma_{k+1}^1(w) = (x_1,\ldots,x_{k+1})$, for all $w \in N$, where

  (a) $x_1 = \sigma_{k+1,1}^1(w) = \sigma_{2,1}^1(w)$,
  (b) $x_i = \sigma_{k+1,i}^1(w) = \sigma_{k,i-1}^1(\sigma_{2,2}^1(w))$ for $2 \leqslant i \leqslant k+1$.

By induction on k, it follows that the Cantor coding scheme (σ_1^k, σ_k^1) between N and N^k is recursive. We will assume these Cantor schemes between N and N^k throughout.

In the following example a different Cantor scheme between N and N^2 is given, which is just as good as the one we have adopted for use.

EXAMPLE 5.1

(1) The function $f: N^2 \to N$, $f = \lambda xy[2^x(2y+1) \dot{-} 1]$, is a recursive bijection from N^2 to N.

(2) Using Cantor coding schemes a function $f: N^n \to N^k$ ($n \geqslant 1$ and $k \geqslant 1$) can be encoded as a function $h: N \to N$. This coding is defined by:

$$cod(f) \triangleq \lambda x[\sigma_1^k(f(\sigma_n^1(x)))],$$
for all functions $f: N^n \to N^k$.

$$cod^{-1}(h) \triangleq \lambda x[\sigma_k^1(h(\sigma_1^n(x)))],$$
for all functions $h: N \to N$.

Note that the coding function cod is a bijection. The fact that such a coding exists implies that all computable functions $N^n \to N^k$ are already 'contained' in the set REC_1^1.

5.2 Gödel numbering

The Cantor schemes of the preceding section can be used to systematically enumerate all k-tuples of natural numbers, for some particular number k. A scheme which can be used to enumerate all finite sequences of natural numbers is called a Gödel numbering scheme. We will use such a scheme to encode SAL(N) programs as natural numbers and to represent terminating computations by natural numbers.

The set of all finite sequences of natural numbers is denoted by N^*, and consists of:

(1) $()$, the sequence of length 0, also denoted by ε.
(2) (n), $n \in N$, all sequences of length 1.
(3) $(n_1, \ldots, n_m) \in N^m$, all sequences of length m, for $m \geqslant 2$.

The *length* of a sequence $x \in N^*$ will be denoted by $L(x)$. If we write $N^0 \triangleq \{()\}$ and $N^1 \triangleq \{(n) | n \in N\}$, then $N^* = \bigcup_{i \geqslant 0} N^i$. In the following two bijections between N^* and N will be discussed.

5.2.1 Using Cantor numberings

Assume given Cantor numbering schemes (σ_1^k, σ_k^1) for all $k \geqslant 1$. Consider the following function from $N^* \to N$.

$$\lambda x[\text{if } L(x) = 0 \text{ then } 0 \text{ else } \sigma_1^2(L(x), \sigma_1^{L(x)}(x))]$$

Clearly this is an injection from N^* to N. It is not a bijection, because $\sigma_1^2(0, x)$ is not the code of a finite sequence for any $x \geqslant 1$. In the rest of the book we will tacitly assume the Gödel scheme as defined in the definition below. We will refer to this numbering as our 'standard numbering'.

Definition 5.1
The function $\langle\,\rangle : N^* \to N$ is defined as follows:

$$\langle\,\rangle \triangleq \lambda x[\text{if } L(x) = 0 \text{ then } 0 \text{ else } 1 + \sigma_1^2(L(x) \dot{-} 1, \sigma_1^{L(x)}(x))]$$

We write $\langle x \rangle$ instead of $\langle\,\rangle(x)$.

The kth element in a sequence y can be obtained from $x = \langle y \rangle$ by the function τ.

$$\tau \triangleq \lambda xk[\text{if } (x = 0 \text{ or } k > L(x)) \text{ then } 0$$
$$\text{else } \sigma_{L(x),k}^1(\sigma_{2,2}^1(x \dot{-} 1))]$$

Given $\langle x \rangle$, the function ω replaces the element x_k of the sequence x by the element z, and determines the Gödel number of the resulting sequence.

$$\omega \triangleq \lambda xkz[\text{if } k > L(x) \text{ then}$$
$$\text{if } L(x) > 0 \text{ then } 1 + \sigma_1^2(k \dot{-} 1, \sigma_1^k(\sigma_{L(x)}^1 \sigma_{2,2}^1((x \dot{-} 1)), 0, \ldots, 0, z))$$
$$\text{else } 1 + \sigma_1^2(k \dot{-} 1, \sigma_1^k(0, \ldots, 0, z))$$

$$\begin{aligned}&\text{else}\\&\quad \text{if } k=0 \text{ then } x\\&\quad \text{else}\\&\quad 1+\sigma_1^2(L(x)\dot{-}1,\sigma_1^{L(x)}(\tau(x,1),\ldots,\tau(x,k-1),z,\\&\qquad\qquad \tau(x,k+1),\ldots,\tau(x,L(x))))]\end{aligned}$$

■

Note that the functions L, τ and ω are SAL(**N**) computable. However, the function $\langle\,\rangle$ and its inverse are *not* SAL(**N**) computable. These functions are computable in the intuitive sense, but they cannot be SAL(**N**) computable, because SAL(**N**) programs have by definition a fixed finite number of input variables and output variables. Thus we cannot have an arbitrary-length sequence of natural numbers as input or output. Hence the function $\langle\,\rangle$ and its inverse $\langle\,\rangle^{-1}$ are only SAL(**N**) computable relative to a coding. We could proceed as follows:

- Given a sequence $(x_1,\ldots,x_n)$ of length n, write x_i in binary, $1\leqslant i\leqslant n$, without superfluous leading zeros. Let z_i be the representation of x_i.
- Let y be the number whose ternary representation is $z_1 2 z_2 2 z_3 2\cdots 2 z_n$.
- Write a SAL(**N**) program which given y computes $\langle x_1,\ldots,x_n\rangle$. Another SAL(**N**) program can compute y, given the code number $\langle x_1,\ldots,x_n\rangle$.

Thus the Gödel scheme is computable relative to the coding sketched above. Note that this amounts to applying another numbering scheme, namely assigning the code number y to the sequence $(x_1,\ldots,x_n)$. (This numbering, as described above, is not a Gödel scheme because there are numbers y which are not valid code numbers; but a Gödel scheme using the above coding idea is possible.)

The above is not a counter-example to Church's thesis that all intuitively computable functions are also SAL(**N**) computable. It is not a problem of computability, but a matter of definition. If a function $f\colon A\to B$ is to be computable, it is necessary that the elements of A and B are manipulable, i.e. belong to the underlying data type. Thus to compute a function $N^*\to N$, our underlying data type must have both N and N^* as sorts of the data type.

In intuitive terms, to compute a function value $f(x)$, where $f\colon N^*\to N$, we must first of all store the given input sequence, for instance in an array. The elements can then be accessed by indexing. Any form of storage combined with fitting access functions to determine the length of the sequence and to fetch and store individual elements will do. Now this is precisely what the Gödel scheme does: it provides a method to store a sequence of natural numbers. Moreover, the access functions are SAL(**N**) computable.

There are many Gödel numberings for N^*, each having its own minor advantages and disadvantages. A well-known scheme uses the decomposition into primes.

5.2.2 Gödel schemes using prime decomposition

The function $c: N^* \to N$ is defined as follows:

$$c(()) = 0$$

$$c(x_0, \ldots, x_n) = 2^{x_0} . 3^{x_1} \cdots p_n^{x_n+1} \dot{-} 1$$

The idea of using the unique decomposition of natural numbers into primes in order to construct a bijection from N^* to N is due to Kurt Gödel. He used it to *arithmetize* metamathematics – that is, to encode statements, theorems and proofs as natural numbers. Once this is done, statements about the provability of number-theoretic assertions can be expressed as number-theoretic statements, i.e. within number theory itself. This has proved a very fruitful idea. Coding programs and computations involving natural numbers by natural numbers is essentially the same idea.

5.2.3 Non-bijective numbering schemes

We have called a pair of bijections $(c: N^* \to N, c^{-1}: N \to N^*)$ a Gödel numbering scheme. Bijectiveness is a little too strict a requirement.

If we have two functions $c: N^* \to N$ and $c^{-1}: N \to N^*$, such that $c^{-1}(c(x)) = x$ for all $x \in N^*$, then the pair (c, c^{-1}) is a valid coding scheme. Under this requirement c is injective and c^{-1} is surjective, but neither has to be bijective. Not every natural number is a valid code number, but if it is a valid code number, then it corresponds to a uniquely determined sequence.

Such schemes are somewhat easier to construct.

EXAMPLE 5.2

Let the injective function $c: N^* \to N$ be defined as follows.

$$c(()) = 0$$

$$c(x, y) = \sigma_1^2(x+1, c(y)), \text{ for any } x \in N \text{ and } y \in N^*.$$

Thus $c(5,3,0,7) = \sigma_1^2(6, \sigma_1^2(4, \sigma_1^2(1, \sigma_1^2(8,0))))$. Assuming our standard Cantor scheme, we have $\sigma_1^2(8,0) = 36$, $\sigma_1^2(1,36) = 739$, $\sigma_1^2(4,739) = 277\,135$, and finally $c(5,3,0,7) = \sigma_1^2(6, 277\,135) = 38\,403\,982\,646$.

The corresponding access functions are computed by the following SAL(**N**) programs.

length(x) *input*: x
output: z

```
        method:
          y:= x;
          while y ≠ 0 do
            if σ¹₂,₁(y) ≠ 0 then
              z:= z + 1;
              y:= σ¹₂,₂(y)
            else y:= 0
            fi
          od

kth term input: x, k
         output: t
         method:
          y:= x;
          while y ≠ 0 ∧ |z − k| ≠ 0 do
            if σ¹₂,₁(y) ≠ 0 then
              z:= z + 1;
              if |z − k| = 0 then t:= σ¹₂,₁(y) ∸ 1 fi;
              y:= σ¹₂,₂(y)
            else y:= 0
            fi
          od
```

5.2.4 Gödel numberings of strings over an alphabet V

V^* denotes the set of all strings over V, i.e. finite sequences of elements of V. Assigning Gödel numbers to strings is a simpler problem than assigning Gödel numbers to finite sequences of numbers, because the alphabet V is by definition finite.

We could, for instance, read strings as representations of natural numbers in a base $|V|+1$ number system, by numbering the elements of $V: v_1, \ldots, v_{|V|}$. In this way, not every number is a valid code.

A second possibility is to number the elements of V^* according to the ordering by length, and within equal lengths ordering alphabetically. If $V = \{a, b, c\}$ and ε denotes the empty string we have:

0	ε
1	a
2	b
3	c
4	aa
5	ab
6	ac
⋮	⋮
28	bca
⋮	⋮

In general, if $V=\{v_1,v_2,\ldots,v_r\}$, this coding is defined as follows:

$$c(\varepsilon)=0$$
$$c(v_i)=i, \text{ for all } i,\ 1\leqslant i\leqslant r$$
$$c(wv_i)=r.c(w)+i, \text{ for all } w\in V^* \text{ and } i,\ 1\leqslant i\leqslant r.$$

Thus

$$c(a_{i_m}\cdots a_{i_0})=i_m.r^m+i_{m-1}.r^{m-1}+\cdots+i_1.r+i_0$$

5.2.5 Gödel numbering of SAL(N) programs

We will now define a Gödel numbering of SAL(**N**) programs using our standard Gödel numbering as defined in Definition 5.1 (page 97).

Definition 5.2

The statements of a given SAL(**N**) program are coded as given in the following table:

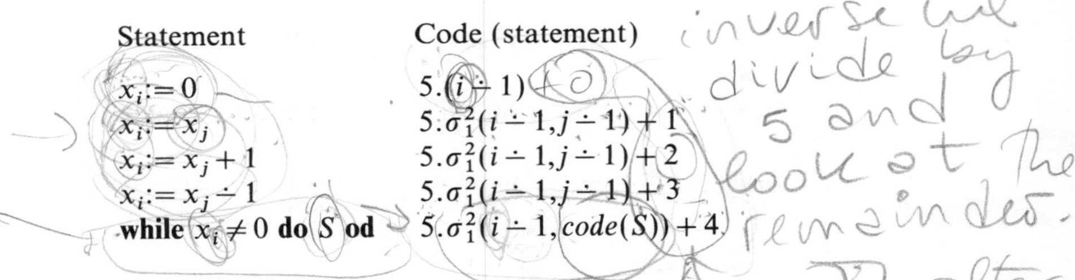

Statement	Code (statement)
$x_i:=0$	$5.(i\dot{-}1)$
$x_i:=x_j$	$5.\sigma_1^2(i\dot{-}1,j\dot{-}1)+1$
$x_i:=x_j+1$	$5.\sigma_1^2(i\dot{-}1,j\dot{-}1)+2$
$x_i:=x_j\dot{-}1$	$5.\sigma_1^2(i\dot{-}1,j\dot{-}1)+3$
while $x_i\neq 0$ **do** S **od**	$5.\sigma_1^2(i\dot{-}1,\mathit{code}(S))+4$

Finally, let $S=S_1,S_2,\ldots,S_m$ be a sequence. Then

$$\mathit{code}(S)\triangleq\langle \mathit{code}(S_1),\mathit{code}(S_2),\ldots,\mathit{code}(S_m)\rangle\dot{-}1$$ ■

It remains to assign codes to programs. A SAL(**N**) program is a 4-tuple (n,k,p,P). We have described above how to find the code number of the sequence P. The use of Gödel numbers of SAL(**N**) programs is restricted to the case where n and k are fixed. That is, for each n and k we will assume a fixed Gödel numbering. In these circumstances n,k, and p are of no concern – n and k are fixed and p can be taken equal to the index of the variable with the highest index occurring in the sequence P. Hence, the code of the program (n,k,p,P) is $\mathit{code}(P)$.

For completeness, consider how to code (n,k,p,P). The numbers n,k, and p are not independent. First, $k\geqslant 1$, and second $p\geqslant\max(n,k,m)$, where m is the index of the variable with the highest index occurring in P. Therefore define

$$\mathit{code}((n,k,p,P))\triangleq\sigma_1^4(n,k\dot{-}1,p\dot{-}\max(n,k,m),\mathit{code}(P)).$$

EXAMPLE 5.3

Consider the following program

```
input: x1, x2
output: x1
method:
  while x2 ≠ 0 do
    x1 := x1 + 1;
    x2 := x2 ∸ 1
  od
```

Assume we are interested in a numbering of the programs computing functions $N^2 \to N$. Thus we do not have to worry about the input and output variables.

$$code(x_1 := x_1 + 1) = 5.\sigma_1^2(0,0) + 2 = 2$$

$$code(x_2 := x_2 \dot{-} 1) = 5.\sigma_1^2(1,1) + 3 = 23$$

Now we combine these two statements into one sequence S:

$$code(S) = \sigma_1^2(1, \sigma_1^2(2, 23)) = \sigma_1^2(1, 348) = 61\,423$$

Now we can determine the code of the while statement

$$code(\textbf{while } x_2 \neq 0 \textbf{ do } S \textbf{ od}) = 5.\sigma_1^2(1, code(S)) + 4$$

$$= 5.\sigma_1^2(1, 61\,423) + 4 = 9\,432\,730\,119$$

The program consists of a one-element sequence, consisting solely of the while statement. Thus

$$code(program) = \sigma_1^2(0, \sigma_1^1(9\,432\,730\,119))$$

$$= 44\,488\,198\,763\,093\,972\,259$$

Do not shiver at the magnitude of this number. You should compare with the text of the program not the number, but the representation of the number.

5.3 A universal program

In Section 5.2 we discussed Gödel numberings of finite sequences of natural numbers and also a Gödel numbering of SAL(**N**) programs. Thus we have:

(1) a natural number g
(2) the corresponding SAL(**N**) program P_g with Gödel number g
(3) the function f_{P_g} computed by this program P_g.

Given a number, we can find the uniquely determined SAL(**N**) program having this number as its Gödel number. Conversely, given any SAL(**N**) program, we can find the uniquely determined Gödel number of this program.

On the other hand, the correspondence between Gödel numbers and REC_k^n is not bijective. Given a Gödel number there is a unique program and a unique function corresponding to this number. But given a function $f \in REC_k^n$ there are many – even infinitely many – programs and thus (infinitely) many Gödel numbers corresponding to this function.

Definition 5.3

Let F be any countable set of functions. An *indexing* of F is a surjective function $h: N \to F$. An *index* of a function $f \in F$ is any natural number i such that $h(i) = f$. ■

The Gödel numbering of SAL(**N**) programs as defined in Definition 5.2, gives us an indexing of REC_k^n for every $n \geq 0$, $k \geq 1$. A particular function of REC_k^n has infinitely many indices under this indexing, as has been noted before.

Definition 5.4

Let F be any countable set of number-theoretic functions from N^n to N^k, and h an indexing of F.

The function $\psi: N^{n+1} \to N^k$ is *universal* for the class F relative to indexing h, if and only if

$$\psi(i, x) = h(i)(x), \text{ for all } i \in N \text{ and } x \in N^n. \quad ■$$

Thus, if ψ is universal, then $h(i) = \lambda x[\psi(i, x)]$. In this case we also write ψ_i instead of $h(i)$. This notation leaves the indexing relative to which ψ is universal for F, implicit.

EXAMPLE 5.4

(1) Let $F = \{C_r^2 | r \in N\}$ be the subset of REC_1^2 consisting of all constant functions. The function $h: N \to F$ defined by

$$h(i) \triangleq \lambda x_1 x_2 [i] \qquad \text{for all } i \in N$$

is an indexing of F. The function $\psi: N^3 \to N$ defined by

$$\psi \triangleq \lambda i x_1 x_2 [i]$$

is universal for F, relative to this indexing.

(2) Let $F = \{\lambda x[ax + b] | a, b \in N\}$, then F is a subset of REC_1^1. Our standard Cantor numbering scheme gives an indexing $h: N \to F$ for this set:

$$h(i) \triangleq \lambda x[\sigma_{2,1}^1(i)x + \sigma_{2,2}^1(i)]$$

The function $\psi: N^2 \to N$ defined below is universal for F relative to the indexing h.

$$\psi \triangleq \lambda i x[\sigma_{2,1}^1(i)x + \sigma_{2,2}^1(i)]$$

The interesting point is, that there exists a SAL(**N**) computable function, which is universal for REC_k^n, relative to our standard indexing defined according to the Gödel numbering of SAL(**N**) programs.

A program computing a function universal for REC_k^n is called a *program universal* for REC_k^n, or a *programming system*. (We also speak of a universal function and a universal program, leaving implicit the set of functions for which the function and the program are universal.)

Suppose you are programming in a Pascal environment. You write a Pascal program and prepare input for it. Your program and its input will then be presented to another program, which will compute the output of your program, provided your program and its input contained no errors.

The program which generates this output for you is a universal program for the Pascal computable functions. Usually this universal program is thought of as being divided into subprograms to read in your program text, compile it, link the necessary subroutines and then load and execute this. But considered as a whole it is simply a universal program. Therefore the term 'programming system' is appropriate.

Let us now turn to the task of writing a SAL(**N**) program, which is universal for REC_k^n, relative to the indexing implied by our standard Gödel numbering of SAL(**N**) programs having n input and k output variables, as given in Definition 5.2 (page 101). To represent the values of the program variables, we use our standard Gödel numbering of N^*, as given in Definition 5.1 (page 97).

Algorithm 5.1 (to compute universal function)

input: $g \in N$ {the Gödel number of a SAL(**N**) program}

$x \in N^n$ {input to program having Gödel number g}

output: $y \in N^k$ {output of the program with Gödel number g}

method:

{M, the memory, is a Gödel number storing the values of the program variables}

initialize(M);
simulate(g, M);
output(M)

■

The macro *initialize* copies the input values x_1 to x_n into the variable M, which is the Gödel number of an initially empty sequence of natural numbers used to store the values of the program variables.

initialize(M)

$M := \langle x_1, x_2, \ldots, x_n \rangle$

Similarly, the macro *output* copies from the M the first k values into the output variables y_1 to y_k.

output(M)

$y_1 := \tau(M, 1)$;
$\vdots$
$y_k := \tau(M, k)$

The crucial part of the above universal program is the macro *simulate*. It operates as follows:

- Initially g is the Gödel number of a non-empty sequence of SAL statements. The macro fetches the first statement S_1.
- If S_1 is an assignment statement, it is executed and g is set equal to the Gödel number of the remaining sequence.
- If S_1 is a while statement, say **while** $x_i \neq 0$ **do** S **od**, where S is a sequence, we proceed as follows:
  - — if x_i is indeed zero, as determined from the memory M, then the Gödel number of the remaining sequence of SAL statements is assigned to g;
  - — if x_i is not zero in the current memory, then g is set equal to the Gödel number of the sequence

    S; **while** $x_i \neq 0$ **do** S **od**; *remaining statements*;

These actions are repeated until the sequence of statements to be executed has become empty.

Algorithm 5.2

simulate(g, M)

$L := 1 + \sigma^1_{2,1}(g)$;

$\{L$ is the length of the sequence of statements that remain to be executed$\}$

while $L \neq 0$ **do**

$S := \sigma^1_{L,1}(\sigma^1_{2,2}(g))$;

$r :=$ remainder of S divided by 5;

if $r \neq 4$ **then**

execute(S, r, M);

pop(g);

$L := L \dot{-} 1$

else

$x := \tau(M, \sigma^1_{2,1}((S \dot{-} 4)/5) + 1)$;

if $x \neq 0$ **then**

$z := \sigma^1_{2,2}((S \dot{-} 4)/5)$;

push(z, g);

$L := L + \sigma^1_{2,1}(z)$

else

pop(g);

$L := L \dot{-} 1$

fi

fi

od ■

The macro *execute* updates the values of the program variables as determined by the statement.

Algorithm 5.3

execute(S, r, M)

if $r = 0$ **then**

$x := 1 + S/5$

else

$x := 1 + \sigma^1_{2,1}((S \dot{-} r)/5)$;

$y := 1 + \sigma^1_{2,2}((S \dot{-} r)/5)$

fi;

if $r = 0$ **then** $M := \omega(M, x, 0)$ **fi**;

if $r = 1$ **then** $M := \omega(M, x, \tau(\mathrm{M}, \mathrm{y}))$ **fi**;

if $r = 2$ **then** $M := \omega(M, x, \tau(\mathrm{M}, \mathrm{y}) + 1)$ **fi**;

if $r = 3$ **then** $M := \omega(M, x, \tau(\mathrm{M}, \mathrm{y}) \dot{-} 1)$ **fi** ■

Finally, the macros *pop* and *push* serve to update the sequence of statements that remain to be executed. They are defined as follows:

$pop(g)$

$$g := \sigma_1^2(L \dot{-} 2, \sigma_1^{L \dot{-} 1}(\sigma_{L,2}^1(\sigma_{2,2}^1(g)), \ldots, \sigma_{L,L}^1(\sigma_{2,2}^1(g))))$$

$push(z, g)$

$$L_z := 1 + \sigma_{2,1}^1(z);$$

$$g := \sigma_1^2(L + L_z \dot{-} 1, \sigma_1^{L+L_z}(\sigma_{L_z}^1(\sigma_{2,2}^1(z)), \sigma_L^1(\sigma_{2,2}^1(g))))$$

Clearly, the above algorithm computes a function which is universal for REC_k^n. The universal function computed by the program is generally denoted by ϕ. The values of n and k are not included in the notation and are determined from the context.

The following phrases are customary:

- Let ϕ be a recursive enumeration of REC_k^n.
- Let $\phi_0, \phi_1, \ldots, \phi_i$ be a recursive enumeration of REC_k^n.

These phrases mean the same, namely that $\phi \in REC_k^{1+n}$ denotes a recursive function which is universal for REC_k^n. The term enumeration hints at the fact that the correspondence $i \mapsto \lambda x[\phi(i, x)]$ is the indexing, and thus enumeration, relative to which ϕ is universal. Often, one writes ϕ_i instead of $\lambda x[\phi(i, x)]$. Finally, ϕ is called the *standard enumeration* of REC_k^n, because it is obtained by the standard indexing of SAL(**N**) programs as defined in Definition 5.2 (page 101).

Many function classes besides *REC* have universal functions. In particular, the class *PRIM* (see Definition 4.6 on page 84), has a recursive universal function. Primitive recursive functions are the functions computed by loop programs, that is, programs using only times statements and no while statements. As for SAL(**N**) programs, Gödel numbers can be assigned to loop programs. Using that Gödel numbering, a program, similar to the one above, can be constructed which is universal for *PRIM*.

It will be shown in the next section that any function, which is universal for *PRIM*, cannot itself be primitive recursive. More can be shown. Let F be any of the subrecursive classes defined in Section 4.5. Then there is a recursive function which is universal for F. Furthermore, if f is a function, universal for F, then f does not belong to F.

These assertions can be proved, using the so-called diagonal method presented in the next section.

5.4 The diagonal method

The diagonal method is due to Georg Cantor, who used it to prove that the real numbers are uncountable.

To illustrate the method we will prove that the set F of all total functions from N to N is not countable.

Let $h:N \rightarrow F$ be any arbitrary function from N to F. We show that h is not surjective. This implies that there does not exist a surjection from N onto F, and thus no bijection between N and F. Therefore F is not countable.

We turn now to the proof that h is not surjective. Let $f_i \triangleq h(i)$, in other words, the function $h(i) \in F$ is denoted by f_i. Define a function $f:N \rightarrow N$ as follows:

$$f \triangleq \lambda x[f_x(x)+1]$$

The function $f:N \rightarrow N$ is total, because all the f_i are total functions. But f cannot be in the range of h. Assume that for some n_0 we have $f_{n_0} = f$. Then

(1) $f(n_0) = f_{n_0}(n_0)$, by the assumption on n_0.
(2) $f(n_0) = f_{n_0}(n_0)+1$, by the definition of f.

Since (1) and (2) must both hold, we have a contradiction, and therefore $f \neq h(n)$ for all $n \in N$. The construction of the function f is illustrated in Figure 5.2.

Above, we have proved that the function $h:N \rightarrow F$ cannot be surjective, by constructing a function f not in $\{h(n) | n \in N\}$. The function f is said to be defined, or constructed, by diagonalizing over $\{h(n) | n \in N\}$. The method cannot be applied if the set of functions F contains partial functions.

Let h be an indexing of a countable subset F of the set of all partial functions from N to N, and let f_i denote the partial function $h(i)$. Define f as follows:

$$f \triangleq \lambda x[f_x(x)+1]$$

The preceding proof that f is not in the range of h now fails. Assume that $f = h(i_0)$ for some particular $i_0 \in N$. Then

(1) $f(i_0) = f_{i_0}(i_0)$, by the assumption that $f = h(i_0)$, and
(2) $f(i_0) = f_{i_0}(i_0)+1$, by the definition of f.

Consequently, $f_{i_0}(i_0) = f_{i_0}(i_0)+1$. This is a contradiction *only if* $f_{i_0}(i_0)$ is defined.

Since F may contain partial functions, it is not guaranteed that $f_{i_0}(i_0)$ is defined, and we have failed to derive a contradiction. Therefore, we have failed to show that f is not in the range of h, i.e. not in F.

Because SAL(**N**) computable functions may be partial, we cannot use the diagonal method directly to define a function which is not SAL(**N**)

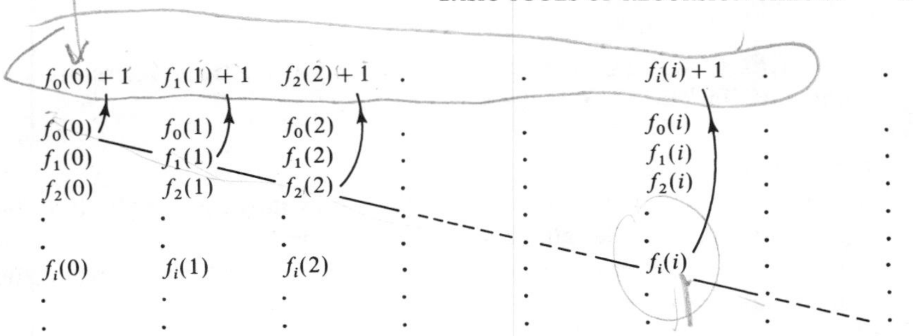

Figure 5.2 Construction of f, using diagonal method.

computable. But it is clear that such noncomputable functions exist. REC_k^n has an indexing and is therefore countable, whereas the set of all partial functions $N^n \to N^k$ is a superset of the set of all total functions $N^n \to N^k$, and is therefore uncountable.

Consider $PRIM_1^1$, the class of all functions $N \to N$ computable by loop programs, i.e. macro programs which may use times statements, but no while statements (see Definitions 4.5 and 4.6 on pages 83 and 84). Clearly, we can assign Gödel numbers to these loop programs, and using these, design a program universal for $PRIM_1^1$. Let ψ be the universal function. Define the function f by diagonalizing over $PRIM_1^1$, using ψ:

$$f \triangleq \lambda x[\psi(x,x)+1]$$

Assume now that there is function ψ that is universal for $PRIM_1^1$, and is itself primitive recursive, i.e. $\psi \in PRIM_1^2$. By diagonalizing over this enumeration ψ, we define the function f as above.

It then follows that $f \in PRIM_1^1$ (from the definition of f and the assumption that ψ is primitive recursive). Also, $f \notin PRIM_1^1$ (from the fact that primitive recursive functions are total and f is defined by diagonalizing over $PRIM_1^1$). Thus we have a contradiction.

Therefore, there is no primitive recursive function which is universal for $PRIM$. (Similarly for the other classes defined in Section 4.5, on subrecursive hierarchies.)

Now consider the class $TREC$.

Theorem 5.1

There is no recursive function which is universal for $TREC$.

Proof

Let ψ be a function which is universal for $TREC_1^1$. Then ψ is a total function. Define the function f by diagonalizing over $TREC_1^1$ using $\psi: f \triangleq \lambda x[\psi(x,x)+1]$. Then $f \notin TREC_1^1$. If ψ is recursive, then

$\psi \in TREC_1^2$, and $f \in TREC_1^1$, which is a contradiction. Therefore the universal function ψ is not recursive. ■

Corollary 5.1

It is not possible to define a programming language such that the programs of this language compute all and only total recursive functions, and such that one can effectively list all the legal programs of the language.

Proof

If the legal programs can be listed effectively, then Gödel numbers can effectively be assigned to these legal programs, and a recursive function, universal for $TREC$ could be constructed. ■

Thus, a model of computation, which avoids the Halting Problem, is not sufficiently powerful to capture the concept of effectiveness in its full generality.

5.5 The *s–n–m* theorem

The s–n–m theorem is a useful technical result about indices of computable functions.

Let $f: N^2 \rightarrow N$ be a SAL(**N**) computable function. For every specific value $a \in N$, the function $f_a \triangleq \lambda x[f(a,x)]$ is also SAL(**N**) computable. Given a program for f, a program computing f_a can be obtained by inserting the statement $x_1 := a$ at the beginning of the program.

The s–n–m theorem reformulates this observation in terms of Gödel numbers of programs.

In the following we will consider functions $N^n \rightarrow N^k$ for various values of n and k. $\phi^{(n,k)}$ denotes the function universal for REC_k^n, according to our standard enumeration. Note that $\phi^{(n,k)} \in REC_k^{n+1}$. If no superscript is given, $(1,1)$ is assumed.

Theorem 5.2

For each $n, m \geqslant 0$ and $k \geqslant 1$, there exists a total recursive function s_m^n such that

$$\phi^{(n,k)}_{s_m^n(i,a)} = \lambda x[\phi_i^{(n+m,k)}(a,x)]$$

for all $i \in N$ and $a \in N^m$.

Proof

Given the program computing $\phi_i^{(n+m,k)}$, we convert it into a program computing $\lambda x[\phi_i^{(n+m,k)}(a,x)]$, by inserting the statements:

(1) $x_{n+m} := x_n;\ x_{n+m-1} := x_{n-1};\ \ldots;\ x_{m+1} := x_1;$

(2) $x_m := a_m;\ x_{m-1} := a_{m-1};\ \ldots;\ x_1 := a_1;$

at the beginning of the program. We must show how to compute the Gödel number of the resulting program.

Since n, m and k are constants, as far as the function s_m^n is concerned, the first sequence of statements is constant and has a particular Gödel number, say z_1. The Gödel number of the second sequence of statements is not constant, it depends on the values of the parameters a_1 to a_m.

A macro $x_i := a_i$ stands for the sequence consisting of a_i statements $x_i := x_i + 1$. We prefix this sequence with the statement $x_i := 0$, which simplifies our programs somewhat. The Gödel number of the assignment $x_i := a_i$ is therefore computed by:

input: $a \in N$, $i \in N$
output: g {the Gödel number of $x_i := a$}
method:

  if $a = 0$ **then**
    $g := 5(i \dot{-} 1)$
  else
    $g := 5\sigma_1^2(i \dot{-} 1, i \dot{-} 1) + 2;$

    do $a \dot{-} 1$ **times**
      $g := \sigma_1^2(5\sigma_1^2(i \dot{-} 1, i \dot{-} 1) + 2, g)$
    od;
    $g := \sigma_1^2(5(i \dot{-} 1), g)$
  fi;
  $g := \sigma_1^2(a, g)$

The following program computes the function s_m^n, which is therefore a total recursive function. It uses the above program as a macro denoted by *Gödel*(a, i), and the macro *push* from Section 5.3.

Algorithm 5.4 (to compute the function s_m^n)
input: $i, a_1, a_2, \ldots, a_m$
output: g {an index of $\lambda x[\phi_i^{(n+m,k)}(a, x)]$}
method:
  $g := i;$
  $x :=$ *Gödel*$(a_1, 1)$; *push*(x, g);
  $x :=$ *Gödel*$(a_2, 2)$; *push*(x, g);
  ⋮
  $x :=$ *Gödel*(a_m, m); *push*(x, g);
  push(z_1, g)

From the program we conclude that the function s_m^n is primitive recursive, because it is computed by a loop program. ∎

The s–n–m theorem is a powerful result; its usefulness will become gradually clear. For now, only an example is given. Other applications of the theorem will be found throughout the book: the first such application is in the next section.

EXAMPLE 5.5

(1) There is a total recursive function f such that $\phi_{f(i,j)} = \lambda x[\phi_i(\phi_j(x))]$. Let

$$\psi \triangleq \lambda ijx[\phi_i(\phi_j(x))]$$

By Church's thesis $\psi \in REC_1^3$. Let z_0 be an index for ψ, i.e. $\psi = \phi_{z_0}^{(3,1)}$. By the s–n–m theorem

$$\phi_{s_2^1(z_0,i,j)} = \lambda x[\psi(i,j,x)].$$

Therefore the function $f \triangleq \lambda ij[s_2^1(z_0,i,j)]$ satisfies the requirements.

(2) For every total recursive function $f: N \to N$, there is an $m \in N$ such that $\phi_m = \phi_{f(m)}$. That is, the programs with Gödel numbers m and $f(m)$ are equivalent – they compute the same function.

Define the function ψ as follows:

$$\psi \triangleq \lambda xy[\text{if } \phi_y(y)\downarrow \text{ then } \phi_{\phi_y(y)}(x) \text{ else } \uparrow]$$

By Church's thesis, ψ is recursive and by the s–n–m theorem there is a total recursive function $h \in TREC_1^1$ such that $\phi_{h(y)} = \lambda x[\psi(x,y)]$. Because $f \in TREC_1^1$, the composition of f and h, $\lambda x[f(h(x))]$ is also recursive and therefore has an index, say n.

The number $m \triangleq h(n)$ is the 'fixed point' we are looking for.

$$\begin{aligned}
\phi_m &= \phi_{h(n)} \\
&= \lambda x[\psi(x,n)] \\
&= \lambda x[\phi_{\phi_n(n)}(x)] \qquad \text{for } \phi_n \in TREC \text{ and thus } \phi_n(n)\downarrow \\
&= \phi_{f(h(n))} \\
&= \phi_{f(m)}
\end{aligned}$$

(3) In Section 4.3 bounded minimization was discussed. We had

an infinite number of bounded minimization operators μ_b, applicable to total functions and defined as follows.

Let $f: N^{n+1} \to N$ be any total function. The function $\mu f: N^{n+1} \to N$ is defined as follows:

$$\mu f \triangleq \lambda xy[\text{if } (f(x,z) \neq y \text{ for all } z \geqslant 0) \text{ then } \uparrow \\ \text{else the smallest } z \text{ such that } f(x,z) = y]$$

As an alternative a single operator $\mu_B: REC_1^n \times N \to REC_1^n$, such that $\mu_B(\phi_i, b) = \mu_b \phi_i$ whenever ϕ_i is total, was suggested. Define the function $\psi: N^{n+3} \to N$ as follows:

$$\psi(i,b,x,y) \triangleq \begin{cases} 0 & \text{if } (\forall z \leqslant b) \\ & [(\phi_i^{(n+1,1)}(x,z)\downarrow \text{ and } \phi_i^{(n+1,1)}(x,z) \neq y] \\ z_0 & \text{if } \phi_i^{(n+1,1)}(x,z_0) = y \text{ and } (\forall z \leqslant z_0) \\ & [(\phi_i^{(n+1,1)}(x,z)\downarrow \text{ and } \phi_i^{(n+1,1)}(x,z) \neq y] \\ \uparrow & \text{in all other cases} \end{cases}$$

By Church's thesis ψ is partial recursive, and by the s–n–m theorem there is a total recursive function μ_B such that

$$\phi_{\mu_B(i,b)}^{(n+1,1)} = \lambda xy[\psi(i,b,x,y)]$$

Thus, this single bounded minimization operator is indeed a recursive operator, in that given any bound b and an index of any SAL(**N**) program P, we can effectively determine an index of a SAL(**N**) program P_b which computes $\mu_b f_P$, if at least f_P is a total function, i.e. P terminates for all its legal inputs.

Item (2) above is not given its full significance if viewed simply as an example. The result is known as 'the recursion theorem', and is a deep and powerful result. But a thorough discussion of this theorem is beyond the scope of this book. Here it serves only as an assertion, far from obvious, which can easily be proved using the s–n–m theorem.

5.6 The universal program and the *s–n–m* theorem: just happenstance?

One might think that the existence of a universal function, and the validity of the s–n–m theorem are peculiarities of our standard Gödel scheme and corresponding indexing π_0 of the partial recursive functions. But this is not the case.

Any model of computation and any effective listing of the finite descriptions of functions according to this model induces a Gödel numbering and an associated indexing of REC^n_k, $n \geqslant 0$ and $k \geqslant 1$. If we have a proof that the model is equivalent to SAL(**N**) as defined in Section 2.2, then the indexing admits a universal function and an s–n–m theorem.

Note that we are discussing indexings of REC^n_k, thus we actually are considering families $\{\pi^{(n,k)}: N \to REC^n_k \mid n \geqslant 0, k \geqslant 1\}$ of indexings, one for each value of n and k. Carefully note that $\pi^{(n,k)}(i)$ is a *function* $N^n \to N^k$. This function has index i relative to π.

Definition 5.5

Let the family $\{\pi^{(n,k)}: N \to REC^n_k \mid n \geqslant 0,\ k \geqslant 1\}$ be any indexing of the partial recursive functions. The indexing π is *acceptable* if and only if it admits a universal function and a corresponding s–n–m theorem. That is, for all n, m and k:

(1) there is a π-universal function, i.e. a function $\psi^{(n,k)} \in REC^{n+1}_k$, such that

$$\pi^{(n,k)}(i) = \lambda x[\psi^{(n,k)}(i,x)]$$

and

(2) the s–n–m theorem is valid. That is, there is a total recursive function $S^n_m \in TREC^{m+1}_1$, such that

$$\pi^{(n,k)}(S^n_m(i,a)) = \lambda x[(\pi^{(n+m,k)}(i))(a,x)].$$

or, using the universal function from point (1) above:

$$\psi^{(n,k)}_{S^n_m(i,a)} = \lambda x[\psi^{(n+m,k)}_i(a,x)]. \qquad \blacksquare$$

We will now prove that an indexing π is acceptable if and only if the indexing can be translated into our standard indexing π_0 using recursive functions. That is, given a π-index we can effectively compute a π_0-index, and vice versa.

Theorem 5.3

Let $\{\pi^{(n,k)}_0: N \to REC^n_k \mid n \geqslant 0, k \geqslant 1\}$ be our standard indexing. An indexing

$$\{\pi^{(n,k)}: N \to REC^n_k \mid n \geqslant 0, k \geqslant 1\}$$

is acceptable if and only if there are families of total recursive functions

$$\{g^{(n,k)} \mid n \geqslant 0, k \geqslant 1\} \text{ and } \{h^{(n,k)} \mid n \geqslant 0, k \geqslant 1\}$$

such that for all n and k:

$$\pi_0(i) = \pi(g(i)) \text{ and } \pi(i) = \pi_0(h(i)), \text{ for all } i \geqslant 0$$

where π, π_0, g and h stand for $\pi^{(n,k)}$, $\pi_0^{(n,k)}$, $g^{(n,k)}$, and $h^{(n,k)}$ respectively. In other words, for each π_0-index we can find a corresponding π-index using g, and for each π-index, we can find a corresponding π_0-index using h. ∎

From the above theorem we conclude that, no matter what model of computation we choose, and no matter what effective listing of the finite descriptions of functions according to this model we choose, the result will always be an acceptable indexing.

Let M be the computation model. If we have a constructive proof that M is equivalent to SAL(**N**), we can exhibit a SAL(**N**) computable function $f_{\text{SAL}(\mathbf{N})}$ which incorporates a given M-computable function f_M. This defines the function g. For the function h, we proceed similarly.

We turn now to the proof of the above theorem. This will be presented by means of two lemmas. In the lemmas and their proofs, we will drop the superscript (n, k) to make the formulae easier to read, as in the above theorem.

Lemma 5.1

An indexing π has a universal function if and only if there is a total recursive function $h \in TREC_1^1$ such that $\pi = \lambda i[\pi_0(h(i))]$.

Proof

- Assume that such a function h exists and that $\phi \in REC_k^{n+1}$ is universal for REC_k^n relative to π_0. Then $\psi \triangleq \lambda ix[\phi(h(i), x)]$ is universal for REC_k^n relative to the indexing π, that is $\pi(i) = \lambda x[\psi(i, x)]$.
- Assume that $\psi \in REC_k^{n+1}$ is universal for REC_k^n relative to the indexing π. Then ψ has a π_0-index, say z_0. By the π_0-s–n–m theorem, we have $\lambda x[\psi(i, x)] = \phi^{(n,k)}_{s_m^n(z_0, i)}$. Thus, there is a total recursive function $h \triangleq \lambda i[s_m^n(z_0, i)]$ which gives a π_0-index for every π-index. In other words, $\pi = \lambda i[\pi_0(h(i))]$. ∎

Lemma 5.2

Let π be an indexing of REC_k^n, and assume that $\pi = \lambda i[\pi_0(h(i))]$ for some particular $h \in TREC_1^1$. An s–n–m theorem relative to π is valid if and only if there exists a total recursive function $g \in TREC_1^1$ such that $\pi_0 = \lambda i[\pi(g(i))]$.

Proof

- Assume that such a function g exists. Let $\psi \in REC_k^{n+m}$ be any partial recursive function, and i a $\pi^{(n+m,k)}$-index for ψ. Then $h(i)$ is a $\pi_0^{(n+m,k)}$-index for ψ. By the π_0-s–n–m theorem, there is a total recursive function s^n_m such that

$$\phi_{s^n_m(h(i),a)} = \lambda x[\psi(a,x)].$$

A $\pi^{(n,k)}$-index for $\lambda x[\psi(a,x)]$ is therefore $g(s^n_m(h(i),a))$.

Define the function S^n_m as follows:

$$S^n_m \triangleq \lambda ia[g(s^n_m(h(i),a))]$$

Then $\pi^{(n,k)}(S^n_m(i,a)) = \lambda x[(\pi^{(n+m,k)}(i))(a,x))]$. In other words, we have an S–n–m theorem.

- Assume now that we have an S–n–m theorem relative to π. That is, there is a total recursive function $S^n_m \in TREC^{m+1}_1$, such that

$$\pi^{(n,k)}(S^n_m(i,a)) = \lambda x[(\pi^{(n+m,k)}(i))(a,x))].$$

We must show how to compute an equivalent π-index, given a π_0-index. Let $\phi \in REC^{n+1}_k$ be universal for REC^n_k relative to our standard π_0 indexing, and let z_0 be a $\pi^{(n+1,k)}$-index for ϕ. By the π-S–n–m theorem we have

$$\lambda x[\phi(i,x)] = \pi(S^n_1(z_0,i)).$$

Therefore, $\pi_0 = \lambda i[\pi(S^n_1(z_0,i))]$, and consequently there is a total recursive function $g \in TREC^1_1$ such that $\pi_0 = \lambda i[\pi(g(i))]$. ■

The above two lemmas together imply the theorem. Therefore the universal function and the s–n–m theorem are not just happenstance. Any model of computation constructively equivalent to SAL(**N**) has a universal program and a valid s–n–m theorem.

This gives added significance to all properties which can be proved from the universal function and the s–n–m-theorem, as for instance the recursion theorem (item (2) of Example 5.5). All these theorems will hold relative to any acceptable indexing, and are therefore independent of the model of computation.

In conclusion, all mainstream results are model independent, and differ only in irrelevant detail, such as whether a particular function has index 113 or 53 971.

Skip to 6

5.7 While statements and procedure calls

Common programming languages, such as Pascal, incorporate a concept which is not included in SAL(**N**), namely procedures.

In this section we show that the while statement can be traded against the if statement plus procedures. We show how to implement the while statement using if statements and procedures, and construct a SAL(**N**) program which is universal for PROCSAL(**N**), a language having procedures and if statements but no while statements.

As procedures are not strictly necessary to understand the following chapters, the subject can be skipped. The result is not a basic tool of recursion theory. It is included here to allow procedures in a later stage, and as a further exercise in manipulating programs and computations using Gödel numbers.

In a program, a procedure may be called from anywhere in the program, or from anywhere within some other procedure, or from anywhere within the procedure itself.

A *recursive* procedure is a procedure which calls itself. This use of the word 'recursive' must be distinguished from the use 'recursive function'. In the latter the word applies to functions and not to programs, and it denotes membership in some particular class.

Definition and usage of a procedure is denoted as follows:

(1) Procedure X is defined by giving the sequence of statements which constitute the procedure body:

$$X[\langle sequence \rangle]$$

(2) The procedure X is called from somewhere else as follows:

$$\langle tuple \rangle := X \langle tuple \rangle$$

In the above, a tuple is a sequence $(x_1, x_2, \ldots, x_n)$.

For example, a procedure to perform addition could be:

$PLUS$[**if** $x_2 \neq 0$ **then**

$x_2 := x_2 \dot{-} 1;$

$x_1 := PLUS(x_1, x_2);$

$x_1 := x_1 + 1$

fi

]

We will define the programming language PROCSAL(**N**), which has if statements and procedures but no while statements, and is otherwise like SAL(**N**). The meaning of the statements, especially the procedure, will be explained, then a Gödel numbering and a universal program for PROCSAL(**N**) will be described.

The universal program is itself a SAL(**N**) program. This then shows that PROCSAL(**N**) is not more powerful than SAL(**N**).

It remains to show the converse. That is done by describing a method to convert an arbitrary SAL(**N**) program to an equivalent PROCSAL(**N**) program.

The syntax of PROCSAL(N)

$$\langle \mathit{identifier} \rangle ::= \{x_i \mid i = 1, 2, 3, \dots\}$$

$$\langle \mathit{procedure} \rangle ::= \{X_i \mid i = 1, 2, 3, \dots\}$$

$$\begin{aligned}\langle \mathit{assignment} \rangle ::= {} & \langle \mathit{identifier} \rangle := 0 \mid \\ & \langle \mathit{identifier} \rangle := \langle \mathit{identifier} \rangle \mid \\ & \langle \mathit{identifier} \rangle := \langle \mathit{identifier} \rangle + 1 \mid \\ & \langle \mathit{identifier} \rangle := \langle \mathit{identifier} \rangle \dot{-} 1\end{aligned}$$

$$\begin{aligned}\langle \mathit{id\text{-}sequence} \rangle ::= {} & \langle \mathit{identifier} \rangle \mid \\ & \langle \mathit{id\text{-}sequence} \rangle, \langle \mathit{identifier} \rangle\end{aligned}$$

$$\langle \mathit{tuple} \rangle ::= (\langle \mathit{id\text{-}sequence} \rangle)$$

$$\langle \mathit{call} \rangle ::= \langle \mathit{tuple} \rangle := \langle \mathit{procedure} \rangle \langle \mathit{tuple} \rangle$$

$$\begin{aligned}\langle \mathit{statement} \rangle ::= {} & \langle \mathit{assignment} \rangle \mid \\ & \langle \mathit{call} \rangle \mid \\ & \mathbf{if}\ \langle \mathit{identifier} \rangle \neq 0\ \mathbf{then}\ \langle \mathit{sequence} \rangle\ \mathbf{fi}\end{aligned}$$

$$\begin{aligned}\langle \mathit{sequence} \rangle ::= {} & \langle \mathit{statement} \rangle \mid \\ & \langle \mathit{sequence} \rangle; \langle \mathit{statement} \rangle\end{aligned}$$

$$\langle \mathit{definition} \rangle ::= \langle \mathit{procedure} \rangle [\langle \mathit{sequence} \rangle]$$

$$\begin{aligned}\langle \mathit{definitions} \rangle ::= {} & \langle \mathit{definition} \rangle \mid \\ & \langle \mathit{definition} \rangle; \langle \mathit{definitions} \rangle\end{aligned}$$

$$\langle \mathit{program} \rangle ::= \{(n, k, p, \langle \mathit{definitions} \rangle) \mid n \geqslant 0, k \geqslant 1 \text{ and } n, k \leqslant p\}$$

In a program (n, k, p, D), n and k specify, as in SAL(**N**), that x_1 to x_n are the input variables, and x_1 to x_k are the output variables. p specifies that x_1 to x_p are the program variables. Thus p must be greater than or equal to the maximum of n, k, and the index of the highest indexed variable which actually occurs in D. To run the program we execute the call:

$$(x_1, x_2, \dots, x_k) := X_1(x_1, x_2, \dots, x_n)$$

The meaning of a procedure call $(x_{i_1}, \dots, x_{i_r}) := X(a_1, \dots, a_s)$ is as follows:

(1) The current values of the program variables x_1 to x_p are temporarily stored in some places not accessible by the program.

(2) The program variables x_1 to x_p are re-initialized:

- x_i is set equal to a_i, for all i, $1 \leqslant i \leqslant s$.
- the remaining variables are set equal to zero.

(3) The sequence S in the definition $X[S]$ of X is executed.

(4) On termination of the computation induced by this sequence S, the old values of the program variables are restored, except for the variables x_{i_1} to x_{i_r}, which keep the value that resulted from steps (2) and (3) above.

(5) The computation, induced by the sequence in which this call statement occurred, is resumed.

The meaning of the other statements is as discussed for SAL(**N**).

Definition 5.6

We assume our standard Gödel numbering as given in Definition 5.1 (page 97). Tuples $(x_{i_1}, x_{i_2}, \ldots, x_{i_r})$ as occurring in the procedure calls in a given PROCSAL(**N**) program are assigned the Gödel number $\langle i_1 \dot{-} 1, i_2 \dot{-} 1, \ldots, i_r \dot{-} 1 \rangle \dot{-} 1$.

The statements of a given PROCSAL(**N**) program are coded as given in Table 5.1.

Let $S = S_1, S_2, \ldots, S_m$ be a sequence. Then

$$code(S) \triangleq \langle code(S_1), code(S_2), \ldots, code(S_m) \rangle \dot{-} 1$$

Finally, let the program consist of the definitions:

$$X_1[S_1];\ X_2[S_2];\ \ldots;\ X_n[S_n]$$

Then the Gödel number of the program is

$$\langle \sigma_1^2(0, code(S_1)), \sigma_1^2(1, code(S_2)), \ldots, \sigma_1^2(n \dot{-} 1, code(S_n)) \rangle \dot{-} 1$$

■

Table 5.1 Gödel numbers of *PROCSAL(N)* statements.

Statement	*code*(statement)
$x_i := 0$	$6(i \dot{-} 1)$
$x_i := x_j$	$6\sigma_1^2(i \dot{-} 1, j \dot{-} 1) + 1$
$x_i := x_j + 1$	$6\sigma_1^2(i \dot{-} 1, j \dot{-} 1) + 2$
$x_i := x_j \dot{-} 1$	$6\sigma_1^2(i \dot{-} 1, j \dot{-} 1) + 3$
$tuple1 := X_i\ tuple2$	$6\sigma_1^3(i \dot{-} 1, code(tuple1), code(tuple2)) + 4$
if $x_i \neq 0$ **then** S **fi**	$6\sigma_1^2(i \dot{-} 1, code(S)) + 5$

Corresponding to this Gödel numbering a universal program is designed similar to Algorithm 5.1. The main difference is in memory management.

To simulate computations of PROCSAL(**N**) programs, we must be able to store the current values of the program variables, and later recall the stored values. Because the called procedure might itself call yet another procedure, we must be able to store a sequence:

(1) current values of program variables

(2) values of the program variables just before the most recent procedure call

(3) values of the program variables just before the most recent but one procedure call

and so forth.

Similarly we must keep track of what statements of which procedures remain to be executed. Thus we must be able to store a sequence:

(1) remaining statements of currently running procedure

(2) remaining statements of procedure which called the currently running procedure

(3) remaining statements of procedure which called the procedure which called the currently running procedure

and so forth.

Thus the memory is interpreted as the Gödel number of a sequence of frames. Each frame consists of three components:

(1) *Vf*: the Gödel number of a sequence $(v_1, v_2, \ldots, v_p)$ representing the values of the programming variables.

(2) *Pf*: the Gödel number of a sequence of PROCSAL(**N**) statements which remain to be executed.

(3) *Lf*: the length of *Pf*.

Below we sketch the corresponding universal program.

Algorithm 5.5 (to compute universal function)
input: $g \in N$ {the Gödel number of a PROCSAL(**N**) program}
$x \in N^n$ {input to the program having Gödel number g}
output: $y \in N^k$ {output of this program}

method:
  $defs := g + 1;$
  $Vf := \langle x_1, x_2, \ldots, x_n \rangle;$
  $Pf := code((x_1, \ldots, x_k) := X_1(x_1, \ldots, x_n));$
  $memory := \langle \sigma_1^3(Vf, Pf, 1) \rangle;$
  simulate;
  $y := (\tau(Vf, 1), \tau(Vf, 2), \ldots, \tau(Vf, k))$ ■

The macro *simulate* is expanded as follows:

Algorithm 5.6
simulate
  *call*_level$:= 0;$
  repeat
    $(Vf, Pf, Lf) := \sigma_3^1(\tau(memory, 1));$
    while $Lf \neq 0$ **do** $S := \sigma_{Lf,1}^1(\sigma_{2,2}^1(Pf));$
      $r :=$ *remainder of S divided by* 6;
      $pop(Pf);$
      $Lf := Lf \dot{-} 1$
      if $r \neq 5$ **then**
        $execute(S, r)$
      else
        $x := \tau(Vf, \sigma_{2,1}^1((S \dot{-} 5)/6) + 1);$
        if $x \neq 0$ **then**
          $z := \sigma_{2,2}^1((S \dot{-} 5)/6);$
          $push(z, Pf);$
          $Lf := Lf + \sigma_{2,1}^1(z)$
        fi
      fi
    od;
    $call\_level := call\_level \dot{-} 1;$
    $pop(memory);$
    $(oldVf, oldPf, oldLf) := \sigma_3^1(\tau(memory, 1));$
    $Lf := oldLf \dot{-} 1;$
    $S := \sigma_{oldLf,1}^1(\sigma_{2,2}^1(oldPf));$
    $out := \sigma_{3,2}^1((S \dot{-} 4)/6);$
    $Vf := restore(oldVf, out);$
    $pop(oldPf);$
    $memory := \omega(memory, 1, \sigma_1^3(Vf, oldPf, Lf))$
  until $call\_level = 0$ ■

In the above the 'repeat–until' construct is used. The macro '**repeat** S **until** $x = 0$', is a shorthand for 'S; **while** $x \neq 0$ **do** S **od**'.

The expansion of the macros *pop* and *push* is the same as in our Algorithm 5.1 (see page 104). The macro *restore* copies the values in *oldVf* back into Vf; except for the variables specified in *out*, these keep their values as specified in Vf.

The expansion of the macro *execute* is largely as specified for Algorithm 5.2 (page 106), so we specify only the part which has to do with the call statement.

```
execute(S, r)
  ⋮
  if r = 4 then
     call_level := call_level + 1;
     memory := ω(memory, 1, σ³₁(Vf, Pf, Lf));
     proc := σ¹₃,₁((S ∸ 4)/6) + 1;
     in := σ¹₃,₃((S ∸ 4)/6);
     Vf := init(Vf, in);
     Pf := τ(defs, proc);
     Lf := L(Pf);
     push(σ³₁(Vf, Pf, Lf), memory)
  fi;
  ⋮
```

The macro *init* re-initializes the program variables. That is, the values of the variables specified in *in*, are copied to the variable $x_1, x_2, \ldots$ and the remaining variables are set to zero.

The above program starts with a single statement which remains to be executed, namely the call $(x_1, \ldots, x_k) := X_1(x_1, \ldots, x_n)$. The simulation then proceeds by executing the statements, every one in its turn.

- If the statement corresponds to an operation of the data type, this operation is executed on the currently valid set of program variables.
- A statement **if** $x \neq 0$ **then** S **fi** is skipped if x happens to be zero in the currently valid set of program variables, otherwise the sequence S is inserted at the beginning of the queue of remaining statements.
- The most complicated is the procedure call. To execute a call, a new set of program variables is created and properly initialized, and the old set is stored in a Gödel number. Computation is resumed by executing the first statement of the called procedure. Once the computation induced by this procedure is finished, the previous set of program variables is restored, and then changed according to the call.

Although the universal program is only sketched, it is clear that it can be expanded into a SAL(**N**) program. Therefore, every function which can be computed by a PROCSAL(**N**) program can also be computed by a SAL(**N**) program. It remains to show the converse.

Let (n, k, p, P) be a SAL(**N**) program. An equivalent PROCSAL(**N**) program (n, k, p, D) is obtained as follows:

(1) Let D be the sequence of procedure definitions, which consists of a single definition: $X_1[P]$.

(2) If D contains definitions of procedures X_1 to X_n, and definition $X_j[S_j]$ contains a while statement **while** $x \neq 0$ **do** S **od**, then replace this while statement by the call $(x_1, \ldots, x_p) := X_{n+1}(x_1, \ldots, x_p)$; and add to D the definition

```
X_{n+1}[
        if x ≠ 0 then
          S;
          (x_1,...,x_p) := X_{n+1}(x_1,...,x_p)
        fi
]
```

(3) Repeat the above step (2) until there is no while statement left.

Clearly, any SAL(**N**) program can be converted into an equivalent PROCSAL(**N**) program, using the above method.

In conclusion, SAL(**N**) and PROCSAL(**N**) are equally powerful models of computation. Thus while statements can be traded against if statements and procedures.

5.8 Exercises

5.1 Let (σ_1^n, σ_n^1) be a Cantor numbering scheme of N^n. A function $f: N^n \to N$ of n variables can be represented by a function of one variable $\hat{f} \triangleq \lambda x[f(\sigma_1^n(x)]$. For the case $n = 3$, prove the following assertions:

(a) $\hat{f}$ is primitive recursive (elementary) if and only if f is primitive recursive (elementary).

(b) Let $h = f \cdot (g_1, g_2, g_3)$. Express the function $\hat{h}$ in the functions $\hat{f}$ and $\hat{g}_i$ $(i = 1, 2, 3)$.

(c) Express the functions $\hat{f}^*$ and $\hat{f}^{\nabla_2}$ in f.

5.2 Consider the following Gödel numbering scheme.

$(x_1, \ldots, x_n)$ is mapped onto $p_0^{x_0+1} \cdot p_1^{x_1+1} \cdots p_n^{x_n+1}$.

This mapping is injective but not surjective; thus not every natural number represents a sequence.

(a) Determine the Gödel number of the sequences $(3, 2, 1, 0)$ and $(7, 1, 0, 0, 0, 0, 0, 0)$.

(b) Determine the sequence with Gödel number 2744.

(c) Which of the natural numbers 404 and 250 are Gödel numbers of sequences?

(d) Prove that the function

$$g \triangleq \lambda x[\text{if } (x \text{ is the Gödel number of a sequence}) \text{ then } 1 \text{ else } 0],$$

is primitive recursive (elementary).

(e) Prove that the function

$$L: N \to N, \qquad L \triangleq \lambda x[\text{if } (x \text{ is the Gödel number of } (x_1, \ldots, x_n)) \text{ then } n \text{ else } 0]$$

is primitive recursive (elementary).

5.3 In this problem we use the standard Gödel numbering as defined in Section 5.2. Prove that the following predicates are primitive recursive (use bounded quantification and logical connectives);

(a) $P(x)$ is 'x is the Gödel number of a sequence'.

(b) $M(x, y)$ is 'x is the Gödel number of a sequence and y is the maximal element in this sequence'.

(c) $P(x)$ is 'x is the Gödel number of a palindrome' (a palindrome is a sequence that reads the same forwards as backwards).

(d) $R(x)$ is 'x is the Gödel number of a sequence in which the same number appears more than once'.

Note that all these predicates are also elementary.

5.4 Let A be an abstract class of algorithms computing total number-theoretic functions. Prove that for each $n \geqslant 0$ and $k \geqslant 1$ there exists a number-theoretic function $f: N^n \to N^k$ which is not computed by any member of A.

5.5 Prove by diagonalization that the number of monotonically increasing functions $f: N \to N$ is uncountably infinite.

5.6 Let P be any property of functions $f: N \to N$. For each indexing $\phi_0, \phi_1, \ldots$ of REC_1^1 the function f is defined as follows:

$$f \triangleq \lambda x[\text{if } (\phi_x \text{ has property } P) \text{ then } 1 \text{ else } 0]$$

If f is not recursive with respect to the standard indexing, then f is not recursive with respect to any acceptable indexing. Prove this.

5.7 Consider the class F of all polynomial functions of one variable, i.e. the set of all functions of the form

$$f \triangleq \lambda x[a_0 + a_1 x + \cdots + a_n x^n]$$

where the coefficients a_i are natural numbers. The function f is assigned the index

$$i \triangleq p_0^{a_i} . p_1^{a_1} \cdots p_{n-1}^{a_n}$$

The polynomial function with index i is denoted by f_i.

(a) Prove that there is a primitive recursive (elementary) function $N^2 \to N$ which is universal for the class F

(b) Prove that there exist primitive recursive functions g and h such that:

(i) $f_{g(u,v)} = \lambda x[f_u(x) + f_v(x)]$

(ii) $f_{h(u,v)} = \lambda x[f_u(x) . f_v(x)]$

5.8 Prove that there exists a SAL(**N**) program $(1, 1, p, P)$ such that

$$f_{(1,1,p,P)}(x) = \text{'the Gödel number of } P\text{'} \qquad \text{for all } x \in N.$$

One might say that P is able to reproduce itself.
(*Hint*: use the s–n–m theorem and the recursion theorem.)

5.9 Let $\phi: N^2 \to N$ be a function universal for REC_1^1. Prove that there exists a primitive recursive function $f: N^2 \to N$ such that

$$range(\phi_{f(i,j)}) = range(\phi_i) \cup range(\phi_j).$$

5.10 Let $\psi_i^{(n,k)}$ be an acceptable indexing of REC_k^n. Prove that there exists a total recursive function $T: N^{n+2} \to \{0, 1\}$ such that for all i and x:

(a) there is an m such that $T(i, x, m) = 1$ if and only if $\psi_i^{(n,k)}\downarrow$, and

(b) if $T(i, x, m) = 1$ then $T(i, x, m + 1) = 1$.

5.11 Prove the following statement. A predicate P which is recursive with respect to any acceptable indexing is recursive with respect to every acceptable indexing.

5.12 Let ϕ_i and ψ_i be two acceptable indexings of REC_1^1. Prove that there exists a function $f: N \to N$ that has the same index in both indexings.

5.13 Prove that the recursion theorem is valid with respect to every acceptable indexing.

5.9 References

The encoding schemes used in this chapter were first used by Gödel (1965). This encoding is now called the 'Gödel numbering' or 'arithmetization' of metamathematics.

Turing (1936) introduced the concept of a universal computing machine. At the same time Kleene (1936) defined the T predicate (see Section 6.2) and later proved that this predicate could be used to construct a function universal for REC. Also the s–n–m theorem and the recursion theorem were first proved by Kleene (1936). Acceptable indexings were extensively studied by Post (1943) and Rice (1953). Indexings or Gödel numberings are now an important subject in recursion theory.

Chapter 6
An Introduction to Recursion Theory, and Solvable and Unsolvable Problems

In this chapter the *Halting Problem* for SAL(**N**) programs is introduced. This is the most basic unsolvable problem of recursion theory.

We also consider recursively solvable approximations of the Halting Problem. Such an approximation is obtained by asking 'does this computation halt within a given number of steps?', instead of 'does this computation ever halt?'. This question is the resource-bounded Halting Problem. This problem is recursively solvable, if the resource cost behaves sufficiently well, as is the case for the resource time.

Finally, we will continue our study of predicates, prove some of the closure properties discussed in Section 4.4, and give examples of non-recursively decidable predicates. The discussion is concluded with a meta-theorem: 'Any nontrivial property is recursively undecidable'.

6.1 The Halting Problem

The Halting Problem for SAL(**N**) programs is the following question:

> Does the computation on a given input x, induced by a given program P, terminate?

We want to show that the problem is not solvable by algorithmic methods. To do so, the problem must be rephrased in terms of natural numbers, thus instead of a program we take its Gödel number. Also, we must be more precise about the program and its inputs, that is we must specify the numbers of input and output variables. This results in the Halting Problem for SAL(**N**) programs with n input and k output variables. This problem will be denoted by HP^n_k.

> HP^n_k: Given an index $i \in N$ and an input value $x \in N^n$, does the computation on input x, induced by the program with Gödel number i, terminate?

Thus the Halting Problem HP^n_k is the question whether given values i_0 and x_0 for i and x satisfy the predicate

$$\lambda ix[\phi_i^{(n,k)}(x)\downarrow]$$

HP^n_k is recursively solvable if and only if the above predicate is recursively decidable; that is, if and only if the function H defined below belongs to $TREC$.

$$H \triangleq \lambda ix[\text{if } \phi_i^{(n,k)}(x)\downarrow \text{ then } 1 \text{ else } 0]$$

Before proceeding to prove that the Halting Problem is recursively unsolvable, we introduce some terminology.

We talk about problems and predicates. These closely resemble one another – so closely in fact that it is not really necessary to discuss problems separately, but we do so because they are so intuitively appealing.

We use the following terminology.

Let P be a predicate having n free variables x_1 to x_n, ranging over N.

- The *problem* P is the following question:

  Given particular values a_1 to a_n, does $(a_1, \ldots, a_n)$ satisfy the predicate; that is, is $P(a_1, \ldots, a_n) = \text{TRUE}$?

- An *instance* of the problem is a sequence of values which can legally be substituted for the variables in the predicate. Thus an instance of the problem P is an n-tuple $(a_1, \ldots, a_n) \in N^n$.
- The problem is called *recursively solvable* if the corresponding predicate is recursively decidable.
- The problem is called *partially recursively solvable* if the corresponding predicate is recursively enumerable.

We turn once more to the Halting Problem. To begin, we restrict ourselves to HP^1_1.

Theorem 6.1

HP_1^1, the Halting Problem for SAL(**N**) programs with one input and one output variable, is recursively unsolvable.

Proof

Assume that HP_1^1 is recursively solvable; that is, the function

$$H \triangleq \lambda ix[\text{if } \phi_i(x)\downarrow \text{ then } 1 \text{ else } 0]$$

is recursive and therefore has an index, say m. In other words, $H = \phi_m^{(2,1)}$. Consider the function $g: N \to N$ defined as follows:

$$g \triangleq \lambda x[\text{if } H(x,x) = 0 \text{ then } 1 \text{ else } \uparrow]$$

This function g is computed by the following program:

```
input: x
output: y
method:
   y:= 1;
   while H(x, x) ≠ 0 do
      y:= 0
   od
```

Therefore, if H is recursive, then g is recursive and has an index, say n. Note that the index n can be effectively determined from an index m of the function H. Then $H(n,n) \neq 0$ and $H(n,n) \neq 1$.

(1) If $H(n,n) = 1$, then $\phi_n(n)\downarrow$, by the definition of H. Then $g(n)\downarrow$, because n is an index for g by the choice of n. Then $H(n,n) \neq 1$, by the structure of the above program. In conclusion, $H(n,n)$ cannot be equal to 1.

(2) If $H(n,n) = 0$, then $\phi_n(n)\uparrow$, by the definition of H. Then $g(n)\uparrow$, because n is an index for g by the choice of n. Then $H(n,n) \neq 0$ by the structure of the above program computing g. In conclusion, $H(n,n)$ cannot be equal to 0.

Thus $H(n,n)$ is unequal to 0 and unequal to 1. Since by definition $H(n,n)$ must be either 0 or 1, we have a contradiction. Therefore the assumption, that HP_1^1 is recursively solvable must be wrong, consequently HP_1^1 is not recursively solvable. ■

In the above proof, we use Church's thesis to infer that the function g has an index. However, we are not able to specify a program which computes

g, because the program, as given in the above proof, contains a subroutine of which we do not know the code. Using Church's thesis in such hypothetical contexts is somewhat dubious. We derive a contradiction from the assumption that H is recursive and therefore conclude that it is not. But if $H \notin REC$, then of course our assumption that H is recursive is wrong. But not only that, also the argument by which we arrive at a contradiction is no longer applicable, because we cannot call Church's thesis.

Also note that Church's thesis is *not* a valid rule of argument. If you give a proof using Church's thesis, you must be able to substitute for all calls on Church's thesis formally acceptable, more detailed, longer and most probably less readable chains of argument.

A more rigorous (though still not completely rigorous) proof of the recursive unsolvability of the Halting Problem for SAL(**N**) programs is given below.

Consider the function

$$\psi \triangleq \lambda zx[\text{if } \phi_z^{(2,1)}(x,x)=0 \text{ then } 1 \text{ else} \uparrow]$$

To evaluate $\psi(z,x)$ proceed as follows:

- Evaluate $\phi_z^{(2,1)}(x,x)$. If $\phi_z^{(2,1)}(x,x)\uparrow$, we can proceed no further. But this is not a problem, because in this case $\psi(z,x)$ is undefined too.
- If the computation of $\phi_z^{(2,1)}(x,x)$ terminates, compare the result with 0. If it is equal to 0 then output 1 and halt, otherwise enter some nonterminating computation.

Thus by Church's thesis, $\psi \in REC_1^2$. Note that this call on Church's thesis is perfectly legal, because we can completely specify a program which computes ψ. By the s–n–m theorem there is a total recursive function h such that $\phi_{h(z)} = \lambda x[\psi(z,x)]$.

Now we have found a uniform way to disprove the assumption that $H = \phi_z^{(2,1)}$, because $H(h(z),h(z)) \neq \phi_z^{(2,1)}(h(z),h(z))$, for all $z \geqslant 0$.

(1) Either $\phi_z^{(2,1)}(h(z),h(z))\uparrow$, in which case H and $\phi_z^{(2,1)}$ differ on the argument $(h(z),h(z))$ and are therefore unequal

(2) or $\phi_z^{(2,1)}(h(z),h(z))\downarrow$. If $H = \phi_z^{(2,1)}$ then we obtain a contradiction as follows:

- If $H(h(z),h(z))=1$, then $\phi_z^{(2,1)}(h(z),h(z))=1$, so $\psi(z,h(z))\uparrow$; hence $\phi_{h(z)}(h(z))\uparrow$, and therefore $H(h(z),h(z)) \neq 1$. In conclusion, $H(h(z),h(z))$ is unequal to 1.
- If $H(h(z),h(z))=0$, then $\phi_z^{(2,1)}(h(z),h(z))=0$, so $\psi(z,h(z))\downarrow$; hence $\phi_{h(z)}(h(z))\downarrow$, and therefore $H(h(z),h(z)) \neq 0$. In conclusion, $H(h(z),h(z))$ is unequal to 0.

Since H is by definition total and assumes no other values than 0 and 1, we have a contradiction. Thus for $\phi_z^{(2,1)}(h(z),h(z))\downarrow$, the functions H and $\phi_z^{(2,1)}$ differ on the argument $(h(z),h(z))$.

This argument is somewhat more effective than the one in the preceding proof. Given any program whatsoever, the above method effectively determines an argument value, for which the function computed by the given program does not have the required value.

Theorem 6.2

The Halting Problem for SAL(**N**) programs with n input and k output variables is recursively unsolvable.

Proof

Let $\psi \triangleq \lambda i x_1,\ldots,x_n[(\phi_i(x_1),1,\ldots,1)]$($k-1$ times '1'). By Church's thesis ψ is in REC_k^{n+1}. By the s–n–m theorem there is a total recursive function h such that $\phi_{h(i)}^{(n,k)} = \lambda x[\psi(i,x)]$. Clearly, $\phi_i(x)\downarrow$ if and only if $\phi_{h(i)}^{(n,k)}(x,0,\ldots,0)\downarrow$. If HP_k^n were recursively solvable, then HP_1^1 would be recursively solvable too. The latter not being the case, we conclude that HP_k^n is not recursively solvable either. ■

The above proof shows an example of a proof method called reduction. In the above proof, HP_1^1 is reduced to HP_k^n, which we denote by $HP_1^1 \leqslant HP_k^n$. This is achieved by giving a total recursive function, which transforms an instance of HP_1^1 into an instance of HP_k^n, such that HP_1^1(instance) = TRUE if and only if HP_k^n(transformed instance) = TRUE. In the above proof, this function was left implicit. It is defined as follows:

$$f \triangleq \lambda i x[(h(i),x,0,\ldots,0)]$$

Definition 6.1

Let A and B be predicates. Predicate A is *reducible* to predicate B ($A \leqslant B$) if and only if there exists a total recursive function f such that

$$(\forall x)[A(x) \text{ if and only if } B(f(x))].$$

To express that the function f is used in the reduction, we say that $A \leqslant B$ via f. ■

Theorem 6.3

Let A and B be predicates and $A \leqslant B$. If B is recursively dedicable, then A is recursively decidable too. If B is recursively enumerable, then A is recursively enumerable too.

Proof

Let $A \leqslant B$ via the total recursive function h.

(1) If B is recursively decidable then $B = \lambda x[sg(f(x)) = 1]$ for some $f \in TREC$. Then $A = \lambda x[sg(f(h(x))) = 1]$. Therefore A is recursively decidable too.

(2) If B is recursively enumerable, then $B = \lambda x[sg(f(x)) = 1]$, for some $f \in REC$. Then $A = \lambda x[sg(f(h(x))) = 1]$. Therefore A is recursively enumerable too. ■

Thus to prove that some problem X is not recursively solvable, find some other problem Y known to be recursively unsolvable, and show that $Y \leqslant X$. This technique will be used frequently.

The reduction concept is very important in structuring the class of non-recursively solvable problems, and will be discussed further in Chapter 7.

Theorem 6.4

The Halting Problem for SAL(**N**) programs having n input and k output variables is partially recursively solvable.

Proof

Let $\psi \triangleq \lambda ix[1 + \pi_1^k(\phi_i^{(n,k)}(x))]$. Then $\psi \in REC_1^{n+1}$ by Church's thesis, and $HP_k^n = \lambda ix[sg(\psi(i, x)) = 1]$. Therefore HP_k^n is a recursively enumerable predicate and the Halting Problem for SAL(**N**) programs having n input and k output variables is partially recursively solvable. ■

A more restricted version of the Halting Problem will be used frequently.

Definition 6.2

Let K be the problem to determine whether the computation on a given input x, induced by the program with Gödel number x, terminates:

$$K \triangleq \lambda x[\phi_x(x)\downarrow]$$

The set $\{x \mid \phi_x(x)\downarrow\}$ will also be denoted by K. ■

Theorem 6.5

The predicate $K = \lambda x[\phi_x(x)\downarrow]$ is recursively enumerable, but not recursively decidable.

Proof

(1) K is recursively enumerable by Church's thesis, for $K = \lambda x[sg(1 + \phi_x(x)) = 1]$.

(2) The problem is not recursively decidable. This follows directly from the proof of Theorem 6.1. There we started with the function

$$H \triangleq \lambda ix[\text{if } \phi_i(x)\downarrow \text{ then } 1 \text{ else } 0],$$

but showed in fact, that the assumption that $\lambda x[H(x,x)]$ is a total recursive function, leads to a contradiction. ■

We can also show that K is not recursively solvable, using the technique of reduction.

Let $\psi \triangleq \lambda xyz[sg(1+\phi_x(y)).z]$.

By the s–n–m theorem there is a total recursive function h such that $\phi_{h(x,y)} = \lambda z[\psi(x,y,z)]$.

(1) If $HP_1^1(x,y) = \text{TRUE}$, and thus $\phi_x(y)\downarrow$, then $\phi_{h(x,y)} = \lambda z[z]$, and therefore $\phi_{h(x,y)}(h(x,y))\downarrow$, whence $K(h(x,y)) = \text{TRUE}$.

(2) If, on the other hand, $HP_1^1(x,y) = \text{FALSE}$, and therefore $\phi_x(y)\uparrow$, then $\phi_{h(x,y)} = \lambda z[\uparrow]$, and therefore $K(h(x,y)) = \text{FALSE}$.

In conclusion, $HP_1^1 \leqslant K$, via the function h defined above. Furthermore, we also have that $K \leqslant HP_1^1$, via the function $\lambda x[(x,x)]$. Thus, K is recursively solvable if and only if HP_1^1 is recursively solvable. In other words, K and HP_1^1 have the same degree of unsolvability. This is not the case for problems in general; there are problems having mutually different degrees of unsolvability, i.e., which cannot be reduced to one another. These points will be discussed further in Chapters 7 and 8.

Recall from Section 4.1 the definition of the operator $\bar{\mu}$.

- Let $f: N^{n+1} \to N$ be any partial function
- Let $V(x,y) \triangleq \{z \mid f(x,z) = y\}$
- Define the function $(\bar{\mu}f)$ as follows:

$$(\bar{\mu}f)(x,y) \triangleq \text{if } V(x,y) \neq \varnothing \text{ then } \min V(x,y) \text{ else } \uparrow.$$

As already stated in Section 4.1, the operator $\bar{\mu}$ does not adequately extend bounded minimization to non-total functions, because there are partial recursive functions f, such that $(\bar{\mu}f)$ is not a partial recursive function. The function ψ defined below, is an example.

$$\psi(x,y) \triangleq \text{if } (\phi_y(y)\downarrow \text{ and } x \leqslant y) \text{ then } x \text{ else } \uparrow$$

By Church's thesis $\psi \in REC_1^2$. We will show that the assumption '$\bar{\mu}\psi$ is partial recursive' leads to the conclusion 'K is a recursively decidable

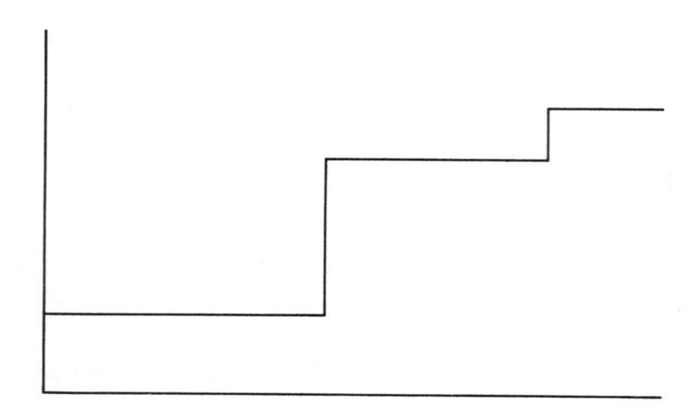

Figure 6.1 The function $f \triangleq \lambda x[\bar{\mu}\psi(x,x)]$.

predicate'. By Theorem 6.5, K is not recursively decidable, and therefore $\bar{\mu}\psi$ is not a partial recursive function.

Assume that $\bar{\mu}\psi$ is partial recursive, then $f \triangleq \lambda x[\bar{\mu}\psi(x,x)]$ is also partial recursive. But we can say more about the function f.

- f is total

  $$f(x) = \bar{\mu}\psi(x,x) = \text{the least } z\text{, if any, such that } \phi_z(z)\downarrow \text{ and } z \geqslant x.$$

  The function f is total because there are infinitely many z such that $\phi_z(z)\downarrow$.
- If $x_1 \leqslant x_2$ then $f(x_1) \leqslant f(x_2)$.
  $f(x_1)$ is the least z such that $\phi_z(z)\downarrow$ and $z \geqslant x_1$. If $x_2 \leqslant z$ then $f(x_1) = f(x_2)$. Otherwise $f(x_1) < f(x_2)$. Thus f is a step function with $range(f) = \{x \mid K(x) = \text{TRUE}\}$: see Figure 6.1.

Using the function f, we can decide whether or not $K(x) = \text{TRUE}$, as follows:

input: x
output: y $\{\text{if } K(x) = \text{TRUE then } y = 1 \text{ else } y = 0\}$
method:

$\{z = 0,\ y = 0\}$

while $f(z) < x$ **do** $z := z + 1$ **od**;

if $f(z) = x$ **then** $y := 1$ **else** $y := 0$ **fi**

Thus if $\bar{\mu}\psi$ is partial recursive, then K must be recursively decidable. Therefore, the operator $\bar{\mu}$ is not an adequate extension of the minimization operator μ to partial functions.

We close this section with some examples of recursively unsolvable problems.

EXAMPLE 6.1

(1) Since the recursively decidable predicates are closed with respect to negation (see Table 4.3 on page 79) it follows that $\neg HP_k^n$ and $\neg K$ are also recursively unsolvable.

(2) Let Γ be the problem to determine whether there is any input value on which the computation induced by a given program terminates. That is,

$$\Gamma \triangleq \lambda x[dom(\phi_x) = \varnothing] = \lambda x[\phi_x(y)\uparrow, \text{ for all } y]$$

We show that this problem is not recursively solvable by reducing the problem $\neg K$ to it. Let

$$\psi \triangleq \lambda xy[\phi_x(x).y]$$

By Church's thesis ψ is recursive, and by the s–n–m theorem there is a total recursive function such that $\phi_{f(x)} = \lambda y[\psi(x, y)]$. Then $\neg K \leqslant \Gamma$ via f, because

(1) If $\neg K(x) = \text{TRUE}$, then $\phi_x(x)\uparrow$, and then $(\forall y)[\psi(x, y)\uparrow]$, and then $(\forall y)[\phi_{f(x)}(y)\uparrow]$, and then $\Gamma(f(x)) = \text{TRUE}$.

(2) If $\neg K(x) = \text{FALSE}$, then $\phi_x(x)\downarrow$, then $(\forall y)[\psi(x, y)\downarrow]$, and then $(\forall y)[\phi_{f(x)}(y)\downarrow]$, and then $\Gamma(f(x)) = \text{FALSE}$.

Thus $\neg K \leqslant \Gamma$ via f. Because $\neg K$ is not recursively solvable, Γ is not recursively solvable either (by Theorem 6.3).

(3) We once more present the example above, now using terminology in accordance with the interpretation of K and Γ as sets.

$$\Gamma = \{x | dom(\phi_x) = \varnothing\}$$

Let ψ and f be as above. Then $\bar{K} \leqslant \Gamma$ via f, where $\bar{K} \triangleq N - K$.

(1) If $x \in \bar{K}$, then $\phi_x(x)\uparrow$, and then $(\forall y)[\psi(x, y)\uparrow]$, and then $(\forall y)[\phi_{f(x)}(y)\uparrow]$, and then $f(x) \in \Gamma$.

(2) If $x \notin \bar{K}$, then $\phi_x(x)\downarrow$, and then $(\forall y)[\psi(x, y)\downarrow]$, and then $(\forall y)[\phi_{f(x)}(y)\downarrow]$, and then $f(x) \notin \Gamma$.

(4) The problem to determine whether two given programs are equivalent, i.e. compute the same function, is recursively unsolvable. That is, the predicate

$$E \triangleq \lambda xy[\phi_x = \phi_y]$$

is not recursively decidable.

Let n be a Gödel number of the function $\lambda x[\uparrow]$, so that $\Gamma(n) = \text{TRUE}$. Let ψ and f be as in (2) above, and define $h \triangleq \lambda x[(f(x), n)]$. Then $\neg K \leqslant E$ via h. Therefore, program equivalence is a recursively unsolvable problem.

6.2 The resource-bounded Halting Problem

We have seen that the Halting Problem *HP* is recursively unsolvable. This is due to the fact that the length of a computation is unbounded. In this section we will study the problem to determine whether there are terminating computations using at most a given amount of some particular computing resource.

First consider the problem:

> Does the computation of a given program P on a given input value x, terminate within a given number t of steps?

This can be expressed as a predicate as follows.

$$T^{(n,k)} \triangleq \lambda ixt[\text{the computation of } \phi_i(x) \text{ requires at most } t \text{ steps}]$$

The above predicate is known as 'Kleene's T-predicate'.

We shall show that the T-predicate is recursively decidable by giving a program which computes the characteristic function of the predicate. This program is obtained by slightly modifying our universal program (Algorithm 5.1).

Algorithm 6.1 (to compute the characteristic function of the T-predicate)

input: $g \in N$ {a Gödel number}
$x \in N^n$ {input to program having Gödel number g}
$bound \in N$ {the maximum number of computation steps}
output: y {$y =$ if $T^{(n,k)}(g, x, bound) = \text{TRUE}$ then 1 else 0}
method:

```
initialize (memory);
cost:= 0;
y:= bounded_simulate (g, memory, bound)
```
■

The macro *bounded_simulate* is similar to *simulate* (Algorithm 5.2) but keeps track of the accumulated cost of the computing resource, in this case time. It stops simulating the computation as soon as this cost exceeds the given bound.

Algorithm 6.2 (for *bounded_simulate*)

input: $g, memory, bound$
output: a
method:

$L := 1 + \sigma^1_{2,1}(g)$;
while $L \neq 0 \wedge cost \leqslant bound$ **do**
  $S := \sigma^1_{L,1}(\sigma^1_{2,2}(g))$;
  $r :=$ *remainder of* S *divided by* 5;
  if $r \neq 4$ **then**
    $execute(S, r, memory)$;
    $pop\ (g)$;
    $L := L \dot{-} 1$
  else
    $x := \tau(memory, \sigma^1_{2,1}((S \dot{-} 4)/5) + 1)$;
    if $x \neq 0$ **then**
      $z := \sigma^1_{2,2}((S \dot{-} 4)/5)$;
      $push\ (z, g)$;
      $L := L + \sigma^1_{2,1}(z)$
    else
      $pop\ (g)$;
      $L := L \dot{-} 1$
    fi
  fi
od;
if $L = 0$ **then** $a := 1$ **else** $a := 0$ **fi** ■

The macro *execute* updates the accumulated cost, but is otherwise the same as used for Algorithm 5.2.

execute $(S, r, memory)$

$update\_cost$;
if $r = 0$ **then**
  $x := 1 + S/5$
else
  $x := 1 + \sigma^1_{2,1}((S \dot{-} r)/5)$;
  $y := 1 + \sigma^1_{2,2}((S \dot{-} r)/5)$
fi;
if $r = 0$ **then** $M := \omega(M, x, 0)$ **fi**;
if $r = 1$ **then** $M := \omega(M, x, \tau(M, y))$ **fi**;
if $r = 2$ **then** $M := \omega(M, x, \tau(M, y) + 1)$ **fi**;
if $r = 3$ **then** $M := \omega(M, x, \tau(M, y) \dot{-} 1)$ **fi**

The macro *update_cost* is

```
update_cost
  cost:= cost + 1
```

The expansion of the other macros is the same as for Algorithm 5.1.

Clearly, Algorithm 6.1 computes the characteristic function of Kleene's T-predicate, and has only terminating computations, because the sequence in the scope of the while statement in *bounded_simulate* is executed no more than specified in the input parameter *bound*. Therefore Kleene's T-predicate is recursively decidable. It can be proved that the characteristic function of this predicate is primitive recursive.

Kleene's T-predicate is very convenient, in circumstances where you have to consider various computations induced by a given program.

EXAMPLE 6.2

Prove the following assertion:

> Let ϕ_z be a recursive function $N \rightarrow N$, which has non-empty range, i.e. there is at least one x such that $\phi_z(x)\downarrow$. Then there exists a total recursive function f which enumerates the range of ϕ_z; that is,
>
> $$\{f(x) | x \in N\} = \{\phi_z(x) | x \in N\}.$$

You cannot simply try and compute $\phi_z(0)$, $\phi_z(1)$, $\phi_z(2), \ldots$, because if for example $\phi_z(13)$ happens to be undefined, the computation will not terminate, and you will never proceed to the computation of $\phi_z(14)$. But you can try all pairs (argument value, length of computation).

Thus we arrive at the following program to compute the function f:

input: x
output: y
method:

$x_1 := \sigma^1_{2,1}(x)$; $t := \sigma^1_{2,2}(x)$;
if $T(z, x_1, t)$ **then**
  $y := \phi_z(x_1)$
else
  $y := a$; $\{$such that $\phi_z(a)\downarrow\}$
  $\{$since $range(\phi_z) \neq \varnothing$,$\}$
  $\{$such a value a must exist$\}$
fi

Kleene's T-predicate is the customary *resource-bounded termination predicate*, but there are many other possibilities. If we expand the macro *update_cost* in some other way to model the use of another computing resource, we are done. The resulting predicate will be denoted by T_r, where r denotes the resource in question.

The predicate T_r will be recursively decidable if the resource cost function behaves properly. Let F_i denote the 'follow function' of the program with Gödel number i, as defined in Definition 3.4 (page 46). Then

$$F_i^0(1,x), F_i^1(1,x), F_i^2(1,x), \ldots, F_i^j(1,x), \ldots$$

where $F_i^k(1,x)$ denotes the state vector reached after k steps, can be identified with the computation induced by the program with Gödel number i on input value x. Then the cost function behaves properly, and therefore T_r is recursively decidable if for every i, x and t, there is a j, such that $cost(F_i^j(1,x)) > t$.

The cost function of the resource *time*, $cost(F_i^j(1,x)) = j$, clearly satisfies this requirement.

Note, however, that the cost function of the resource *space* does *not* satisfy this requirement. Although we do not have a space measure defined on our compound data type **N**, it is obvious that the following program requires only a bounded amount of space, in any reasonable sense of the word space, because the space used does not depend on the length of the computation, but only on the value of x.

input: x
output: y
method:
  while $x \neq 0$ **do** $y := x$ **od**

Nevertheless, a recursively decidable space-bounded termination predicate can be defined, because there is a maximum to the magnitude of a number which can be represented using no more than a given amount of space.

Let m be the maximum number which can be represented within a given amount s of space. Let (n, k, p, P) be a SAL(**N**) program having its lines numbered from 1 to f. (Numbering the lines of a program and the state vector concept are discussed in Section 3.4.) The number of different state vectors for which no more space than s is used is at most $(1+f).m^p$. The numbers f and p can be computed from the Gödel number of the program. If a terminating computation using no more space than s exists, then the length of this computation is at most equal to the number of mutually different state vectors using no more space than s, that is, the length is at most $(1+f).m^p$.

Thus, the algorithms discussed earlier can be adapted so as to compute the characteristic function of a space-bounded termination predicate, by appropriately expanding the macro *update_cost*, and changing the while condition in the expansion of *bounded_simulate* (Algorithm 6.2) to '$length \neq 0 \wedge cost \leqslant bound \wedge number\_of\_steps \leqslant (1+f).m^p$'.

The recursive decidability of Kleene's termination predicate depends on the Gödel numbering of programs. As can be seen from the proof of Lemma 5.1 (page 115), any acceptable indexing admits a Kleene T-predicate. That is, there is a recursively decidable predicate T, such that

(1) if $\phi_i(x)\uparrow$, then $T(i,x,y) = \text{FALSE}$ for all $y \geqslant 0$

(2) if $\phi_i(x)\downarrow$, then there is some y_0 such that $T(i,x,y) = \text{if } y < y_0 \text{ then}$ FALSE else TRUE

Maybe interpreting the value of the parameter y as the time used is not very credible, but the recursively decidable predicate T exists. This predicate can be used as Kleene's T-predicate is used with respect to our model SAL(**N**).

6.3 Recursively decidable and enumerable predicates

The classes of recursively decidable, and of recursively enumerable, predicates have been defined in Section 4.4. Also, some of their closure properties with respect to propositional connectives and quantification have been discussed. In this section we will take a closer look at predicates, especially at recursively enumerable predicates. Finally, some predicates, $\neg K$ among them, will be shown to be not recursively enumerable.

Recursively enumerable predicates were defined to be those predicates which can be expressed in the form $\lambda x[sg(f(x))) = 1]$, for some function $f \in REC$. The name 'recursively enumerable' was explained by the remark that it is possible to list effectively the TRUE instances ($\{x | P(x) = \text{TRUE}\}$) of a recursively enumerable predicate. We will now prove that this is indeed the case.

Theorem 6.6

The predicate P is recursively enumerable if and only if either P is always FALSE, or there is a total recursive function f, such that

$$\{f(m) | m \in N\} = \{x | P(x) = \text{TRUE}\}.$$

Proof

- Let $\psi \in REC_1^n$ and $P = \lambda x[sg(\psi(x)) = 1]$.
  If P is always FALSE, there is nothing to prove. So let us assume that $P(a) = \text{TRUE}$, where a is a particular value in N^n. The total recursive function f effectively listing the TRUE instances of P is computed by the program:

input: $m \in N$
output: $x \in N^n$
method:

$s := \sigma^1_{2,1}(m)$; $t := \sigma^1_{2,2}(m)$;
{assume that z is an index for ψ}
if $T^{(n,1)}(z, \sigma^1_n(s), t)$ **then**
  if $sg(\psi(\sigma^1_n(s))) = 1$ **then**
    $x := \sigma^1_n(s)$
  else $x := a$
  fi
else $x := a$
fi

- For the converse:
  - — If P is always FALSE, then $P = \lambda x[sg(C^n_0) = 1]$. Thus P is recursively decidable and recursively enumerable.
  - — Assume that P is not always FALSE, and the set of all TRUE instances is enumerated by $f \in TREC^1_n$. Define $\psi(x) \triangleq 1 + \mu m[f(m) = x]$, for all $x \in N^n$. Then $\psi \in REC^n_1$ and $P = \lambda x[sg(\psi(x)) = 1]$, and therefore P is recursively enumerable. ■

If a predicate is recursively decidable, then its negation is also recursively decidable. Therefore, both the set of all TRUE instances and the set of all FALSE instances can be listed effectively. The converse is also true.

Theorem 6.7

A predicate P is recursively decidable if and only if both P and $\neg P$ are recursively enumerable.

Proof

- If P is recursively decidable, then $\neg P$ is recursively decidable too, and both P and $\neg P$ are recursively enumerable.
- Since the always FALSE predicate is recursively decidable, we only have to consider the case where both P and $\neg P$ are recursively enumerable, and neither is always FALSE. Then there are functions $f, g \in TREC^1_n$, enumerating the TRUE instances of P and $\neg P$ respectively. The following program computes the characteristic function of P:

input: $x \in N^n$
output: y {if $P(x)$ then $y = 1$ else $y = 0$}
method:

$m := 0$;

while $f(m) \neq x \wedge g(m) \neq x$ **do** $m := m + 1$ **od**;

if $f(m) = x$ **then** $y := 1$ **else** $y := 0$ **fi**

Thus P is recursively decidable. ∎

If a problem is recursively solvable, and you wish to determine what the solution is for a given instance x, then you can systematically search the set of all yes-instances, and simultaneously the set of all no-instances. You are guaranteed to find your instance x in one of these, at some moment of time. There might be a better solution method, but this systematic search will at least work.

If, on the other hand, the problem is not recursively solvable, but only partially recursively solvable, then you can do no better than search the set of all yes-instances, and simultaneously a proper subset (maybe the empty set) of the set of all no-instances. If yours is a yes-instance, it will be found to be so, by this search procedure. But if it is a no-instance, the search might not terminate. There may of course be a subset of instances which can easily be solved, but a general procedure cannot be better than the above-described search method.

Every recursively enumerable predicate can be written in a so-called *normal form*. The theorem below gives three such normal forms.

Theorem 6.8

The following statements are equivalent:

(1) $Q \in PRED(REC_1^n)$, i.e. Q is recursively enumerable

(2) There is a $\phi_i^{(n,1)}$ such that $Q = \lambda x[\phi_i^{(n,1)}(x)\downarrow]$

(3) There is an $m \in N$ such that $Q = \lambda x[(\exists t)[T^{(n,1)}(m, x, t)]]$.

Proof

- (1) implies (2).

  If Q is recursively enumerable, then there is an $f \in REC$, such that

  $$Q = \lambda x[sg(f(x)) = 1].$$

  Define the function ψ as follows:

  $$\psi \triangleq \lambda x[\text{if}(f(x)\downarrow \text{ and } sg(f(x)) = 1) \text{ then } 1 \text{ else } \uparrow]$$

By Church's thesis $\psi \in REC$, and clearly $Q = \lambda x[\psi(x)\downarrow]$.

- (2) implies (3)
  Thus, $\psi(x)\downarrow$ means the same thing as $(\exists t)[T^{(n,1)}(m, x, t)]$, where m is an index for the recursive function ψ.
- (3) implies (1)
  Let $Q = \lambda x[(\exists t)[T^{(n,1)}(m, x, t)]]$, and define the function ψ as follows:

$$\psi \triangleq \lambda x[1 + \phi_m(x)]$$

Then, clearly $\psi \in REC_1^n$ and $Q = P_\psi \triangleq \lambda x[sg(\psi(x)) = 1]$. ■

In the above theorem, (1) is our definition of recursive enumerability. When discussing sets, form (2) is used. It states that a set is recursively enumerable if and only if it is the domain of a partial recursive function. This immediately provides a recursive indexing of the recursively enumerable sets.

Definition 6.3

$$W_i^n \triangleq dom(\phi_i^{(n,1)})$$ ■

Number (3) in the above theorem is the so-called *Kleene normal form*. It directly gives an indexing of the recursively enumerable predicates.

Definition 6.4

$$Q_i^{(n)} \triangleq \lambda x[(\exists t)[T^{(n,1)}(i, x, t)]]$$ ■

Note that there does not exist an indexing of the recursively decidable predicates.

We turn now to the closure properties of the recursively enumerable and the recursively decidable predicates with respect to propositional connectives and quantification. These are summarized in Table 6.1.

The crucial property in Table 6.1 is that the set of all recursively enumerable predicates is not closed with respect to negation.

Theorem 6.9

The class of all recursively enumerable predicates is not closed with respect to negation. The predicate K as given in Definition 6.2 (page 132) is recursively enumerable but not recursively decidable. The predicate $\neg K$ is not recursively enumerable.

Proof

We already known from Theorem 6.5 (page 132) that K is recursively enumerable but not recursively decidable. If the recursively enumerable predicates were closed with respect to negation, then $\neg K$ would be

Table 6.1 Closure properties of recursively decidable and recursively enumerable predicates with respect to propositional connectives and quantification.

Operator	Recursively decidable predicates	Recursively enumerable predicates
$\neg$	closed	not closed
$\rightarrow$	closed	not closed
$\wedge$	closed	closed
$\vee$	closed	closed
$\forall$	not closed	not closed
bounded $\forall$	closed	closed
$\exists$	not closed	closed
bounded $\exists$	closed	closed

recursively enumerable. If K and $\neg K$ are both recursively enumerable, then by Theorem 6.7 K is recursively decidable. Therefore, $\neg K$ is not recursively enumerable, and the class of recursively enumerable predicates is not closed with respect to negation. ■

We turn now to the closure properties of the recursively decidable predicates. Apart from unbounded quantification, they have all been proved in Section 4.4. Since by Theorem 6.8 every recursively enumerable predicate can be obtained by existential quantification from the recursively decidable T-predicate, it follows that the recursively decidable predicates are not closed with respect to existential quantification. Non-closure with respect to universal quantification is now immediate, $(\exists x)P(x)$ is equivalent to $\neg((\forall x)(\neg P(x)))$.

We consider the closure properties of the recursively enumerable predicates. We only show closure with respect to $\vee$ and $\exists$. The proofs of the remaining ones are similar, and are left as an exercise. Kleene's T-predicate and normal form are crucial in these proofs.

Theorem 6.10

Let $Q_i^{(n)}$ and $Q_j^{(n)}$ be recursively enumerable predicates having n free variables. The predicate $Q \triangleq Q_i^{(n)} \vee Q_j^{(n)}$ is also recursively enumerable.

Proof

Let ψ be the function computed by the following program:

input: $x \in N^n$

output: y {output is immaterial here}
method:
$t := 0;$
while $\neg T^{(n,1)}(i,x,t) \wedge \neg T^{(n,1)}(j,x,t)$ **do** $t := t+1$ **od**

Then, by Church's thesis, ψ is recursive. Let m be an index of ψ. Then $Q_i^{(n)} \vee Q_j^{(n)} = Q_m^{(n)}$. ■

The above proof shows somewhat more than is stated in the text of the theorem, namely that the recursively enumerable predicates are *effectively closed* with respect to $\vee$. That is, there is a total recursive function f such that $Q_i^{(n)} \vee Q_j^{(n)} = Q_{f(i,j)}^{(n)}$. This follows directly by the s–n–m theorem. All the closures given in Table 6.1 are effective closures.

Theorem 6.11

Let $Q_i^{(n)}$ be a recursively enumerable predicate. Then $(\exists x_k)Q_i^{(n)}$ is recursively enumerable too.

Proof

Consider the following program:

input: $x_1, \ldots, x_{k-1}, x_{k+1}, \ldots, x_n$
output: y {value of y is immaterial}
method:
$t := 0;$
while $\neg T^{(n,1)}(i, x_1, \ldots, x_{k-1}, \sigma_{2,1}^1(t), x_{k+1}, \ldots, x_n, \sigma_{2,2}^1(t))$ **do**

$t := t+1$

od

Let m be an index for the function $\psi \in REC_1^{n-1}$ computed by the above program. Then clearly $Q_m^{(n-1)} = (\exists x_k)Q_i^{(n)}$. ■

We close this section with a meta-theorem stating that only trivial properties of recursive functions can be effectively decided.

Let P be any nontrivial property of partial recursive functions $N \rightarrow N$, where nontrivial means that there are functions having property P and also functions which do not have property P. Let Q be the predicate stating that the function computed by a given program has property P, that is

$$Q \triangleq \lambda i[P(\phi_i)]$$

Since P is a nontrivial property, Q is neither always TRUE nor always FALSE. Even more so: there are infinitely many TRUE instances and infinitely many FALSE instances.

Assume for the time being that the everywhere undefined function does not have property P and that the partial recursive function f does have property P. Define

$$\psi \triangleq \lambda xy[sg(1+\phi_x(x)).f(y)]$$

By Church's thesis $\psi \in REC_1^2$, and by the s–n–m theorem there exists a total recursive function h such that $\phi_{h(x)} = \lambda y[\psi(x, y)]$. Then $K \leqslant Q$ via h because

(1) If $K(x) = \text{TRUE}$, then $\phi_x(x)\downarrow$, and then $\phi_{h(x)} = f$. Therefore $\phi_{h(x)}$ has property P and $Q(h(x)) = \text{TRUE}$.

(2) If $K(x) = \text{FALSE}$, then $\phi_x(x)\uparrow$, and then $\phi_{h(x)}$ is equal to the everywhere undefined function, and therefore does not have property P. Thus $Q(h(x)) = \text{FALSE}$.

Also $\neg K \leqslant \neg Q$, as can easily be checked.

Therefore Q is not recursively decidable and $\neg Q$ is not recursively enumerable. No conclusion can be drawn from the above about whether or not Q is recursively enumerable.

If the everywhere divergent function happens to have property P, then the above argument fails. But in that case we can consider the property not-P, for which the above argument is valid.

In general:

> if P is a nontrivial property of partial recursive functions, and $Q \triangleq \lambda i[P(\phi_i)]$, then Q is not recursively decidable. If the everywhere undefined function has property P, then Q is not recursively enumerable, and otherwise $\neg Q$ is not recursively enumerable.

Note that in the above it is crucial that P is a property of functions: if $\phi_n = \phi_m$ then $Q(n)$ iff $Q(m)$. There are many predicates which are not of this type, for instance the predicate K.

EXAMPLE 6.3

(1) $A_1 \triangleq \lambda i[\phi_i \text{ is total}]$
A_1 is not recursively decidable, and $\neg A_1$ is not recursively enumerable.

(2) $A_2 \triangleq \lambda i[\text{range } (\phi_i) \text{ is finite}]$
A_2 is not recursively decidable, and not recursively enumerable.

(3) The predicate A_1 is not recursively enumerable. This can be shown for instance by diagonalizing over the TRUE instances of A_1, or by a reduction which is radically different from the one above. We will follow this course. Define

$$\psi \triangleq \lambda xy[\text{if} \neg T(x, x, y) \text{ then } y \text{ else } \uparrow]$$

By Church's thesis this ψ is recursive, and by the s–n–m theorem there exists a total recursive function h, such that $\phi_{h(x)} = \lambda y[\psi(x, y)]$. Then $\neg K \leqslant A_1$ via the function h, because

(a) If $\neg K(x) = \text{TRUE}$ then $\phi_x(x)\uparrow$, then $(\forall y)(\neg T(x, x, y))$, and then $(\forall y)[\phi_{h(x)}(y)\downarrow]$, and then $A_1(h(x)) = \text{TRUE}$.

(b) If $\neg K(x) = \text{FALSE}$ then $\phi_x(x)\downarrow$, then $(\exists z)(\forall y)(\text{if } y > z$ then $T(x, x, y))$, and then $(\exists z)(\forall y)[\text{if } y > z \text{ then } \phi_{h(x)}(y)\uparrow]$, and then $A_1(h(x)) = \text{FALSE}$.

Therefore A_1 is not recursively enumerable.

(4) When designing a program to solve a complicated problem, it is useful and customary to split the problem into smaller, easier to solve problems. The task of writing programs to solve these subproblems is assigned to different individuals. These individuals are given specifications of what their programs are supposed to do. We then are confronted with the problem of determining whether these programs are correct; that is, whether all pairs (input value, output value) satisfy a given predicate P, in other words, whether the program behaves according to the given specification. Since this is a property of functions, the problem to determine whether a given program satisfies a given constraint P is recursively unsolvable.

In Example 6.1 we showed that the problem of determining whether two programs are equivalent is recursively unsolvable. This is an immediate consequence of the fact that the problem of determining whether a program satisfies a particular constraint is recursively unsolvable.

In (3) above we have seen that the predicate $\lambda x[\phi_x$ is a total function$]$ is not recursively enumerable. This result was mentioned before in Corollary 5.1 on page 110, which stated that $TREC$ cannot be characterized by a programming language. That corollary was proved by diagonalization, the alternative hinted at in (3) above.

6.4 Exercises

6.1 Prove that the Halting Problem HP_1^1 is not recursively decidable and the predicate $\lambda x[\phi_x(x)\downarrow]$ is a recursively enumerable predicate.

6.2 Discuss whether the following predicates (and their negations) are recursively decidable or recursively enumerable.

(a) $\lambda x[\exists y)(\phi_x(13) = y)]$.

(b) $\lambda x[(\exists y)(\phi_x(y) = 13)]$.

(c) $\lambda x[(\exists y)(\phi_y(x) = 13)]$.

(d) $\lambda x[\phi_x(x)\downarrow]$.

(e) $\lambda xy[\phi_x = \phi_y]$.

(f) $\lambda x[dom(\phi_x)$ is a finite set$]$.

(g) $\lambda x[(\forall y)($if $P(y)$ then $\phi_x(y)\downarrow$ and $R(\phi_x(y))]$, where P and R are arbitrary recursively decidable predicates.

(h) $\lambda x[dom(\phi_x) \subseteq \{p \in N \mid p$ is prime$\}]$.

6.3 Prove that a predicate P is recursively decidable if and only if both P and $\neg P$ are recursively enumerable.

6.4 Prove that the class of recursively decidable predicates is not closed with respect to unbounded quantification.

6.5 Is the space-bounded Halting Problem recursively decidable?

6.6 Discuss whether the predicate

$$P \triangleq \lambda x[\text{the computation on input } x \text{ of the program with Gödel number } x \text{ takes precisely } x \text{ steps}]$$

is recursively decidable, or recursively enumerable, or not recursively enumerable.

6.7 Prove or disprove the following assertions.

(a) Every infinite recursively decidable set contains an infinite recursively enumerable subset.

(b) Every infinite recursively enumerable set contains an infinite recursively decidable subset.

6.8 Prove that there exists a total recursive function $p: N \to N$ such that for every predicate P in $PRED(PRIM_1^n)$ there is a natural number k such that $P(x)$ if and only if $p(\sigma_1^n(x)) \neq 0$, for all $x \in N^n$.

6.9 Let ψ_i be an acceptable indexing of REC_1^1 and V a subset of REC_1^1.

Prove that the set $A = \{x \mid \psi_x \in V\}$ is recursively decidable if and only if V is empty or $V = REC_1^1$ (Rice's theorem).

6.10 Let ϕ_i be the standard indexing of REC_1^1, and P the predicate

$$P \triangleq \lambda x[(\exists y)T(x, x, y)],$$

where T is Kleene's T-predicate.

(a) Prove that the predicate P is recursively enumerable but not recursively decidable.

(b) Prove using the predicate P that the Halting Problem for REC_1^1 is recursively unsolvable.

6.11 Prove that there exists an indexing of all recursively enumerable subsets of N. Why does such an indexing not exist for all recursively decidable subsets of N?

6.5 References

The earliest references to the Halting Problem in relation to recursively decidable and recursively enumerable sets can be found in Kleene (1936), Turing (1936/1937) and Church (1936). The basic properties of recursively enumerable sets were investigated by Kleene (1943) and Post (1943, 1944). Theorems like 'A set A is recursively decidable if and only if A and its complement are both recursively enumerable' were first proved by Post. The method to obtain undecidability results using function index sets is due to Rice (1953); related results are proved by Dekker and Myhill (1958).

Chapter 7
Reducibility

In preceding chapters various predicates have been shown to be non-recursively decidable or non-recursively enumerable. Two methods of proof have been employed:

- Show that the assumption that predicate X is recursively decidable (recursively enumerable) leads to a contradcition. This method was used to show that the Halting Problem HP_1^1 for SAL(**N**) programs is not recursively decidable. (See Theorem 6.1 on page 129, its proof and the subsequent discussion.)
- Show that if the predicate X is recursively decidable (recursively enumerable), then so is some other predicate which we already know not to be so, e.g. HP_1^1, or K or maybe $\neg K$. This method was used to show that the Halting Problem HP_n^k is recursively unsolvable (Theorem 6.2), and also to show that program equivalence is a recursively undecidable property (item (4) of Example 6.1 on page 135).

The latter method is called *reduction*. The concept has been defined in Definition 6.1 on page 131, and was used in subsequent theorems and examples.

In this chapter we will study various reducibilities, all generalizations of the one defined in Section 6.1. We will study the question of whether or not it is possible to prove any statement of the form 'predicate X is not recursively decidable' by showing that the predicate K, or equivalently HP_1^1, can be reduced to the predicate X. Similarly, we wish to know whether we can prove any statement 'X is not recursively enumerable', by reducing the predicate $\neg K$, or $\neg HP_1^1$, to the predicate X.

The method of reduction is not only useful for proving negative results, such as non-decidability or non-enumerability. Reduction is a general means to study whether, and if so, how, the difficulties of two problems are related to one another. These latter aspects will be discussed in later chapters, covering complexity issues.

To simplify matters, attention will be restricted to membership problems. That is, the discussion is phrased in terms of sets, and not in terms of predicates. This brings no loss of generality, because there is a one-to-one correspondence between sets and predicates, as discussed in Section 4.4. Furthermore, only subsets of N will be discussed, subsets of N^k can then be taken care of using our standard Cantor scheme (see Section 5.1). In the following we will often use $\bar{A}$ instead of $N - A$.

Two concepts of reducibility are distinguished: Karp (or strong) reducibility and Cook (or Turing) reducibility. In either case the idea is to determine whether or not the characteristic function of a set can be obtained from the characteristic function of another set using some computational scheme. The way these schemes are restricted determines the type of reducibility. In this chapter we discuss Karp reducibility; Cook reducibility is the subject of the next chapter.

7.1 Karp reducibility

Definition 7.1

Let L be any subset of the set of all functions $f:N \to N$ which contains the identity function and is closed with respect to composition; let A and B be subsets of N. The set A is L-reducible to set B, notation $A \leqslant_L B$, if and only if there is a function f in L, such that $x \in A$ iff $f(x) \in B$. That is,

$$A \leqslant_L B \qquad \text{iff } (\exists f \in L)(\forall x \in N)[x \in A \text{ iff } f(x) \in B]$$

The expression '$A \leqslant_L B$ via h' means that for this particular function h, we have

$$(\forall x \in N)[x \in A \text{ iff } h(x) \in B].$$

The relation $\leqslant_L$ is called a *Karp reducibility relation.* ■

According to the definition above $\chi_A = \chi_B \cdot f$. Thus, the computation scheme to compute the characteristic function of A from the characteristic function of B is restricted in two ways:

- just one question about the set B may be asked, that is, the characteristic function of B may be computed just once.

- only functions in L may be used to determine what membership question to ask about the set B.

The relation $\leqslant_L$ is reflexive and transitive because the identity function is an element of L and L is closed w.r.t. functional composition. Thus $\leqslant_L$ induces an equivalence relation $\equiv_L$ defined as follows:

$$A \equiv_L B \qquad \text{iff } A \leqslant_L B \text{ and } B \leqslant_L A$$

The equivalence classes induced by this relation are called *degrees*. They contain all and only those sets which have the same degree of difficulty modulo L. That is, assuming that the functions in L can be computed free of charge, or at least at negligible cost, functions in the same equivalence class have the same degree of difficulty.

Equivalence classes are ordered by the reducibility relation as follows:

$$[A]_{\equiv_L} \leqslant_L [B]_{\equiv_L} \qquad \text{iff } A \leqslant_L B$$

Although reducibility among equivalence classes is not the same binary relation as reducibility among sets, they are so strongly related that they will be denoted by the same symbol.

The following example illustrates the definition of a Karp-type reducibility relation, as defined above.

EXAMPLE 7.1

Let L be the set of all injective functions from N to N. According to the definition above, $A \leqslant_L B$ if and only if there is an injective function $f: N \to N$ such that $f(A) \subseteq B$ and $f(N - A) \subseteq N - B$. Thus $A \equiv_L B$ if and only if A and B have the same cardinality, and $N - A$ and $N - B$ also have the same cardinality. We have the classes:

ZERO containing all sets having zero elements; that is, the empty set $\varnothing$, and no others.

co-ZERO containing all sets whose complements have zero elements; that is, the set N and no others.

ONE containing all sets having a single element; for example the sets $\{0\}, \{1\}, \ldots, \{123\}, \ldots$.

co-ONE containing all sets whose complements have exactly one element, for example the sets $N - \{0\}$, $N - \{1\}$, $N - \{123\}$.

⋮

I containing all sets having precisely i elements.

co-I containing all sets whose complements have precisely *i* elements.

⋮

ALEPH containing all subsets of *N* which are infinite, and whose complements are infinite too.

The order induced on this set of equivalence classes is:

$$ZERO \leqslant_L ONE \leqslant_L \cdots \leqslant_L I \leqslant_L \cdots \leqslant_L ALEPH,$$

$$co\text{-}ZERO \leqslant_L co\text{-}ONE \leqslant_L \cdots \leqslant_L co\text{-}I \leqslant_L \cdots \leqslant_L ALEPH,$$

whereas the classes *I* and *co-I* are incomparable.

Note that the class *ALEPH*, the greatest element in the ordering given above, contains sets which are not recursively enumerable, e.g. the set $\bar{K}$, and also recursively decidable sets of very low complexity, e.g. the set of all even numbers. All of these sets have the same degree of difficulty, assuming that all injective functions from *N* to *N* can be computed at negligible cost. This assumption is very unrealistic, because there are injective functions $N \rightarrow N$ which are not effectively computable, let alone at negligible cost.

If a reducibility relation is to be useful, it must discriminate between sets that are recursively decidable and those that are not. The set *L* must be chosen accordingly.

We will discuss a number of possibilities:

- *L* is the set of all total recursive functions: many-to-one reducibility
- *L* is the set of all functions computable in polynomial time
- *L* is the set of all functions computable in logarithmic space.

The first of these corresponds to one of the reducibility relations studied in recursive function theory. The most important question studied is how the recursively enumerable sets are ordered by this reducibility relation. The reducibility is called *Karp reducibility*, or *strong reducibility*, to distinguish it from other types, in particular from Cook reducibility, which is to be defined later.

The recursively decidable sets, apart from $\varnothing$ and *N*, are all mutually equivalent with respect to Karp and Cook reducibility relations. The sets $\varnothing$ and *N* are pathological cases, $[\varnothing]_\equiv = \{\varnothing\}$, and $[N]_\equiv = \{N\}$. In complexity theory, we will require more discriminating power. That is, the set of all recursively decidable sets should be partitioned into many $\equiv_L$-classes. The last two of the above list prove to be suitable in complexity theory.

The following theorem states some basic facts about $\leqslant_L$ which are independent of the choice of L.

Theorem 7.1

(1) If $L \subseteq M$ then $\leqslant_L$ is finer than $\leqslant_M$, i.e. $\leqslant_L \subseteq \leqslant_M$ and $\equiv_L \subseteq \equiv_M$.

(2) If $A \leqslant_L B$ then $\bar{A} \leqslant_L \bar{B}$.

(3) If L is a subset of $TREC$ then the properties of being recursively decidable or recursively enumerable are hereditary downwards:

  (a) If B is recursively decidable and $A \leqslant_L B$ then A is recursively decidable.

  (b) If B is recursively enumerable and $A \leqslant_L B$ then A is recursively enumerable.

Proof

Parts (1) and (2) are obvious. As to (3):

- Let $A \leqslant_L B$ via $f \in L$. Hence $\chi_A = \chi_B \cdot f$. Thus if $\chi_B \in TREC$, then $\chi_A \in TREC$. Thus, if B is recursively decidable, then so is A.
- If B is recursively enumerable, then, by Theorem 6.8 (page 142), there is some $\phi \in REC$ such that $B = \{x \mid \phi(x)\downarrow\}$. But then $A = \{x \mid \phi(f(x))\downarrow\}$. Since $\phi \cdot f \in REC$, we have that A is recursively enumerable. ■

In Theorem 6.3 (page 131) we have already seen a special case of (3) above, namely the case $L = TREC$.

From part (3) it follows that if a degree contains a recursively decidable set, it contains recursively decidable sets only. The same holds for recursively enumerable sets. It therefore makes sense to speak about recursively decidable and recursively enumerable degrees. Although customary, this is a slight abuse of terminology. It can be, and generally is, impossible to determine whether a given set belongs to a given recursively decidable degree. That is to say, the predicate $\lambda x[W_x \in [W_m]_\equiv]$ may be recursively undecidable, although $[W_m]_\equiv$ is a recursively decidable degree. (W_x is the recursively enumerable set with Gödel number x, as defined in Definition 6.3 on page 143.)

7.2 Many-to-one reducibility

The reducibility and equivalence relations determined by $L = TREC$ are called many-to-one reducibility and many-to-one equivalence and are denoted by $\leqslant_m$ and $\equiv_m$, or by $\leqslant$ and $\equiv$ if it is clear from the context that we mean many-to-one reducibility and equivalence.

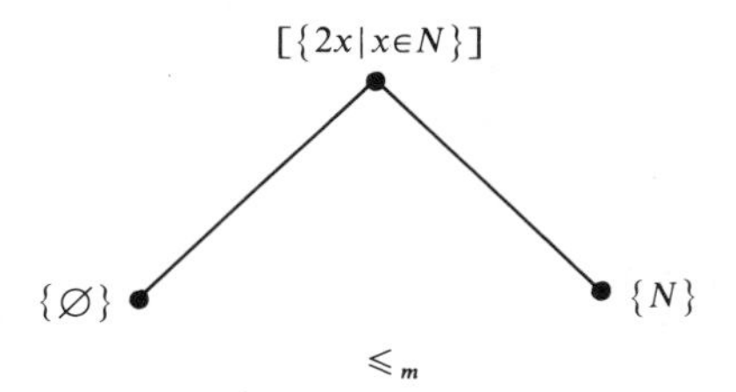

Figure 7.1 Recursive sets with respect to $\leqslant_m$.

Theorem 7.2

The Karp reducibility $\leqslant_m$ is an upper semi-lattice, i.e. any two degrees have a unique least upper bound. Furthermore the least upper bound of two recursively decidable degrees is a recursively decidable degree, and the least upper bound of two recursively enumerable degrees is a recursively enumerable degree.

Proof

For sets A and B, we define a disjoint union of A and B as follows:

$$A \sqcup B \triangleq \{2x \mid x \in A\} \cup \{2x+1 \mid x \in B\}.$$

Then $A \leqslant_m A \sqcup B$ via $\lambda x[2x]$ and $B \leqslant_m A \sqcup B$ via $\lambda x[2x+1]$, and hence $[A \sqcup B]_{\equiv_m}$ is an upper bound of $[A]_{\equiv_m}$ and $[B]_{\equiv_m}$.

It is also a least upper bound, for if $A \leqslant_m C$ via f and $B \leqslant_m C$ via g, then $A \sqcup B \leqslant_m C$ via $h \triangleq \lambda x[\text{if}(x \text{ is even}) \text{ then } f(x/2) \text{ else } g((x-1)/2)]$. ■

The structure of the recursive sets with respect to $\leqslant_m$ is depicted in Figure 7.1.

The sets $\varnothing$ and N are incomparable with respect to $\leqslant_m$, and constitute classes by themselves. The remaining recursively decidable subsets of N constitute a single $\equiv_m$-class. Let A and B be recursively decidable, proper subsets of N. Let $b \in B$ and $c \in N - B$. Then $A \leqslant_m B$ via the function $\lambda x[\text{if } x \in A \text{ then } b \text{ else } c]$, which is recursive because A is recursively decidable. Thus all recursively decidable proper subsets of N are $\equiv_m$-equivalent. Note that $[\varnothing]_{\equiv_m}$ is a recursively decidable degree, but the predicate $\lambda x[W_x \in [\varnothing]_{\equiv_m}]$ is not recursively decidable by items (2) and (3) of Example 6.1 (page 135).

Theorem 7.3

There exist non-recursively decidable sets which are incomparable with respect to $\leqslant_m$.

Proof

Let A be a non-recursively decidable subset of N such that A is recursively enumerable and $\bar{A}$ is not recursively enumerable. Then A and $\bar{A}$ are incomparable with respect to $\leqslant_m$.

- Assume that $\bar{A} \leqslant_m A$. Since A is recursively enumerable it follows by Theorem 7.1 that $\bar{A}$ is recursively enumerable, which, by assumption, it is not. Thus $\bar{A}$ does not precede A with respect to $\leqslant_m$.
- Assume now that $A \leqslant_m \bar{A}$. By Theorem 7.1 this implies that $\bar{A} \leqslant_m \bar{\bar{A}} = A$. Thus A does not precede $\bar{A}$.

It remains to show that such sets A exist. But we already know that. The set $K = \{x \mid \phi_x(x)\downarrow\}$ is recursively enumerable by Theorem 6.5 (page 132), and its complement $\bar{K}$ is not recursively enumerable by Theorem 6.9 (page 143). Thus K and $\bar{K}$ are incomparable with respect to $\leqslant_m$. ■

We will see later, that there also exist sets A and B, both recursively enumerable, which are incomparable with respect to $\leqslant_m$.

7.3 Completeness

Definition 7.2

Let M be any set of subsets of N and let $\leqslant$ be any reducibility ordering. A set $A \subseteq N$ is called $\leqslant$-complete for M iff

(1) $A \in M$

(2) $(\forall B)[\text{if } B \in M \text{ then } B \leqslant A]$

If set A satisfies (2) but not necessarily (1), then A is called $\leqslant$-hard for M. ■

Thus A is $\leqslant$-complete for M iff $[A]_\equiv$ is the greatest element in the reducibility ordering among the classes $[B]_\equiv$, $B \in M$. If A is $\leqslant$-complete for M, then A itself belongs to the class M. In other words, the membership problem of A is the most difficult problem in M.

The set A is $\leqslant$-hard for M if $[A]_\equiv$ is an upper bound of the classes $[B]_\equiv$, $B \in M$. It does not matter whether or not A belongs to the class M. In other words, the membership problem of A is at least as hard a problem as any problem in M.

EXAMPLE 7.2

(1) The set $\{0\}$ is $\leqslant_m$-complete for the class of all recursively decidable sets, as can be seen from Figure 7.1 (page 155).

(2) Consider once again Example 7.1 (page 152).

(a) The class $ALEPH$ is $\leqslant_L$-complete for the set of all subsets of N relative to this particular reducibility $\leqslant_L$, where L is the set of all injective functions $N \to N$.

(b) The class $ALEPH$ is also $\leqslant_L$-hard for the set of all finite subsets of N.

(c) There is no $\equiv_L$-class which is $\leqslant_L$-complete for the set of all finite subsets of N.

Theorem 7.4

There is a recursively enumerable set which is $\leqslant_m$-complete for the set of all recursively enumerable sets.

Proof

Consider the set $K_0 \triangleq \{\sigma_1^2(x, y) | \phi_x(y)\downarrow\} = \{\sigma_1^2(x, y) | (\exists z)\, T(x, y, z)\}$ where T is Kleene's termination predicate, as defined in Section 6.2.

(1) Because Kleene's termination predicate T is recursively decidable, and thus recursively enumerable, and recursively enumerable predicates are closed with respect to existential quantification, it follows that K_0 is recursively enumerable (see Section 6.2, and Theorems 6.7 and 6.11).

(2) Let A be any recursively enumerable set. Then A has an index according to our Gödel numbering of recursively enumerable sets, as given in Definition 6.3 (page 143). Let this index be a, that is $A = W_a$. Then

$$A = \{y | (\exists z)\, T(a, y, z)\}.$$

Thus $A \leqslant_m K_0$ via the function $\lambda x[\sigma_1^2(a, y)]$. ■

By the above theorem, the membership problem of any recursively enumerable set can be reduced to the membership problem of the set K_0. Moreover, this reduction can be done via an injective function.

The above set K_0 is a prime example of a complete set. Later, we will meet resource-bounded versions of the set K_0 as sets $\leqslant$-complete for particular complexity classes and associated resource-bounded reducibility relations.

There are many sets which are $\leqslant_m$-complete for the class of all recursively enumerable sets. Among these is the set K, which we have used in many proofs.

Theorem 7.5

The set $K = \{x | \phi_x(x)\downarrow\}$ is $\leqslant_m$-complete for the class of all recursively enumerable sets.

Proof

By Theorem 6.5, K is recursively enumerable. It remains to show that for all recursively enumerable sets A, $A \leqslant_m K$.

We will show that $K_0 \leqslant_m K$. Since K_0 is $\leqslant_m$-complete, and $\leqslant_m$ is transitive, this implies that $A \leqslant_m K$ for all recursively enumerable sets A.

Define the function ψ as follows:

$$\psi \triangleq \lambda xy[sg(1 + \phi_{\sigma^1_{2,1}(x)}(\sigma^1_{2,2}(x)))].$$

By Church's thesis this function ψ is recursive. By the s–n–m theorem, there is a total recursive function f such that $\phi_{f(x)} = \lambda y[\psi(x, y)]$. $K_0 \leqslant_m K$ via this function f, because

(1) If $x \in K_0$, then $\phi_{\sigma^1_{2,1}(x)}(\sigma^1_{2,2}(x))\downarrow$, and then $\phi_{f(x)}$ is the constant function $\lambda y[1]$, and therefore $f(x) \in K$.

(2) If $x \notin K_0$, then $\phi_{\sigma^1_{2,1}(x)}(\sigma^1_{2,2}(x))\uparrow$, and then $\phi_{f(x)}$ is the everywhere divergent function $\lambda y[\uparrow]$, and therefore $f(x) \notin K$. ■

The function f, via which a recursively enumerable set A is reduced to K in the above proof, is not necessarily injective, as was the case in the reduction to K_0 in Theorem 7.4. We can show, however, that the membership problem of any recursively enumerable set can be reduced to the membership problem of the set K via an injective function. This injective function can be obtained from the function f constructed in the above proof, using the technique of 'padding', that is, constructing successively larger Gödel numbers describing the same function.

There is no special interest in obtaining an injective function, but the padding technique is interesting in itself, and will be used later in the book.

Lemma 7.1

There exists a total recursive function $p: N^2 \rightarrow N$ such that

(1) $(\forall x)(\forall y)[\phi_x = \phi_{p(x,y)}]$

(2) $(\forall x)(\forall y)[x < p(x, y) < p(x, y + 1)]$.

Proof

A program computing p can be constructed according to the following computation scheme

input: x, y
output: z
method:

(1) *compute the program text corresponding to Gödel number* x;

(2) *append '; **while** $v \neq 0$ **do**' to this text, where v is a variable which does not already occur in the program text;*
(3) *append y copies of '$x_1 := x_1$;' to this text;*
(4) *append '$x_1 := x_1$ **od**' to this text;*
(5) *compute z, the Gödel number of the newly constructed program text*

The reader is invited to expand the above computation scheme to a macro program which performs the necessary manipulations on Gödel numbers.

It is clear that the function computed according to the above computation scheme, is total. It satisfies condition (1) since the statements added have no net effect. Whether or not the function also satisfies condition (2) depends on the Gödel numbering used. In our standard Gödel numbering the Gödel numbers increase with program lengths and hence the function computed by the program above satisfies (2) as well.

Note that the computations induced by a program with Gödel number $p(x)$ are almost the same as those induced by the program with Gödel number x. The space and time complexities are almost the same. The state vectors corresponding to the program having Gödel number $p(x)$ have one extra component, representing the value of v, which is always zero. Furthermore, the computation is one step longer. In this final computation step, the while loop is skipped, because the value of the variable v is zero. The variable v was initialized to zero, and it was never assigned any other value. ∎

The proof given above works only for some particular kind of Gödel numbering. The lemma however is true for any acceptable Gödel numbering because given x and y it is always possible to append as many copies of '$x_1 := x_1$;' as is required to make $p(x, y)$ larger than the maximum of the Gödel numbers x, $p(x, 0)$, $p(x, 1), \ldots, p(x, y-1)$ obtained thus far.

We turn now to the problem of reducing the membership problem of any recursively enumerable set to the set K via an injective function.

Let $K_0 \leqslant_m K$ via f as constructed in Theorem 7.5. Let p be the padding function as defined in the preceding lemma. Define the function $\bar{f}$ as follows:

$$\bar{f}(0) = f(0)$$

$$\bar{f}(x) = p(f(x), \mu y[(\forall z < x)[p(f(x), y) \neq \bar{f}(z)]]), \qquad x > 0.$$

Thus to compute $\bar{f}(x)$, we first compute $f(x)$, and then, using Lemma 7.1, find successively larger Gödel numbers for the same function $\phi_{f(x)}$, until we have found one different from any $\bar{f}(z)$, $0 \leqslant z < x$. Clearly, $\bar{f}$ is an injective total recursive function.

Recall from the proof of Theorem 7.5 that $\phi_{f(x)}$ is the constant function $\lambda y[1]$ if $x \in K_0$ and therefore $f(x) \in K$, and is otherwise the everywhere divergent function $\lambda y[\uparrow]$. Therefore $\phi_{f(x)}(f(x))\downarrow$ if and only if $\phi_{p(f(x),y)}(p(f(x),y))\downarrow$.

Thus the membership problem of K_0 can be reduced to the membership problem of K via an injective function. Because the membership problem of any recursively enumerable set can be reduced to the membership problem of the set K_0 via an injective function, we are done.

A similar technique is applicable to any set A which is $\leqslant_m$-complete for the class of all recursively enumerable sets. If A is $\leqslant_m$-complete for the class of all recursively enumerable sets, then $K \leqslant_m A$ via some total recursive function f. From this f we can derive an injective total recursive function $\bar{f}$, via which $K \leqslant_m A$. Since every recursively enumerable set can be reduced to K via an injective function, every recursively enumerable set can be reduced to A via an injective function. In Section 7.5 we will discuss the matter of reducing via injective functions further.

7.4 Productive and creative sets

In this section we will discuss sets which are $\leqslant_m$-complete for the class of all recursively enumerable sets. These complete sets can be characterized in terms of a property of their complements: productiveness. Productiveness asserts that there is a mechanical procedure to falsify the statement: the set in question is equal to W_x, the recursively enumerable set with Gödel number x. ($W_x = dom(\phi_x)$: see Definition 6.3 on page 143.)

Definition 7.3

The set A is *productive* if there exists a $\kappa \in TREC_1^1$ such that

$$(\forall x)[\kappa(x) \in A \text{ if and only if } \kappa(x) \notin W_x]$$

The function κ is called productive for A. ■

Thus, if κ is a productive for A, then $\kappa(x) \in ((A - W_x) \cup (W_x - A))$, for all $x \in N$, and $\kappa(x)$ is a counter-example to the statement '$A = W_x$'.

It is possible – and customary – to allow productive functions to be partial. The definition then reads:

The function $f \in REC_1^1$ is productive for the set A if

$(\forall x)[\text{if } W_x \subseteq A \text{ then } f(x)\downarrow \text{ and } f(x) \in A - W_x]$.

It can be shown that this definition is equivalent to the one given earlier.

That is, if there is a partial recursive function f which is productive for a set A, then there also exists a total recursive function κ which is productive for the set A, according to Definition 7.3 above.

EXAMPLE 7.3

The set $\bar{K}$ is productive and the identify function I^1 is productive for $\bar{K}$. We have to show that $(\forall x)[I^1(x) \in \bar{K}$ if and only if $I^1(x) \notin W_x]$. Because $I^1(x) = x$, this amounts to $x \in \bar{K}$ if and only if $x \notin W_x$ if and only if $\phi_x(x)\uparrow$, which is true by definition.

Theorem 7.6

If A is productive and $A \leqslant_m B$ then B is productive

Proof

Let ζ be a total recursive function which is productive for A and let $A \leqslant_m B$ via f. Define the function ψ as follows:

$$\psi \triangleq \lambda xyz[\phi_x(\phi_y(z))]$$

By Church's thesis, ψ is recursive, and by the s–n–m theorem there is a total recursive function g such that $\phi_{g(x,y)} = \lambda z[\psi(x, y, z)]$.

Let m be an index for f and define the function ξ as follows:

$$\xi \triangleq \lambda x[f(\zeta(g(x,m)))]$$

The function ξ is productive for B:

- If $\xi(x) \in B$ then $f(\zeta(g(x,m))) \in B$. Because $A \leqslant_m B$ via f, it follows that $\zeta(g(x,m)) \in A$. Because ζ is productive for A, $\zeta(g(x,m)) \notin W_{g(x,m)}$. Therefore,
  - — $\phi_{g(x,m)}(\zeta(g(x,m)))\uparrow$
  - — $\phi_x(\phi_m(\zeta(g(x,m))))\uparrow$
  - — $\phi_x(f(\zeta(g(x,m))))\uparrow$ (by the choice of m, $\phi_m = f$)
  - — $\phi_x(\xi(x))\uparrow$
  - — $\xi(x) \notin W_x$
- Similarly, if $\xi(x) \notin B$ then $f(\zeta(g(x,m))) \notin B$, and then $\zeta(g(x,m)) \notin A$. Because ζ is productive for A, $\zeta(g(x,m)) \in W_{g(x,m)}$. Therefore,
  - — $\phi_{g(x,m)}(\zeta(g(x,m)))\downarrow$
  - — $\phi_x(\phi_m(\zeta(g(x,m))))\downarrow$
  - — $\phi_x(f(\zeta(g(x,m))))\downarrow$ (by the choice of m, $\phi_m = f$)
  - — $\phi_x(\xi(x))\downarrow$
  - — $\xi(x) \in W_x$ ■

The preceding theorem shows that productiveness is hereditary upwards in the $\leqslant_m$-reducibility ordering. Thus the familiar reduction technique can also be used to show that a particular set is productive.

EXAMPLE 7.4

(1) In item (3) of Example 6.1 (page 135) we have seen that $\bar{K} \leqslant_m \Gamma$, where $\Gamma \triangleq \{x \mid W_x = \varnothing\}$. Because $\bar{K}$ is productive, we may conclude that Γ is productive too.

(2) In Example 6.3 we considered the predicate $A_1 \triangleq \lambda i[\phi_i$ is total$]$, and thus equivalently the set $\{x \mid W_x = N\}$. In item (1) of that example we saw that $\bar{K} \leqslant_m \bar{A}_1$, and in item (3) that $\bar{K} \leqslant_m A_1$. Therefore both A_1 and $\bar{A}_1$ are productive sets.

Definition 7.4

The set A is *creative* if A is recursively enumerable and $\bar{A}$ is productive.

■

The term 'creative' is due to Emil Post. It is intended to convey the idea that creative sets capture some aspect of the creative quality of scientific theories.

A theory is essentially a set of well-formed formulae, which is built up as follows.

- There is a finite alphabet of symbols, constants such as 0, 1, 2, etc., indexed variables, operation symbols, predicate symbols, and so forth.
- Formulae can be written using these symbols. The well-formed formulae are those strings over the alphabet which conform to given syntax rules, asserting correct use of parentheses, supplying functions with the correct number of arguments, and so on. For example, the expression $5*(4+3)$ would be considered well-formed, whereas $*=)4-x_7$ would not.
- A theory then, is a set of well-formed formulae. Thus a theory is just a body of factual knowledge, phrased in some particular way. Usually one wants not just factual knowledge, but explanations. These come in the form of proofs. A proof is finite sequence of well-formed formulae such that each formula of the sequence is derived from formulae earlier in the sequence, or is an axiom, that is, a formula which belongs to the theory by definition. Formulae can be derived from other formulae by using given rules of proof, such as *modus ponens*. Given formulae A and $A \rightarrow B$, the rule of *modus ponens* allows you to conclude the formula B.

For example, in elementary number theory we have:

- a single constant, 0.
- three operation symbols, $'$ (successor), $+$ (sum), $*$ (product).
- a single predicate symbol, $=$, for equality.
- axioms, for example $(\forall x)(\forall y)[x*(y') = x*y + x]$.

  Also, for every well-formed formula $A(x)$ the following is an axiom:

  $$(A(0) \wedge (\forall x)[A(x) \rightarrow A(x')]) \rightarrow (\forall x)A(x).$$

  These infinitely many axioms together express the principle of induction.
- rules of proof are *modus ponens* and generalization: $(\forall x)A$ follows from A.

Examples of formulae are

- $(\forall x)(\exists y)[x = y*y]$ (this formula is well-formed, though not valid).
- $(\forall x)(\exists y)[(x*x)' = (x + x) + y]$ (this formula is well-formed and valid).

In Section 5.2 we saw how to put the strings over some finite alphabet into a one-to-one correspondence with the natural numbers. Thus, recursion-theoretic properties (decidability, enumerability, productiveness, creativeness) are meaningful with respect to formal theories.

The following statements regarding elementary number theory can be proved:

- the set of all valid well-formed formulae is productive
- the set of all non-valid well-formed formulae is productive
- the set of all provable well-formed formulae is recursively enumerable
- the set $\{F | \neg F \text{ is provable}\}$ is recursively enumerable
- the set of all unprovable well-formed formulae is productive.

If we make the assumption that our theory is consistent, that is, $F \wedge \neg F$ is not provable, it follows that $\{F | \neg F \text{ is provable}\}$ is a recursively enumerable subset of the productive set of unprovable formulae. Therefore we can find a formula F such that neither F nor $\neg F$ is provable.

In summary, no axiomatization of mathematics can exactly capture all valid number-theoretic assertions, and for any axiomatization which is not contradictory a formula can be found which is valid but not provable in this axiomatization, namely one of F or $\neg F$. The formula F is found using the productiveness of the set of unprovable formulae.

The above considerations give special significance to recursively enumerable sets having a productive complement, hence the name creative set.

We will now proceed to show that the creative sets are precisely the sets which are $\leqslant_m$-complete for the class of all recursively enumerable sets.

Lemma 7.2

If the set A is $\leqslant_m$-hard for the class of all recursively enumerable sets, then $\bar{A}$ is productive.

Proof

Since A is $\leqslant_m$-hard for the class of all recursively enumerable sets, we have $K \leqslant_m A$ and thus $\bar{K} \leqslant_m \bar{A}$ by Theorem 7.1. Since $\bar{K}$ is productive (Example 7.3 on page 161), Theorem 7.6 implies that $\bar{A}$ is productive. ■

Lemma 7.3

If the set $\bar{A}$ is productive, then A is $\leqslant_m$-hard for the class of all recursively enumerable sets.

Proof

We must show that $X \leqslant_m A$ whenever X is a recursively enumerable set. Let $h: N \to N$ be a total recursive function, productive for $\bar{A}$. Define the function ψ as follows:

$$\psi \triangleq \lambda xy[sg(1 + \phi_x(x))]$$

Clearly, ψ is recursive. Thus, by the $s-n-m$ theorem, there is a total recursive function f such that $\phi_{f(x)} = \lambda y[\psi(x, y)]$.
Then $K \leqslant_m A$ via the function $\lambda x[h(f(x))]$.

(1) If $x \in K$, then $\phi_x(x)\downarrow$ and $\phi_{f(x)} = \lambda y[1]$. Therefore $h(f(x)) \in W_{f(x)}$. Because h is productive for $\bar{A}$, we have $h(f(x)) \notin \bar{A}$, that is $h(f(x)) \in A$.

(2) If $x \notin K$, then $\phi_x(x)\uparrow$ and $\phi_{f(x)} = \lambda y[\uparrow]$. Therefore $h(f(x)) \notin W_{f(x)}$. Because h is productive for $\bar{A}$, we have $h(f(x)) \in \bar{A}$, that is $h(f(x)) \notin A$.

Therefore $K \leqslant_m A$ via $\lambda x[h(f(x))]$. Because K is $\leqslant_m$-complete for the class of all recursively enumerable sets, we are done. ■

The preceding two lemmas imply the following theorem:

Theorem 7.7

A set is $\leqslant_m$-complete for the class of all recursively enumerable sets if and only if it is creative.

Proof

- Assume that A is $\leqslant_m$-complete for the class of all recursively enumerable sets. Then A is recursively enumerable and $\leqslant_m$-hard for the class of all recursively enumerable sets. By Lemma 7.2 $\bar{A}$ is productive. Consequently A is creative.
- Assume that A is creative. By definition, A is recursively enumerable and $\bar{A}$ is productive. By Lemma 7.3, A is $\leqslant_m$-hard for the class of all recursively enumerable sets. Consequently, A is $\leqslant_m$-complete for the class of all recursively enumerable sets.

■

In the above discussion, we have seen just two sets, K_0 and K, $\leqslant_m$-complete for the class of all recursively enumerable sets. There are infinitely many such complete sets. However, all of these are essentially identical. That is, for any pair A, B of sets, $\leqslant_m$-complete for the class of all recursively enumerable sets, there is a recursive permutation $f: N \rightarrow N$ such that $f(A) = B$.

7.5 Injective reducibility, cylinders and simple sets

In Section 7.3, we have seen that for any recursively enumerable set A there is an injective function f via which $A \leqslant_m K_0$. In this section we will discuss the reducibility relation $\leqslant_1$, obtained from our general Definition 7.1 (page 151) by setting

$$L = \{f \in TREC_1^1 \mid f \text{ is an injective function}\}.$$

The reducibility relation $\leqslant_1$ is called *injective reducibility*, or *one-to-one reducibility*. It is not so much this restricted reducibility that matters, but the properties of $\leqslant_m$ that come as by-products of studying $\leqslant_1$. As the material is not strictly necessary for what follows, this section can be skipped.

We will see that $\leqslant_1$ indeed produces a more intricate hierarchy than does $\leqslant_m$, and that a set A is $\leqslant_m$-complete for the class of all recursively enumerable sets if and only if it is $\leqslant_1$-complete for this class. In general, however, a $\equiv_m$-class consists of more than one $\equiv_1$-class. Every $\equiv_m$-class has a $\leqslant_1$-greatest element, a set M which is $\leqslant_1$-complete for the $\equiv_m$-class. This greatest element can be obtained from any other set in the $\equiv_m$-class by a process called *cylindrification.*

Finally, we will see that there is an $\equiv_m$-class $[S]_{\equiv_m}$ such that

$$[N]_{\equiv_m} <_m [S]_{\equiv_m} <_m [K]_{\equiv_m}.$$

That is, there is a recursively enumerable set which is not recursively decidable, but this latter fact cannot be proved by reduction from K. In the next chapter, on Turing reducibility, we will discuss a stronger result, namely that there are recursively enumerable degrees D_1 and D_2 such that $[N]_{\equiv} < D_1 < [K]_{\equiv}$, and $[N]_{\equiv} < D_2 < [K]_{\equiv}$, and D_1 and D_2 are incomparable, that is, neither $D_1 \leqslant D_2$ nor $D_2 \leqslant D_1$.

In the following, we will need the following property of the equivalence relation $\equiv_1$.

Lemma 7.4

$A \equiv_1 B$ if and only if there is a recursive bijection $p:N \rightarrow N$ such that $p(A) \equiv B$.

Proof

Assume that there is recursive bijection $p:N \rightarrow N$, such that $p(A) = B$. The inverse p^{-1} of p is a recursive bijection too, as can easily be verified. Clearly, $p^{-1}(B) = A$. Therefore $A \leqslant_1 B$ via p, and $B \leqslant_1 A$ via p^{-1}, and thus $A \equiv_1 B$.

Assume now that $A \equiv_1 B$. Then there are recursive injective functions f and g, such that $A \leqslant_1 B$ via f, and $B \leqslant_1 A$ via g. Using f and g, we will construct a list of pairs $L \triangleq ((a_0, b_0), (a_1, b_1), \ldots)$, such that L is a finite approximation of a bijection $p:N \rightarrow N$, defined by $p(a_i) \triangleq b_i$, and that for all pairs (x, y) on the list L we have $x \in A$ if and only if $y \in B$.

To find the image $p(m)$ of a given element m, we only have to construct the list until we arrive at a pair (m, y). Then $p(m) = y$. Thus p is a computable, and therefore recursive, bijection.

Initializing the list L:

- Let $L = ((0, f(0))$. This clearly is a finite approximation of a bijection, and for all pairs (x, y) on the list, we have $x \in A$ if and only if $y \in B$.

Adding a pair to a list L having k elements:

- Let $L = ((a_0, b_0), (a_1, b_1) \ldots, (a_{k-1}, b_{k-1}))$; $D = \{a_0, a_1, \ldots, a_{k-1}\}$, and $C = \{b_0, b_1, \ldots, b_{k-1}\}$. Find the least i such that $i \notin D$ or $i \notin C$.
  - Assume that $i \notin D$. Let $y = f(i)$. If y does not already occur on the list L, we are done. Otherwise, we must choose another y-value. Suppose $y = b_j$. Then $i \in A$ if and only if $y \in B$, because $A \leqslant_1 B$ via f; and $a_j \in A$ if and only if $b_j \in B$ by the construction of L. Also $f(a_j) \neq y$, because f is injective and $a_j \neq i$. Finally, $i \in A$ if and only if $y = b_j \in B$ if and only if $a_j \in A$ if and only if $f(a_j) \in B$. Therefore, try the value $f(a_j)$ for y.

    Repeat this process until a value for y is found which does not occur in the set C. This is bound to happen, because every new value $f(a_j)$ tried differs from all the values tried earlier. Once a value for y is found which does not occur in C, add the pair (i, y) as the kth element of the list L.

— Assume that $i \notin C$. Proceed similarly to find an x such that (x, i) can be added to the list L. Let $x = g(i)$. Then $x \in A$ if and only if $i \in B$, because $B \leqslant_1 A$ via g. If this x already occurs in the set D, say $x = a_j$, then try the value $x = g(b_j)$. Repeat this process until a value is found which does not occur in D. Add the pair (x, i) as the kth element of the list L.

Clearly, the above process gives a list L of pairs (x, y), such that $x \in A$ if and only if $y \in B$, and the list is a finite approximation of a bijection $p: N \rightarrow N$. For each k, the pairs (k, y) and (x, k) will occur on L for some x and y, before L has length $2k + 2$. Thus if $A \equiv_1 B$ then there is a recursive bijection p such that $p(A) = B$. ■

Obviously, the above lemma is false if we read $\equiv_m$ instead of $\equiv_1$. This already shows that $\leqslant_1$ and $\leqslant_m$ are different reducibility relations.

Definition 7.5

(1) Let A be any subset of N. The *cylindrification* of A is a set denoted by $A \times N$, and defined as follows:

$$A \times N \triangleq \{\sigma_1^2(x, y) \mid x \in A, y \in N\}.$$

(2) The set A is a *cylinder* if there exists a set B such that A is recursively isomorphic to $B \times N$; that is, if there exist a set B and a bijection $p \in TREC_1^1$, such that $A = p(B \times N)$. ■

Theorem 7.8

Let A be any subset of N, then the following properties hold:

(1) $A \leqslant_1 A \times N$

(2) $A \equiv_m A \times N$

(3) A is a cylinder if and only if $(\forall B)[\text{if } B \leqslant_m A \text{ then } B \leqslant_1 A]$

Proof

(1) $A \leqslant_1 A \times N$ via $\lambda x[\sigma_1^2(x, 0)]$

(2) By (1), $A \leqslant_1 A \times N$. Since $\leqslant_1$ is a refinement of $\leqslant_m$ we have $A \leqslant_m A \times N$. Furthermore $A \times N \leqslant_m A$ via $\lambda x[\sigma_{2,1}^1(x)]$. Hence $A \equiv_m A \times N$.

(3) (a) Assume that A is a cylinder, $A = p(C \times N)$ for some set C and recursive bijection $p: N \rightarrow N$. Then $C \times N \leqslant_1 A$ via

p, and $A \leqslant_1 C \times N$ via the inverse bijection p^{-1} of p. Assume that $B \leqslant_m A$. Then

$$B \leqslant_m A \leqslant_m C \times N \leqslant_m C,$$

by (2) above. Let $B \leqslant_m C$ via f, and define the function g as follows: $g \triangleq \lambda x[\sigma_1^2(f(x), x)]$. Clearly g is injective, and $x \in B$ if and only if $f(x) \in C$ if and only if $g(x) \in C \times N$. Thus $B \leqslant_1 C \times N$ via g. Because $C \times N \leqslant_1 A$, we have $B \leqslant_1 A$.

(b) Assume now that $(\forall B)[\text{if } B \leqslant_m A \text{ then } B \leqslant_1 A]$. By (1), $A \leqslant_1 A \times N$. By (2), $A \times N \leqslant_m A$. Then it follows from our assumptions that $A \times N \leqslant_1 A$. Therefore $A \equiv_1 A \times N$, and by the preceding lemma there is a recursive bijection p such that $A = p(A \times N)$. Thus A is a cylinder. ∎

The above theorem shows that there is a $\leqslant_1$-greatest element among the $\equiv_1$-classes contained in a given $\equiv_m$-class. Let A be any set belonging to a given $\equiv_m$-class $Q = [A]_{\equiv_m}$. Then $A \times N$ belongs to Q, and is a cylinder. Thus, for any $B \in Q$, $B \leqslant_m A \times N$, and therefore $B \leqslant_1 A \times N$. Therefore $A \times N$ is $\leqslant_1$-complete for Q. In other words, $[A \times N]_{\equiv_1}$ is a $\leqslant_1$-greatest element among the $\equiv_1$-classes contained in the given $\equiv_m$-class Q.

Consider for example the class of all recursively decidable subsets of N. The structure of this class with respect to the $\leqslant_m$ and $\leqslant_1$ reducibility relations is depicted in Figure 7.2.

The structure of the recursively decidable sets with respect to $\leqslant_m$, has been discussed in Section 7.2. As regards $\leqslant_1$, this structure is very similar to the structure discussed in Example 7.1 (page 152). If A and B are recursively decidable sets, then there is an injective function f such that $x \in A$ if and only if $f(x) \in B$, if and only if there is a recursive injective function f_r such that $x \in A$ if and only if $f_r(x) \in B$.

- The class $[\varnothing]_{\equiv_m}$ consists of a single $\equiv_1$-class, namely $[\varnothing]_{\equiv_1} = \{\varnothing\}$. The cylindrification of $\varnothing$ is indeed empty.

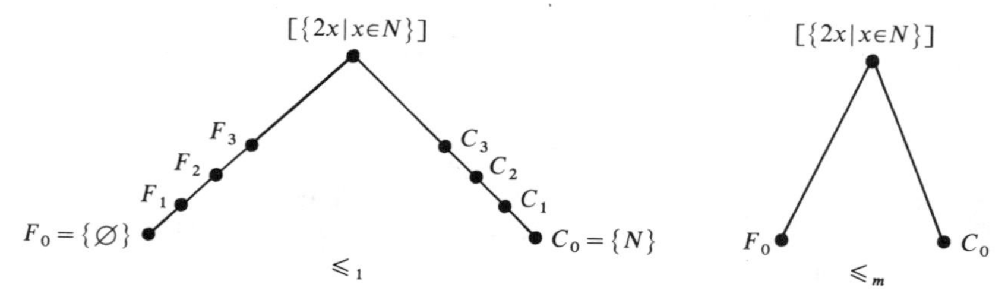

$F_i \triangleq \{X \subseteq N \mid |X| = i\}$, $C_i \triangleq \{X \subseteq N \mid |\overline{X}| = i\}$, $i \geqslant 0$.

Figure 7.2 Recursive sets with respect to $\leqslant_1$ and $\leqslant_m$.

- Similarly, $[N]_{\equiv_m} = [N]_{\equiv_1}$, and the cylindrification of N is N.

  The $\equiv_m$-class $[\{2x|x\in N\}]_{\equiv_m}$ consists of an infinite number of $\equiv_1$-classes. The greatest element is the class $[\{2x|x\in N\}]_{\equiv_1}$. The set $\{2x|x\in N\}$ can be obtained from any finite or co-finite set by cylindrification. For example,

$$\{2x|x\in N\} = p(\{0,1,2\} \times N),$$

  where $p:N\to N$ is the bijection

$$\lambda x[\text{if } 0 \leqslant \sigma^1_{2,1}(x) < 3 \text{ then } 2(\sigma^1_{2,1}(x) + 3\sigma^1_{2,2}(x)) \\ \text{else } 2(\sigma^2_1(\sigma^1_{2,1}(x) \dot{-} 3, \sigma^1_{2,2}(x))) + 1].$$

We will now turn to the construction of recursively enumerable sets which are not cylinders. The existence of such sets shows that there are recursively enumerable $\equiv_m$-classes which consist of more than one $\equiv_1$-class.

Definition 7.6

The set A is *simple* if and only if

(1) A is recursively enumerable and

(2) $\bar{A}$ is infinite and does not contain an infinite recursively enumerable subset. ■

Since the complement of a simple set does not contain any recursively enumerable subset, this complement is itself certainly not recursively enumerable. Thus simple sets are not recursively decidable.

The term 'simple' is from Emil Post. Such a set is simple in that its membership problem is strictly easier than the membership problem of K, i.e. for a simple set S we have $S <_m K$, that is, $S \leqslant_m K$ and not $S \equiv_m K$. To prove this, we will proceed as follows:

- show that simple sets exist,
- show that the $\equiv_m$-class of simple set consists of more than one $\equiv_1$-class,
- show that the $\equiv_m$-class of a set which is $\leqslant_m$-complete for the class of all recursively enumerable sets consists of a single $\equiv_1$-class.

Thus $S \leqslant_m C$ and not $S \equiv_m C$ whenever S is a simple set and C is $\leqslant_m$-complete for the class of all recursively enumerable sets.

Theorem 7.9

There exists a simple set.

Proof

Consider the function f defined as follows:

$$f \triangleq \lambda x[\sigma^1_{2,1}(\mu y[\sigma^1_{2,1}(y) > 2x \wedge T(x, \sigma^1_{2,1}(y), \sigma^1_{2,2}(y))])],$$

where T is Kleene's T-predicate.

Evidently f is a partial recursive function. The value $f(x)$ is some element in the domain of ϕ_x which is larger than $2x$, if such an element exists. If such an element does not exist, then $f(x)$ is undefined. The range, $S \triangleq f(N)$, of f is a simple set:

(1) S is recursively enumerable because it is the range of a partial recursive function. Let $\phi_x : N \to N$ be any partial recursive function. Then

$$\lambda y[\mu z[T(x, \sigma^1_{2,1}(z), \sigma^1_{2,2}(z)) \wedge \phi_x(\sigma^1_{2,1}(z)) = y]]$$

converges exactly on the range of ϕ_x. Therefore, the range of a partial recursive function is recursively enumerable.

(2) To see that the complement of S is infinite, note that in any set $\{0, 1, 2, \ldots, 2k\}$ there are at most k elements belonging to $S = f(N)$ because if $f(x)$ is defined, then $f(x) > 2x$. Therefore, at least k elements belong to $\bar{S} \triangleq N - f(N)$, and therefore $\bar{S}$ is infinite.

(3) $\bar{S}$ does not contain any infinite recursively enumerable subset. Consider W_x, the recursively enumerable set with Gödel number x. If W_x is infinite, the set $\{y \mid y \in W_x \text{ and } y > 2x\}$ is infinite as well. Therefore $f(x)$ is defined and $f(x) \in W_x$. Hence $W_x \cap S \neq \varnothing$ and thus W_x is not a subset of $\bar{S}$. ∎

It will now be shown that simple sets are not cylinders. This implies that a simple set cannot belong to the greatest $\equiv_1$-class contained in the $\equiv_m$-class of this simple set. Therefore this $\equiv_m$-class must consist of more than one $\equiv_1$-class.

Theorem 7.10

If a set is simple, then it is not a cylinder.

Proof

Assume that A is a simple cylinder. Then $A = p(B \times N)$ for some set B and recursive bijection p, and $\bar{A}$ is infinite and does not contain an infinite recursively enumerable subset. Because $\bar{A} = p(\bar{B} \times N)$, it follows that $\bar{B}$ is non-empty. Let $s \in \bar{B}$. The set $\{s\} \times N \subseteq \bar{B} \times N$ is recursively enumerable and infinite. Also $p(\{s\} \times N) \subseteq \bar{A}$ is recursively enumerable and infinite. Therefore $\bar{A}$ contains an infinite recursively enumerable subset. Thus A is not simple and we have a contradiction. ∎

Theorem 7.11

If S is a simple set, then $[S]_{\equiv_m} \neq [S]_{\equiv_1}$. The $\equiv_m$-class of S contains at least two $\equiv_1$-classes, namely $[S]_{\equiv_1}$ and $[S \times N]_{\equiv_1}$.

Proof

If S is a simple set, then by Theorem 7.10 S is not a cylinder, in particular S is not equivalent to $S \times N$, relative to $\equiv_1$. But for any set and hence for S we have $S \equiv_m S \times N$ by Theorem 7.8. Thus $S, S \times N \in [S]_{\equiv_m}$, and $[S]_{\equiv_1}, [S \times N]_{\equiv_1} \subseteq [S]_{\equiv_m}$, but $[S]_{\equiv_1} \neq [S \times N]_{\equiv_1}$. ∎

By the above theorem $\equiv_1$ and $\equiv_m$ do not coincide on the non-recursive, recursively enumerable sets. Thus the reducibilities $\leqslant_1$ and $\leqslant_m$ are actually different partial ordering relations both on the recursively decidable and the recursively enumerable sets.

The preceding theorem shows that there are recursively enumerable $\equiv_m$-classes, classes $[S]_{\equiv_m}$, where S is simple, which consist of at least two $\equiv_1$-classes. It can be shown that the $\equiv_m$-class of a simple set consists of infinitely many $\equiv_1$-classes.

Finally, we want to show that if A is $\leqslant_m$-complete for the class of all recursively enumerable sets, then $[A]_{\equiv_m} = [A]_{\equiv_1}$; that is, the concepts $\leqslant_m$- and $\leqslant_1$-completeness for the class of all recursively enumerable sets coincide.

We know from Section 7.3 that K is $\leqslant_1$-complete for the class of all recursively enumerable sets. We want to show that $K \leqslant_m A$ implies $K \leqslant_1 A$. Given a function f via which $K \leqslant_m A$, we want to construct a function f', via which $K \leqslant_1 A$. This can be done using a technique similar to padding, discussed in Section 7.3.

A tentative definition of f' is:

$$f'(0) = f(0)$$

$$f'(x) = \text{some } y \text{ such that } y \in A \text{ if and only if } f(x) \in A, \text{ and } y \neq f'(x') \text{ for all } x',\ 0 \leqslant x' < x.$$

The 'echo lemma' (Lemma 7.6, below) provides a method to find values y as required in the tentative definition above.

To find a value different from a finite set of given values, we must specify this finite set. This can be done using our Gödel numbering scheme. Thus, we represent the finite set $F_x = \{x_1, x_2, \ldots, x_n\}$ by the Gödel number $\langle x_1, x_2, \ldots, x_n \rangle$. The representation is not unique; the Gödel number determines a list, that is, an ordered set of numbers. This, however, is easily hidden by using appropriate access functions.

In the proof of the echo lemma, we need the following technical result.

Lemma 7.5

Let f be any total recursive function, such that $f(K) \cap f(\bar{K}) = \emptyset$. There exists a recursive injective function $r: N \to N$ enumerating $f(K)$, that is $f(K) = \{r(n) | n \in N\}$.

Proof

$K \leqslant_m f(K)$ via f, because $f(K) \cap f(\bar{K}) = \emptyset$. Therefore $f(K)$ is not recursively decidable, and thus infinite.

The set $f(K)$ is enumerated by the total recursive injective function $r: N \to N$ defined as follows:

- to compute $r(0)$:

  find the least t such that $T(\sigma^1_{2,1}(t), \sigma^1_{2,1}(t), \sigma^1_{2,2}(t))$, where T is Kleene's termination predicate. Set $r(0) = f(\sigma^1_{2,1}(t))$.

- to compute $r(k)$

  find the least t such that

  $$T(\sigma^1_{2,1}(t), \sigma^1_{2,1}(t), \sigma^1_{2,2}(t)) \wedge (\forall i < k)[f(\sigma^1_{2,1}(t)) \neq r(i)].$$

  Set $r(k) = f(\sigma^1_{2,1}(t))$.

Clearly, this function r is recursive, total, and injective. By the s–n–m theorem, an index for r can be obtained from an index for f. ∎

We turn now to the echo lemma.

Lemma 7.6

If A is $\leqslant_m$-complete for the class of all recursively enumerable sets, then there is a total recursive function $e: N \to N$, such that

(1) if $F_x \cap A \neq \emptyset$ then $e(x) \in A - F_x$,

(2) otherwise, if $F_x \cap A = \emptyset$, then $e(x) \in \bar{A} - F_x$.

(F_x is the finite set with code number x, as defined above.)

Proof

Let $K \leqslant_m A$ via f, and $A = W_a$, i.e. a is an index for A according to Definition 6.3 (page 143).

Define the function ψ as follows:

$$\psi \triangleq \lambda xy[\text{if } f(y) \in F_x \text{ then } 0$$
$$\text{else } \mu t[\sigma^1_{2,1}(t) \in F_x \wedge T(a, \sigma^1_{2,1}(t), \sigma^1_{2,2}(t))]]$$

Note that $\psi(x, y)\downarrow$ if and only if $f(y) \in F_x$ or $F_x \cap A \neq \emptyset$. The function ψ is clearly recursive. Therefore, by the s–n–m theorem, there is a total recursive function h such that $\phi_{h(x)} = \lambda y[\psi(x, y)]$. Regarding h we have the following properties.

- $F_x \cap A \neq \varnothing$.

  Then $\phi_{h(x)}(y)\downarrow$ for all $y \in N$. Hence, in this case, $h(x) \in K$.

- $F_x \cap A = \varnothing$.

  — Then $\phi_{h(x)}(y)\downarrow$ if and only if $f(y) \in F_x$. Assume that $h(x) \in K$. Then $\phi_{h(x)}(h(x))\downarrow$, and thus $f(h(x)) \in F_x$. Also $f(h(x)) \in A$, because $K \leqslant_m A$ via f. Thus $f(h(x)) \in F_x \cap A$, which contradicts our assumption $F_x \cap A = \varnothing$. Therefore $h(x) \notin K$.

  — Also $f(h(x)) \notin F_x$, because if $f(h(x)) \in F_x$, then $\psi(x, h(x))\downarrow$, and thus $\phi_{h(x)}(h(x))\downarrow$, and thus $h(x) \in K$.

Define the echo function e as follows:

$$e \triangleq \lambda x[\text{if } f(h(x)) \notin F_x \text{ then } f(h(x)) \text{ else } r(\mu t[r(t) \notin F_x])],$$

where r is an injective total recursive function enumerating $f(K)$; by Lemma 7.5 such a function r exists.

Clearly, e is a total recursive function. Also, for all x, $e(x) \notin F_x$ and $e(x)$ is in the range of f. To see that e satisfies the other requirements:

(1) If $F_x \cap A \neq \varnothing$, then $h(x) \in K$. Therefore $f(h(x)) \in f(K) \subseteq A$, and consequently $e(x) \in A - F_x$.

(2) If $F_x \cap A = \varnothing$, then $f(h(x)) \notin F_x$ and $h(x) \notin K$, and consequently $f(h(x)) \in \bar{A}$. Therefore $e(x) \in \bar{A} - F_x$. ■

The intended use of the echo function e is as follows. An element p is given, and it is required to find new elements $q_1, q_2, \ldots$ such that these are mutually different and also different from p, and also such that $p \in A$ if and only if $q_i \in A$ for all i.

This can be achieved as follows:

(1) $q_1 = e(\langle p \rangle)$

(2) $q_2 = e(\langle p, q_1 \rangle)$

(3) $q_3 = e(\langle p, q_1, q_2 \rangle)$

and so on.

Thus the echo function and the padding function as discussed in Section 7.3 serve similar purposes.

Theorem 7.12

A set is $\leqslant_m$-complete for the class of all recursively enumerable sets if and only if it is $\leqslant_1$-complete for the class of all recursively enumerable sets.

Proof

(1) A set which is $\leqslant_1$-complete, is $\leqslant_m$-complete by definition.

(2) Let A be an $\leqslant_m$-complete set and $K \leqslant_m A$ via f. Using the echo function e, define the function f as follows:

(a) $f'(0) = f(0)$

(b) $f'(k) = z$, as determined by the following program:

$z := f(k);\ x := \varnothing;$

while $(\exists i < k)[z = f'(i)]$ **do**

$x := x \cup \{z\};$
$z := e(x)$

od

Then $K \leqslant_1 A$ via the function f'. Since K is $\leqslant_1$-complete for the class of all recursively enumerable sets (see Section 7.3), it follows that A is $\leqslant_1$-complete for the class of all recursively enumerable sets too. ■

If the set A is complete, then by the above theorem $[A]_{\equiv_1} = [A]_{\equiv_m}$. As we have seen in Theorem 7.11, $[S]_{\equiv_1} \neq [S]_{\equiv_m}$ if S is a simple set. In conclusion, simple sets cannot be $\leqslant_m$-complete for the class of all recursively enumerable sets. Therefore, there exists a set S such that $[\{2x \mid x \in N\}]_{\equiv_m} <_m [S]_{\equiv_m} <_m [K]_{\equiv_m}$, namely the simple set constructed in Theorem 7.9. Therefore, the fact that S is not recursively decidable cannot be proved by reduction from K.

Also note that by the above theorem all sets $\leqslant_m$-complete for the class of all recursively enumerable sets constitute a single $\equiv_1$-class, and are therefore (by Lemma 7.4 on page 166) recursively isomorphic. That is, for any pair A, B of such complete sets there is a recursive bijection $p: N \to N$, such that $p(A) = B$.

7.6 Exercises

7.1 Give an example of an infinite set A such that $A \leqslant_m \bar{A}$. Give also an example of an infinite set B such that $B \leqslant_m \bar{B}$ does not hold. Prove that $C \leqslant_m \bar{C}$ for every finite non-empty set C.

7.2 Give examples of sets which are $\leqslant_m$-hard but not $\leqslant_m$-complete for the class of all recursively decidable sets.

7.3 Prove that the set K is $\leqslant_m$-complete for the class of all recursively enumerable sets.

7.4 Prove that $\bar{K}_0$ is a productive set.

7.5 Investigate the relationship $\leqslant_m$ between the five sets defined below.

(a) $\{x \mid x \text{ is prime}\}$

(b) $\{x \mid x \text{ is even}\}$

(c) $\{x \mid W_x = \varnothing\}$

(d) $\{x \mid W_x \text{ is infinite}\}$

(e) $\{x \mid \phi_x \text{ is total}\}$

7.6 Prove that $A \leqslant_m B$ if A is recursively decidable and B is not.

7.7 A partial recursive function ψ is called productive for the set A if

$$(\forall x)[\text{if } W_x \subseteq A \text{ then } (\psi(x)\downarrow \text{ and } \psi(x) \in A - W_x)].$$

Prove that there is a total recursive function which is productive for A in the sense defined above.

7.8 Let A be a set for which there exists a partial recursive function which is productive for A in the sense defined in the preceding exercise. Prove that A is productive in the sense of Definition 7.3 on page 160.

7.9 If B is recursively enumerable and $A \cap B$ is productive, then A is productive too. Prove this.

7.10 Show that the sets $\{x \mid \phi_x \text{ is total}\}$ and $\{x \mid \phi_x \text{ is not total}\}$ are incomparable with respect to $\leqslant_m$.

7.11 Let $X(f) \triangleq \{A \mid A \text{ is productive with } f \text{ as productive function}\}$. Give examples of total recursive functions f such that $X(f) = \varnothing$ and such that $X(f) \neq \varnothing$.

7.12 Prove that there exist countably infinitely many creative sets.

7.13 Prove by direct construction that the sets $\{x \mid \phi_x \text{ is a bijection}\}$ and $\{x \mid \phi_x \text{ is an injection}\}$ are productive.

7.14 Prove the following assertions.

(a) $A \times B$ is a cylinder if A is a cylinder, or B is a cylinder.

(b) $A \sqcup B$ is a cylinder if both A and B are cylinders.

7.15 Is $K \times K$ a cylinder? Is $K \sqcup \bar{K}$ a cylinder?

7.7 References

For an introduction to the literature the reader is referred to the references in Chapter 8.

Chapter 8
Cook/Turing Reducibility

In Chapter 7, we discussed the Karp reducibility $\leqslant_m$. In an intuitive sense, this relation is a little too discriminative, in that a recursively enumerable set A and its complement $\bar{A}$ are only in the same $\equiv_m$ equivalence class, if the set A is recursively decidable.

Let A be a recursively enumerable set. If $A \equiv_m \bar{A}$, then $\bar{A}$ is recursively enumerable by Theorem 7.1 (page 154). Consequently, A is recursively decidable. Thus the complement of a non-decidable, recursively enumerable set, does not belong to the same degree as the set itself. This is reasonable in that the elements of the set can be effectively listed, whereas the elements of the complement cannot. Intuitively however, the characteristic function χ_A of the set A has the same complexity as the characteristic function $\chi_{\bar{A}}$ of $\bar{A}$, since $\chi_{\bar{A}} = \lambda x[1 \dot{-} \chi_A(x)]$.

The Turing or Cook reducibility relation, to be discussed in this chapter, assigns the same degree to a set and its complement. The idea is the following:

$A \leqslant B$ if χ_A is computable whenever χ_B is computable.

To show that $A \leqslant B$, we write a macro program computing χ_A. This macro program may comprise assignment statements of the form $x := \chi_B(y)$. In general, we cannot expand this macro. Instead we assume that $\chi_B(y)$ is evaluated in unit time by some oracle. In other words, we temporarily add the function χ_B to the underlying data type **N**. Or, in terms of recursive functions, we show that χ_A belongs to $C(B_{rec} \cup \{\chi_B\}, F)$, where $C(B_{rec}, F) = REC$.

Algorithms expressed as such macro programs are called *relative algorithms*. The adjective *relative* expresses that the functions computed by these programs are not necessarily effectively computable. They are effectively computable only relative to the effective computability of the function temporarily added to the underlying data type **N**, or, in terms of recursive functions, temporarily added to the set of base functions.

Relative algorithms give rise to a *relativized theory* concerning relative computability, relative enumerability, relative decidability, etc. In what follows, the basics of this relativized theory will be discussed. As regards the Cook/Turing reducibility relation, we will show that $\equiv_m$ is a refinement of the equivalence relation induced by the Cook/Turing reducibility relation, that there exist recursively enumerable sets which are incomparable with respect to Cook/Turing reducibility and that K is Cook/Turing complete for the class of all recursively enumerable sets.

8.1 Relative algorithms

Relative algorithms describe non-effective computations. The non-effectiveness is caused by the underlying data type. The operations specified in this underlying data type are the actions which can be performed in single computation steps. By adding an operation which is not effectively executable, we get computations in which some of the steps are effectively executable, and others are not effectively executable. The latter consist of the evaluation of a function which is not computable. To stress the non-effectiveness, we say that the data type has been equipped with an *oracle*. The oracle evaluates in some mysterious way the noncomputable function added to the data type as an operation.

EXAMPLE 8.1

Let us add a function $\chi: N \to N$ to the data type **N**, and let $^{\chi}$**N** be the resulting data type. An example of a SAL($^{\chi}$**N**) program is:

input: x
output: y
method:

$p := \sigma^1_{2,1}(x)$; $q := \sigma^1_{2,2}(x)$;

if $\chi(p) = 1 \wedge \chi(q) = 1$ **then** $y := 1$ **else** $y := 0$ **fi**

The function f_P computed by this program depends of course on the function χ.

- Let $\chi = \lambda p[\text{if}\,(p \text{ is prime}) \text{ then } 1 \text{ else } 0]$ for example. Then f_P is the characteristic function of the set $\{\sigma_1^2(p,q) | p \text{ and } q \text{ primes}\}$.
- Now let $\chi = \lambda p[\text{if } \phi_p(p)\downarrow \text{ then } 1 \text{ else } 0]$. Then f_P is the characteristic function of the set $\{\sigma_1^2(p,q) | \phi_p(p)\downarrow \text{ and } \phi_q(q)\downarrow\}$.

In the next section we define the oracle data type ${}^\chi\mathbf{N}$, and the syntax and semantics of the programming language SAL(${}^\chi\mathbf{N}$). These definitions are extensions of those for SAL($\mathbf{N}$) and are therefore largely copies of earlier definitions. Places where the definitions differ from the corresponding oracle-free definitions are marked with the symbol ☞.

We show that there is a universal function, which is computable if the function evaluated by the oracle is computable, and that the s–n–m theorem holds. In consequence, many of the results of Chapter 5 carry over. To understand Cook/Turing reducibility, discussed in Section 8.3, it suffices to comprehend the above example. Then the next section may be skipped.

8.2 Definitions and theorems concerning relative computability

Definition 8.1

The oracle type ${}^\chi\mathbf{N}$ is the many-sorted algebra ${}^\chi\mathbf{N} \triangleq (N, B; 0, +_1, \dot{-}_1, \chi, \neq_0)$ consisting of:

(1) Two sorts:

  (a) $N = \{0, 1, 2, 3, \dots\}$, the set of all natural numbers, and
  (b) $B = \{\text{TRUE}, \text{FALSE}\}$, the set of truth values.

(2) and five operations:

  (a) a nullary operation 0, i.e. a given particular element of N, namely 0, zero
  (b) $+_1 : N \to N$, the successor function.
  (c) $\dot{-}_1 : N \to N$, the predecessor function.
☞ (d) $\chi : N \to N$, the function mysteriously evaluated by the oracle.
  (e) $\neq_0 : N \to B$, a test for zero. ■

Apart from the added oracle, the above data type is the same as our standard type $\mathbf{N}$ (Definition 3.1).

Over this oracle data type the language SAL($^\chi$**N**) is defined. The definition of the syntax and the semantics of SAL(**X**) depends on the data type **X**, but the structure of the definitions is the same for all data types. The definitions given below are extensions of those in Sections 3.2 and 3.4. The extensions are necessary to handle the oracle.

The syntax of SAL($^\chi$N)

⟨*identifier*⟩::= $\{x_i | i = 1, 2, 3, \dots\}$.

⟨*assignment*⟩::= ⟨*identifier*⟩:= 0|
⟨*identifier*⟩:= ⟨*identifier*⟩|
⟨*identifier*⟩:= ⟨*identifier*⟩ + 1|
⟨*identifier*⟩:= ⟨*identifier*⟩ $\dot{-}$ 1
☞ ⟨*identifier*⟩:= χ(⟨*identifier*⟩).

⟨*statement*⟩::= ⟨*assignment*⟩|
while ⟨*identifier*⟩ $\neq 0$ **do** ⟨*sequence*⟩ **od**.

⟨*sequence*⟩::= ⟨*statement*⟩|
⟨*sequence*⟩;⟨*statement*⟩.

⟨*program*⟩::= {($n, k, p,$⟨*sequence*⟩)| $n \geq 0, k \geq 1$ and $n, k \leq p$}.

The computations induced by a SAL($^\chi$**N**) program are defined as sequences of single steps, each step transforming the state vector, as in Section 3.4. The single steps are defined as follows.

Let V_Q be the set of all state vectors of a SAL($^\chi$**N**) program $Q = (n, k, p, P)$ having its lines numbered from 1 to f.

A state vector $v_1 = (s, x) \in V_Q$ is transformed into the state vector $v_2 = (t, y) \in V_Q$ in a single computation step, denoted by $v_1 \vdash_Q v_2$, if and only if:

(1) s: ⟨*assignment*⟩ occurs in the labelled version of Q, $t = s + 1$, $y_r = x_r$ for all r such that $1 \leq r \leq p$ and $r \neq i$, and the value of y_i depends on the assignment statement as specified in Table 8.1.

(2) Otherwise, the line with number s contains a while-head or a while-tail. In that case $y = x$, and the value of t is determined as follows:

   (a) if s: **while** $x_i \neq 0$ **do** :s' occurs in the labelled version of Q, then $t = s + 1$ if $\neq_0(x_i)$ evaluates to TRUE and $t = s' + 1$ otherwise;

   (b) if s: **od** :s' and s': **while** $x_i \neq 0$ **do** :s occur in the labelled version of Q, then $t = s' + 1$ if $\neq_0(x_i)$ evaluates to TRUE and $t = s + 1$ otherwise.

Table 8.1 Semantics of assignment statements.

⟨*assignment*⟩	Value of y_i
$x_i := 0$	$y_i = 0$
$x_i := x_j$	$y_i = x_j$
$x_i := x_j + 1$	$y_i = +_1(x_j)$
$x_i := x_j \dot{-} 1$	$y_i = \dot{-}_1(x_j)$
☞ $x_i := \chi(x_j)$	$y_i = \chi(x_j)$

From this, the meaning ${}^{\chi}M_Q : N^p \to N^p$ of a SAL(${}^{\chi}$**N**) program $Q = (n, k, p, P)$ and the function ${}^{\chi}f_Q : N^n \to N^k$ computed by this program are defined.

Let $a \in N^p$ and $(1, a), s_1, s_2, \ldots, (t, b)$ be a computation of Q.

(1) If the computation terminates with (t, b), i.e. (t, b) is a final state vector and thus $t = f + 1$, then ${}^{\chi}M_Q(a) = b$.

(2) If no terminating computation starting with state vector $(1, a)$ exists, then ${}^{\chi}M_Q(a) = \uparrow$, i.e. undefined.

The function ${}^{\chi}f_Q : N^n \to N^k$ computed by the program, is determined as follows:

$${}^{\chi}f_Q \triangleq \lambda x[\pi^p_{1\ldots k}({}^{\chi}M_Q(x_1, x_2, \ldots, x_n, 0, 0, \ldots, 0))]$$

Finally, we define the set of all SAL(${}^{\chi}$**N**) computable functions:

(1) F^n_k(SAL(${}^{\chi}$**N**)) is the set of all functions $f : N^n \to N^k$ such that there exists a SAL(${}^{\chi}$**N**) program (n, k, p, P) computing f.

(2) F(SAL(${}^{\chi}$**N**)) is the union of all F^n_k(SAL(${}^{\chi}$**N**)), $n \geqslant 0$, $k > 0$.

Gödel numbers can be assigned to SAL(${}^{\chi}$**N**) programs, such that this numbering admits a universal function and a corresponding s–n–m theorem. All of this is achieved by slightly modifying the definitions and algorithms in Chapter 5.

To assign Gödel numbers to programs, first assign code numbers to statements as defined Table 8.2.

A sequence $S = S_1, S_2, \ldots, S_m$ is assigned the number

$$code(S) \triangleq \langle code(S_1), code(S_2), \ldots, code(S_m)\rangle \dot{-} 1$$

and the SAL(${}^{\chi}$**N**) program (n, k, p, P) is assigned the Gödel number $code(P)$.

Theorem 8.1

For every $n \geqslant 0$ and $k \geqslant 1$, there is a SAL(${}^{\chi}$**N**) computable function

Table 8.2 Gödel numbers of SAL($^{\chi}$**N**) statements.

Statement	*code*(statement)
$x_i := 0$	$6.(i \dot{-} 1)$
$x_i := x_j$	$6.\sigma_1^2(i \dot{-} 1, j \dot{-} 1) + 1$
$x_i := x_j + 1$	$6.\sigma_1^2(i \dot{-} 1, j \dot{-} 1) + 2$
$x_i := x_j \dot{-} 1$	$6.\sigma_1^2(i \dot{-} 1, j \dot{-} 1) + 3$
☞ $x_i := \chi(x_j)$	$6.\sigma_1^2(i \dot{-} 1, j \dot{-} 1) + 4$
while $x_i \neq 0$ **do** S **od**	$6.\sigma_1^2(i \dot{-} 1, code(S)) + 5$

$^{\chi}\phi: N^{n+1} \rightarrow N^k$ which is universal for F_k^n(SAL($^{\chi}$**N**)), that is, for every $f \in F_k^n$(SAL($^{\chi}$**N**)) there is an i such that $f = {}^{\chi}\phi_i$.

Proof

Consider Algorithm 5.1. This algorithm can be made into an algorithm to compute a function universal for F_k^n(SAL($^{\chi}$**N**)) by modifying the expansion of the macros *simulate* (Algorithm 5.2) and *execute* (Algorithm 5.3). The modifications consist of updating the way of determining what statement is involved, i.e. taking the remainder of dividing by 6 instead of by 5, and adding a statement to call the oracle.

```
simulate(g, M)
  L := 1 + σ^1_{2,1}(g);
  {L is the length of the sequence of statements that remain to
  be executed}

  while L ≠ 0 do

    S := σ^1_{L,1}(σ^1_{2,2}(g));
    r := remainder of S divided by 6;
    if r ≠ 5 then
      execute(S, r, M);
      pop(g);
      L := L ∸ 1
    else
      x := τ(M, σ^1_{2,1}((S ∸ 5)/6) + 1);
      if x ≠ 0 then
        z := σ^1_{2,2}((S ∸ 5)/6);
        push(z, g);
        L := L + σ^1_{2,1}(z)
      else
        pop(g);
        L := L ∸ 1
      fi
    fi
  od
```

$execute(S, r, M)$

if $r = 0$ **then**
  $x := 1 + S/6$
else
  $x := 1 + \sigma^1_{2,1}((S \dot{-} r)/6);$
  $y := 1 + \sigma^1_{2,2}((S \dot{-} r)/6)$
fi;
if $r = 0$ **then** $M := \omega(M, x, 0)$ **fi**;
if $r = 1$ **then** $M := \omega(M, x, \tau(M, y))$ **fi**;
if $r = 2$ **then** $M := \omega(M, x, \tau(M, y) + 1)$ **fi**;
if $r = 3$ **then** $M := \omega(M, x, \tau(M, y) \dot{-} 1)$ **fi**;
if $r = 4$ **then** $M := \omega(M, x, \chi(\tau(M, y)))$ **fi**; ■

Corollary 8.1

For every $n \geqslant 0$ and $k \geqslant 1$ there is a total SAL($^\chi$**N**) computable function $^\chi T: N^{n+2} \to N$, such that $^\chi T(i, x, t) =$ if ($^\chi\phi_i(x)\downarrow$ within t steps) then 1 else 0.

Proof

The algorithm computing Kleene's termination predicate T sketched in Section 6.2 can be modified so as to compute this function $^\chi T$, by appropriately modifying the expansion of the macros *simulate* and *execute* as has been done in the proof of Theorem 8.1. ■

Beware that, despite the suggestive algorithmic notation, the universal function $^\chi\phi$ and termination function $^\chi T$ are not necessarily effectively computable. These functions are computable only if the function χ is computable.

Theorem 8.2

For every $n, m \geqslant 0$ and $k \geqslant 1$ there is a total recursive function s^n_m such that

$$^\chi\phi^{(n,k)}_{s^n_m(i,a)} = \lambda x[^\chi\phi^{(n+m,k)}_i(a, x)]$$

for all $i \in N$ and $a \in N^m$.

Proof

The proof of this relativized version of the s–n–m theorem is almost identical to the proof of the s–n–m theorem. (In the macro evaluating the Gödel number of $x_i := a$, 5 must be changed into 6, but that is all the modification required.) The evaluation of the function s^n_m requires only the manipulation of Gödel numbers, and does not call for any computations induced by the programs having these Gödel numbers.

Consequently, the s_m^n function is a primitive recursive function, no matter what function is evaluated by the oracle χ. ■

We can now proceed to relativize the concepts of enumerability and decidability.

(1) A predicate P is SAL($^{\chi}$**N**) enumerable if there is a partial SAL($^{\chi}$**N**) computable function f such that $P = \lambda x[sg(f(x)) = 1]$.

(2) A predicate P is SAL($^{\chi}$**N**) decidable if there is a total SAL($^{\chi}$**N**) computable function f such that $P = \lambda x[sg(f(x)) = 1]$.

Many properties of enumerability and decidability carry over. To be specific, all the properties given in Section 6.3 hold for the relativized concepts too.

In the above, relative computability was defined in terms of the programming language SAL($^{\chi}$**N**). An alternative is to define 'χ-recursiveness' and the set $^{\chi}REC$ of all χ-recursive functions, as is done below.

The set $^{\chi}REC$ of all χ-recursive partial functions is the closure $C(B, F)$, where:

(1) The set B of base functions consists of

 (a) the successor function $+_1: N \to N$,
 (b) the predecessor function $\dot{-}_1: N \to N$,
 (c) the identity functions $I^n: N^n \to N^n$, for all $n \geq 0$,
 (d) the projection functions $\pi_i^n: N^n \to N$, for all $n \geq 1$ and i, $1 \leq i \leq n$,
 (e) the zero function $z: N \to N$, $z = \lambda x[0]$,
☞ (f) the function $\chi: N \to N$ evaluated by the oracle.

(2) The set F of operators consists of

 (a) composition of functions,
 (b) Cartesian product of functions,
 (c) exponentiation of functions,
 (d) iteration of functions.

The fact that these two definitions are indeed equivalent is expressed by the equivalence theorem 'F(SAL($^{\chi}$**N**)) = $^{\chi}REC$'.

It is not always a trivial matter to proceed from a theorem and its proof to the relativized version of this theorem and proof, or vice versa. There are problems, among which is one of the most prominent problems in complexity theory, the $\mathbf{P} \overset{?}{=} \mathbf{NP}$ question, which are open in the classical theory, and are solved in a relativized version. That is, there is an oracle f such that $\mathbf{P} = \mathbf{NP}$ in terms of SAL(^{f}**N**) computability, and there also is an oracle g such that $\mathbf{P} \neq \mathbf{NP}$ in terms of SAL(^{g}**N**) computability. (The $\mathbf{P} \overset{?}{=} \mathbf{NP}$ question is discussed at length in Chapter 13.)

8.3 Properties of the Turing reducibility relation

Oracle computations present an unrestricted form of reducing a given problem to some other problem. If we have shown that f is SAL(^{g}N) computable, for some functions f and g, we may conclude: 'if g is computable then f is computable too'.

Definition 8.2

Let A and B be subsets of N.

$A \leqslant_T B$ iff there is a SAL($^{\chi_B}$N) program P, such that $f_P = \chi_A$. In other words, if χ_A is χ_B-recursive.

$A \leqslant_T B$ is pronounced as: 'A is Turing reducible to B'. ■

Clearly, the Turing reducibility relation $\leqslant_T$ is reflexive and transitive, and therefore induces an equivalence relation defined as follows:

$$A \equiv_T B \quad \text{iff} \quad \text{both } A \leqslant_T B \text{ and } B \leqslant_T A$$

The classes induced by this equivalence relation are ordered by the reducibility relation as follows:

$$[A]_{\equiv_T} \leqslant_T [B]_{\equiv_T} \quad \text{iff} \quad A \leqslant_T B$$

The above reducibility relation can also be interpreted as stating that the cost of computing χ_A is not greater than the cost of computing χ_B, provided that the cost of computing the function f_P of the definition above is approximately equal to the cost of computing χ_B. Cook introduced this type of reducibility in complexity theory. To make the cost of evaluating f_P approximately equal to the cost of evaluating χ_B, it will be required that the program P used in the reduction is of low complexity, e.g. has a polynomial running time. The reducibility relation obtained in this way can be used to compare the computing cost of sufficiently complex problems, e.g. problems requiring more than polynomial running time.

The Turing reducibility relation is a limiting case of the Cook resource-bounded reducibility relation. With respect to Turing reducibility all recursively decidable problems are in a single equivalence class, and we compare the computing costs of noncomputable problems, that is, problems having an infinite complexity.

Theorem 8.3

(1) $A \equiv_T \bar{A}$

(2) $A \leqslant_m B$ implies $A \leqslant_T B$

(3) The classes $[A]_{\equiv_T}$ constitute an upper semi-lattice with respect to $\leqslant_T$

(4) ($A \leqslant_T B$ and B recursive) implies A recursive

(5) ($A \leqslant_T B$ and B recursively enumerable) does *not* imply that A is recursively enumerable.

Proof

(1) $\chi_{\bar{A}} = \lambda x[1 \dot{-} \chi_A(x)]$, and thus $\chi_{\bar{A}}$ is χ_A-recursive. Similarly, χ_A is $\chi_{\bar{A}}$-recursive. Thus $A \leqslant_T \bar{A}$ and $\bar{A} \leqslant_T A$, and therefore $A \equiv_T \bar{A}$.

(2) Let $A \leqslant_m B$ via $f \in REC_1^1$. Then $\chi_A = \chi_B \cdot f$, and thus χ_A is χ_B-recursive.

(3) In Theorem 7.2 (page 155) we saw that the $\equiv_m$-classes constitute an upper semi-lattice with respect to $\leqslant_m$, and that $[A \sqcup B]_{\equiv_m}$ is the least upper bound of $[A]_{\equiv_m}$ and $[B]_{\equiv_m}$. By part (2) we have that $[A \sqcup B]_{\equiv_T}$ is an upper bound of $[A]_{\equiv_T}$ and $[B]_{\equiv_T}$. Let $[C]_{\equiv_T}$ be any upper bound, that is $[A]_{\equiv_T} \leqslant_T [C]_{\equiv_T}$ and $[B]_{\equiv_T} \leqslant_T [C]_{\equiv_T}$. Then χ_A is χ_C-recursive and χ_B is χ_C-recursive. Therefore

$$\chi_{A \sqcup B} = \lambda x\left[\text{if } (x \text{ is even}) \text{ then } \chi_A\left(\frac{x}{2}\right) \text{ else } \chi_B\left(\frac{x \dot{-} 1}{2}\right)\right]$$

is χ_C-recursive. This implies that $[A \sqcup B]_{\equiv_T} \leqslant_T [C]_{\equiv_T}$. Consequently, $[A \sqcup B]_{\equiv_T}$ is the least upper bound of $[A]_{\equiv_T}$ and $[B]_{\equiv_T}$.

(4) Immediate.

(5) By part (1) we have $K \equiv_T \bar{K}$, and thus $\bar{K} \leqslant_T K$. Since K is recursively enumerable and $\bar{K}$ is not, we are done. ■

From part (2) of the above theorem it follows that $\equiv_m$ is a refinement of $\equiv_T$, i.e. every Turing degree consists of one or more m-degrees. The class of all recursively decidable subsets of N for example is a single Turing degree, which consists of three m-degrees (see Figure 7.1)

In Definition 7.2 (page 156) the concepts *completeness* and *hardness* were defined for an arbitrary class M of subsets of N and for an arbitrary reducibility ordering. Hence this definition is valid for the Turing reducibility relation. Thus,

(1) a set $A \subseteq N$ is $\leqslant_T$-hard for a class M of subsets of N if $B \leqslant_T A$ whenever $B \in M$,

(2) A is $\leqslant_T$-complete for M if A is $\leqslant_T$-hard for M and furthermore $A \in M$.

Theorem 8.4

Any set which is $\leqslant_m$-hard for the class of all recursively enumerable sets is also $\leqslant_T$-hard for this class.

Proof

If A is $\leqslant_m$-hard for the class of all recursively enumerable subsets of N, then $X \leqslant_m A$ whenever X is recursively enumerable. By part (2) of Theorem 8.3 above, $X \leqslant_T A$ for all recursively enumerable sets $X \subseteq N$. Consequently, A is $\leqslant_T$-hard for the class of all recursively enumerable subsets of N. ■

Corollary 8.2

The sets $K = \{x | \phi_x(x)\downarrow\}$ and $K_0 = \{\sigma_1^2(x, y) | \phi_x(y)\downarrow\}$ are $\leqslant_T$-complete for the class of all recursively enumerable subsets of N. ■

Note that the converse of Theorem 8.4 is false; there are sets which are $\leqslant_T$-hard for the class of recursively enumerable subsets of N, but are not $\leqslant_m$-hard for this class. An example is the set $\bar{K}$. Since $K \leqslant_T \bar{K}$, and K is $\leqslant_T$-complete for the class of all recursively enumerable subsets of N, we have that $\bar{K}$ is $\leqslant_T$-hard for this class. On the other hand, K and $\bar{K}$ are incomparable with respect to $\leqslant_m$, by Theorem 7.3. Thus K does not precede $\bar{K}$ with respect to $\leqslant_m$, and therefore $\bar{K}$ is not $\leqslant_m$-hard for the class of all recursively enumerable subsets of N.

An $\equiv_m$-equivalence class which contains a recursively enumerable set contains recursively enumerable sets only (see Section 7.1). This is not true for $\equiv_T$-classes. In particular K and $\bar{K}$ both belong to $[K]_{\equiv_T}$. Therefore the Turing degree of K consists of more than one m-degree.

In the next section, we show that the $\equiv_T$-class of K contains more than one recursively enumerable $\equiv_m$-class. The proof of this statement requires knowledge of the material in Section 7.5.

One might get the impression that the Turing reducibility relation is so powerful that all recursively enumerable sets belong to one and the same Turing degree. The question of whether this is indeed the case is known as 'Post's problem'. The answer is 'no'. The proof of this is rather involved and will be presented in the final section of this chapter.

Although important methods are shown there, the remaining two sections can be skipped. They are not strictly necessary to understand what follows, and are somewhat laborious.

8.4 The Turing degree of K

Reading this section requires knowledge of some of the material in Section 7.5, namely:

(1) a simple set is a recursively enumerable set, whose complement is itself infinite, but does not contain an infinite recursively enumerable subset.

(2) the membership problem of simple sets is significantly easier than the membership problem of K, that is, simple sets precede K with respect to $\leqslant_m$, but are not equivalent to K with respect to $\equiv_m$.

In this section we construct a simple set S which is Turing-equivalent to K, that is, $S \in [K]_{\equiv_T}$. Thus the Turing degree $[K]_{\equiv_T}$ of K contains at least two recursively enumerable $\equiv_m$-classes, namely $[K]_{\equiv_m}$ and $[S]_{\equiv_m}$. The construction proceeds as follows:

(1) Given a total recursive function $f: N \to N$, we define the set B_f as follows:

$$B_f \triangleq \{x \mid (\exists y)[y > x \wedge f(y) \leqslant x]\}$$

(2) we then show that $B_f \equiv_T f(N)$,

(3) finally, we show that B_f is simple if $f(N)$ is not recursively decidable.

Lemma 8.1

Let $f: N \to N$ be recursive, total and injective, and let

$$B_f \triangleq \{x \mid (\exists y)[y > x \wedge f(y) \leqslant x]\}.$$

Then $f(N) \leqslant_T B_f$.

Proof

To determine whether a given element x belongs to $f(N)$, proceed as follows.

- find z, the $(x+1)$th element not in B_f.

```
z:= 0; t:= 0;
while t ⩽ x do
   if χ_{B_f}(z) = 0 then t:= t + 1 fi;
   z:= z + 1
od;
z:= z ∸ 1
```

- This element z does not belong to B_f. Thus, $\neg(\exists y)[y > z \wedge f(y) \leqslant z]$, in other words $(\forall y)[\text{if } y > z \text{ then } f(y) > z]$. Because $z \geqslant x$, we also have $(\forall y)[\text{if } y > z \text{ then } f(y) > x]$. Therefore $x \in f(N)$ if and only if $x \in f([0, z])$. To see whether $x \in f(N)$, we construct the finite set $f([0, z])$, and see whether x belongs to this set.

Thus $\chi_{f(N)}$ is χ_{B_f}-recursive, and consequently $f(N) \leqslant_T B_f$. ■

Lemma 8.2

Let $f:N \to N$ be recursive, total and injective, and let

$$B_f \triangleq \{x|(\exists y)[y > x \wedge f(y) \leqslant x]\}.$$

Then $B_f \leqslant_T f(N)$.

Proof

To see whether $x \in B_f$, we must check whether there is a $y > x$ such that $f(y) \leqslant x$.

$$(\exists y)[f(y) \leqslant x \wedge y > x] \text{ if and only if}$$
$$(\exists z)[z \in [0, x] \cap \{f(n)|n > x\}]$$

Because f is an injective function

$$\{f(n)|n > x\} = f(N) - \{f(n)|0 \leqslant n \leqslant x\}$$

Therefore,

$$(\exists y)[f(y) \leqslant x \wedge y > x] \text{ if and only if}$$
$$(\exists z)[z \in ([0, x] - \{f(n)|0 \leqslant n \leqslant x\}) \cap f(N)]$$

Since we can effectively construct the finite set $[0, x] - \{f(n)|0 \leqslant n \leqslant x\}$, we have that χ_{B_f} is $\chi_{f(N)}$-recursive. ■

The above two lemmas are summarized in the theorem below.

Theorem 8.5

Let $f:N \to N$ be recursive, total and injective, and let

$$B_f \triangleq \{x|(\exists y)[y > x \wedge f(y) \leqslant x]\}.$$

Then $B_f \equiv_T f(N)$. ■

For the following, we shall require the following property of infinite recursively enumerable sets.

Lemma 8.3

Let A be an infinite recursively enumerable subset of N.

(1) There is a recursive, total, injective function $f:N \to N$, such that $A = f(N)$.

(2) A has an infinite recursively decidable subset B.

Proof

Let A be an infinite recursively enumerable subset of N.

(1) By Theorem 6.6 (page 140), there is a total recursive function $h:N \to N$ such that $h(N)=A$. Using this function h, a recursive, total, injective function $f:N \to N$ such that $f(N)=h(N)=A$, is defined as follows:

$$f(0)=h(0)$$

$$f(n+1)=h(\mu z[(\forall m \leqslant n)[h(z) \neq f(m)]]) \qquad n \geqslant 0.$$

(2) Define the function $g:N \to N$ as follows:

$$g(0)=f(0)$$

$$g(n+1)=f(\mu z[(\forall m \leqslant n)[f(z) > g(m)]]) \qquad n \geqslant 0.$$

The function g is recursive, total, injective and strictly increasing. To see whether $x \in g(N)$, generate $g(0), g(1), \ldots, g(i), \ldots$ until for some i either $g(i)=x$ or $g(i)>x$. In the latter case $x \notin g(N)$, because g is strictly increasing. Thus $g(N)$ is an infinite recursively decidable subset of $f(N)=A$. ∎

We now return to our set B_f defined earlier.

Lemma 8.4

Let $f:N \to N$ be recursive, total and injective, and let

$$B_f \triangleq \{x \mid (\exists y)[y > x \wedge f(y) \leqslant x]\}$$

If $f(N)$ is undecidable, then B_f is a simple set.

Proof

B_f is recursively enumerable, because $\lambda xy[y > x \wedge f(y) \leqslant x]$ is a recursively decidable predicate.

Assume that B_f is not simple. Then $\bar{B}_f$ contains an infinite, recursively enumerable subset. By the preceding lemma, $\bar{B}_f$ then also contains an infinite recursive subset A. Then $f(N) \leqslant_T A$. The proof of this is a slightly modified version of the proof of Lemma 8.1 above.

To determine whether a given element x belongs to $f(N)$, proceed as follows.

- find z, the $(x+1)$th element in A.

```
z:= 0; t:= 0;

while t ⩽ x do
   if χ_A(z) = 1 then t:= t + 1 fi
od;
z:= z ∸ 1
```

- This element z does not belong to B_f, because A is a subset of

$\bar{B}_f$. Thus,

$$\neg(\exists y)[y > z \wedge f(y) \leqslant z],$$

in other words $(\forall y)$[if $y > z$ then $f(y) > z$]. Because $z \geqslant x$, we also have $(\forall y)$[if $y > z$ then $f(y) > x$]. Therefore $x \in f(N)$ if and only if $x \in f([0, z])$. To see whether $x \in f(N)$, we construct the finite set $f([0, z])$, and see whether x belongs to this set.

Thus $\chi_{f(N)}$ is χ_A-recursive. Since χ_A is a recursive function, it follows that $f(N)$ is a recursively decidable set.

Consequently, if $f(N)$ is not recursively decidable, then B_f is simple. ■

We have now acquired all the tools necessary to prove the final theorem of this section.

Theorem 8.6

$[K]_{\equiv_T}$, the Turing degree of K, contains more than one recursively enumerable $\equiv_m$-class.

Proof

K is recursively enumerable. By Lemma 8.3 there is a recursive, total, injective function f, such that $f(N) = K$. The corresponding set B_f is Turing equivalent to K, and simple, because K is not recursively decidable. Thus $[K]_{\equiv_T}$ contains at least two recursively enumerable m-degrees, namely $[K]_{\equiv_m}$ and $[B_f]_{\equiv_m}$. ■

8.5 Post's problem

One might get the impression that Turing reducibility is so powerful that all recursively enumerable sets belong to one and the same Turing degree, i.e. to the class $[K]_{\equiv_T}$. Post's problem is the question of whether this is indeed the case.

The answer to the question is 'no'. There exist infinitely many recursively enumerable sets whose Turing degrees are mutually incomparable.

To solve Post's problem, we have to construct two recursively enumerable sets A and B whose Turing degrees are incomparable. To show that A does not precede B with respect to $\leqslant_T$, we have to show that χ_A is not χ_B-recursive. Because A is recursively enumerable, this means that we must show that $\bar{A}$ is not equal to any χ_B-recursively enumerable set; that is, $\bar{A} \neq {}^{\chi_B}W_i$ for every χ_B-recursively enumerable set ${}^{\chi_B}W_i$. In other words, for every i there is an x such that $x \in (\bar{A} - {}^{\chi_B}W_i) \cup ({}^{\chi_B}W_i - \bar{A})$, that is, such that $x \in A$ if and only if $x \in {}^{\chi_B}W_i$.

The assignment $i \mapsto x$, such that $x \in (\bar{A} - {}^{\chi_B}W_i) \cup ({}^{\chi_B}W_i - \bar{A})$, defines a function $f: N \to N$. This function cannot be recursive.

Assume to the contrary, that this function f is a total recursive function. Define a function ψ as follows:

$$\psi \triangleq \lambda xy[\text{if } \chi_B(x) = 1 \text{ then } \uparrow \text{ else } 1]$$

Then ψ is χ_B-recursive. By the relativized s–n–m theorem (Theorem 8.2), there is a total recursive function $g \in TREC_1^1$ such that $\phi_{g(x)} = \lambda y[\psi(x, y)]$.

- $f(g(x)) \in A$ if and only if
- $f(g(x)) \in {}^{\chi_B}W_x$, by the assumption that $f(x) \in A$ iff $f(x) \in {}^{\chi_B}W_x$, if and only if
- ${}^{\chi_B}W_{g(x)} \neq \varnothing$, because ${}^{\chi_B}W_{g(x)}$ is either empty, or equal to N, if and only if
- $\chi_B(x) \neq 1$, that is, $x \notin B$.

Accordingly, $\bar{B} \leqslant_m A$, and thus $B \leqslant_T A$. Since A and B are to be mutually incomparable with respect to Turing reducibility, the function f cannot be recursive.

Thus, although we must show that for each i there is an x, such that $x \in (\bar{A} - {}^{\chi_B}W_i) \cup ({}^{\chi_B}W_i - \bar{A})$, we cannot actually produce this value x.

The following procedure simulates the computations of a given SAL($^{\chi}$**N**) program assuming that the oracle χ evaluates a given finite set.

input: $k, j \in N$; F {a finite subset of N.
 A finite set $\{x_1, x_2, \ldots, x_n\}$ is represented by the Gödel number $\langle x_1, x_2, \ldots, x_n \rangle$, as in Section 7.5.}
output: E, F_0 {finite subsets of N represented in the same way.}
method:

(1) replace all statements of the form $y := \chi(x)$ in the *SAL*($^{\chi}$**N**) program with index j by

 if $x \in F$ **then** $y := 1$ **else** $y := 0$; $G := G \cup \{x\}$ **fi**;

(2) Let P be the resulting program. Consider the finite set F, whose characteristic function is evaluated by the 'oracle', as an additional input variable, and the set G as an additional output variable of this program P.

(3) $E := \varnothing$; $F_0 := \varnothing$; $i := 0$;

 while $i \leqslant k$ **do**

  run program P on input $(\sigma_{2,1}^1(i), F)$ *for* $\sigma_{2,2}^1(i)$ steps;

  if *P halts within this number of steps* **then**

  $E := E \cup \{\sigma_{2,1}^1(i)\}$;

```
            F_0:= F_0 ∪ G
        fi;

        i:= i + 1
    od
```

The function computed by this procedure will be denoted by *simulate*, e.g.

$$(E, F_0) := \textit{simulate}\ (j, F, k)$$

The purpose of this procedure is to enumerate the set ${}^{\chi_V}W_j$, where V is a set under construction, and only a finite subset F of the set V is as yet known. The procedure determines which elements are enumerated within a certain number of steps, and it records in the set F_0 which elements must be left out of V in order to make the enumeration valid for V as well as for its finite approximation F.

This enumeration is used in the following procedure, which computes finite sets A and B, and partial functions f and g defined on an initial interval $[0, c]$ of N, for some $c \geqslant 0$.

Two infinite recursively enumerable sets A^∞ and B^∞, incomparable with respect to $\leqslant_T$, are obtained by running this procedure on ever growing values of its input variable. The partial functions f and g converge to functions f^∞ and g^∞, which can be used to show that A^∞ and B^∞ are incomparable with respect to $\leqslant_T$. That is, $f^\infty(i) \in A^\infty$ iff $f^\infty(i) \in {}^{\chi_{B^\infty}}W_i$, and similarly $g^\infty(i) \in B^\infty$ iff $g^\infty(i) \in {}^{\chi_{A^\infty}}W_i$.

```
input: bound ∈ N
output: A, B {finite subsets of N};
        f, g {finite approximations of functions N → N}
method:

  f(0):= 0; g(0):= 0;
  a:= 1; b:= 1;
  A:= ∅; B:= ∅;
  k:= 0;

  while k ≤ bound do

    k:= k + 1; f(k):= a; a:= a + 1;
    j:= 0; no_change:= true;

    while j ≤ k and no_change do

      (E, F):= simulate (j, B, k);
      if f(j) ∈ E − A then
        no_change:= false;
        A:= A ∪ {f(j)};
        b:= max {b, 1 + max F};
```

```
            for i:=j to k ∸ 1 do
               g(i):= b;
               b:= b + 1
            od
         fi;
         j:=j + 1
      od;

      g(k):= b; b:= b + 1;
      j:= 0; no_change:= true;

      while j ⩽ k and no_change do

         (E, F):= simulate (j, A, k);
         if g(j)∈E − B then
            no_change:= false;
            B:= B ∪ {g(j)};
            a:= max{a, 1 + max F};
            for i:= j + 1 to k do
               f(i):= a;
               a:= a + 1
            od
         fi;
         j:= j + 1
      od

   od
```

Let R be the preceding program and r the function computed by this program, i.e. $r(bound) = (A, B, f, g)$. Recursively enumerable sets A^∞ and B^∞ are obtained by running program R on ever growing values of the input variable *bound*. More formally, the set A^∞ is the domain of the function computed by the following program.

```
input: x∈N
output: y {value is immaterial}
method:
   bound:= 0; A:= leftmost component of r(bound);
   while x∉A do
      bound:= bound + 1;
      A:= leftmost component of r(bound)
   od
```

The set B^∞ is defined similarly.

The sequence of partial functions f obtained by running program R on ever increasing inputs converges to a function f^∞. Similarly the sequence of partial functions g converges to a function $g^\infty : N \to N$.

That these sequences of partial functions converge to functions $N \to N$ follows from the observation that the number of times that a value is assigned to $f(i)$ or $g(i)$ is finite. Thus $f(i)$ and $g(i)$ must have reached their final values for some sufficiently large value of R's input variable *bound*. For larger values $f(i)$ and $g(i)$ will remain the same. Thus,

$$f^{\infty} \triangleq \lambda i[\text{final value of } f(i)]$$

Similarly,

$$g^{\infty} \triangleq \lambda i[\text{final value of } g(i)]$$

Let $n_f(i)$ be the number of times a value is assigned to $f(i)$ and similarly $n_g(i)$ the number of times a value is assigned to the variable $g(i)$. Then:

$$n_f(0) = 1 \text{ and } n_f(i) \leqslant 1 + n_g(0) + n_g(1) + \cdots + n_g(i-1)$$

$$n_g(0) \leqslant 2 \text{ and } n_g(i) \leqslant 1 + n_f(0) + n_f(1) + \cdots + n_f(i)$$

These relations follow from inspection of the program R

$f(0)$ is assigned a value exactly once

$f(i)$ the value of $f(i)$ can only change if $j < i$ and $g(j)$ is added to B. For each j this can happen at most $n_g(j)$ times. Hence, a value is assigned to $f(i)$ at most $1 + \Sigma_{j<i}\, n_g(j)$ times.

The relation for n_g is derived similarly.

$g(0)$ is initially set to zero. The value can be changed only if $f(0)$ is added to A, and this can happen only once.

$g(i)$ the value of $g(i)$ can only change if $j \leqslant i$ and $f(j)$ is added to A. For each j this can happen at most $n_f(j)$ times. Thus, a value is assigned to $g(i)$ at most $1 + \Sigma_{j \leqslant i}\, n_f(j)$ times.

Consequently, the number of times a value is assigned to $f(i)$ or $g(i)$ is finite, and the sequences of partial functions converge to functions f^{∞}, $g^{\infty}: N \to N$.

These functions are not computable and are not computed by program R. The problem is that although we can prove that at some finite time the value of $f(i)$ is its final value, there is no algorithmic way to decide whether or not the current value already is this final value. That is, we cannot decide whether the values of the current partial function f are the final values, the values of f^{∞}.

It remains to show that A^∞ and B^∞ are indeed incomparable with respect to $\leqslant_T$, and that $f^\infty(i)$ and $g^\infty(i)$ supply the counter-examples to the statements $A^\infty = {}^{\chi_{B^\infty}}W_i$ and $B^\infty = {}^{\chi_{A^\infty}}W_i$.

- Assume $f^\infty(i)\in A$. Then $f^\infty(i)$ was added to A at some stage and $f^\infty(i)\in{}^{\chi_B}W_i$ for some finite subset B of B^∞. The variable b will then get a value large enough to ensure that elements which were tested during the enumeration of ${}^{\chi_B}W_i$, and found not to be in B, will not be assigned to $g(m)$, and therefore not added to B at some later stage, for any $m \geqslant i$. On the other hand, no $g(m)$, $m < i$, can be added to B, otherwise $f(i)$ would not have attained its final value $f^\infty(i)$. Therefore $f^\infty(i)\in{}^{\chi_{B^\infty}}W_i$ whenever $f^\infty(i)\in A^\infty$.
- Assume now that $f^\infty(i)\in{}^{\chi_{B^\infty}}W_i$. Then at some instant of time after $f(i)$ has reached its final value $f^\infty(i)$, B will have acquired all the elements necessary to show that $f^\infty(i)\in{}^{\chi_B}W_i$, and hence $f^\infty(i)$ will be added to A.

A similar argument shows that $g^\infty(i)\in B^\infty$ if and only if $g^\infty(i)\in{}^{\chi_{A^\infty}}W_i$. Thus we have proved the following theorem.

Theorem 8.7

There exist recursively enumerable sets which are incomparable with respect to Turing reducibility. ■

Theorem 8.8

There are at least four different Turing degrees.

Proof

We have seen the degree $[\varnothing]_{\equiv_T}$ of all recursively decidable sets and the degree $[K]_{\equiv_T}$ of K. These two degrees are obviously different. Let A and B be the recursively enumerable sets constructed just now. Then $\varnothing \leqslant_T A$, $\varnothing \leqslant_T B$ and $[\varnothing]_{\equiv_T} \neq [A]_{\equiv_T}$, $[\varnothing]_{\equiv_T} \neq [B]_{\equiv_T}$ and $[A]_{\equiv_T} \neq [B]_{\equiv_T}$. Furthermore $A \leqslant_T K$ and $B \leqslant_T K$, because K is $\leqslant_T$-complete for the class of all recursively enumerable sets. But not $K \leqslant_T A$, otherwise $B \leqslant_T K \leqslant_T A$, similarly not $K \leqslant_T B$. Hence $[\varnothing]_{\equiv_T}$, $[A]_{\equiv_T}$, $[B]_{\equiv_T}$ and $[K]_{\equiv_T}$ are all different. ■

The preceding construction can be generalized to produce not 2 but m recursively enumerable sets which are mutually incomparable. Since there is no bound on m there exist infinitely many mutually incomparable Turing degrees, each of which contains, but does not necessarily consist of, recursively enumerable sets.

The above sets A^∞ and B^∞ are also incomparable with respect to $\leqslant_m$. Thus, there are recursively enumerable $\equiv_m$-classes D_1 and D_2, such that $[N]_\equiv < D_1 < [K]_\equiv$, $[N]_\equiv < D_2 < [K]_\equiv$, and D_1 and D_2 are incomparable, that is, neither $D_1 \leqslant D_2$ nor $D_2 \leqslant D_1$. This strengthens the results of Section 7.5.

8.6 Exercises

8.1 A and B are subsets of N, and A is non-recursive. Discuss the following statement. A is recursive in B if and only if A is recursively enumerable in B.

8.2 Prove that every Turing degree contains a countably infinite number of sets.

8.3 Prove that every Turing degree has an infinite number of Turing degrees above it.

8.4 Prove that the set $\{A | A \leqslant_T B\}$, where B is an arbitrary set, is closed with respect to $\cup$, $\cap$ and complement with respect to B.

8.5 Let $A = \{x | x \in dom(\phi_x)\}$ and $B = \{x | x \in range(\phi_x)\}$.

(a) Prove that A is recursively enumerable and non-recursively decidable.

(b) Which of the following assertions is true?

(i) $A \equiv_T B$
(ii) $A \equiv_m B$
(iii) $A \equiv_1 B$

8.6 Let $A \subseteq N$ be a recursively enumerable set. Prove that A is recursively decidable if $A \leqslant_m N - A$.

8.7 Prove that the complement of a creative set is not recursively enumerable and does not always contain an infinite recursively enumerable set.

8.7 References

The most important reference for reducibility is Post (1944). In this paper the basic concepts of reducibility, Turing reducibility, many-to-one reducibility, and complete and simple sets are introduced. In the same paper the so-called Post problem is posed; see also Friedberg (1957) and Muchnik (1956). The characterization of the hierarchy of the arithmetical predicates using the T-predicate is due to Kleene (1943). For an extensive discussion the reader is referred to the excellent books of Rogers (1967) and Shoenfield (1971).

tue 18 jul 95

Chapter 9
Introduction to Complexity Theory

In the preceding chapters we have concentrated on studying recursively partially solvable problems, i.e. problems on the boundary between recursively solvable and recursively unsolvable problems. This is natural and necessary in order to study the power of algorithmic methods.

Focusing on effectiveness, we almost completely ignored other aspects of algorithmic methods, in particular, the cost involved with applying these methods. (We also ignored every other aspect of practical importance, as for instance robustness, maintainability, elegance, development cost and so forth.)

However, establishing the existence of an algorithmic method to solve a given problem is not our only task. We must develop a specific algorithm which solves the problem at hand, and moreover solves it efficiently.

Thus we must compare different algorithms solving the given problem, and select the best one. An important criterion, though not the only one, is the cost of executing the algorithm.

It is not always obvious what to charge for executing the algorithm. When considering matrix multiplication algorithms, the number of performed multiplications/divisions and additions/subtractions, is a reasonable measure of cost, but this measure is of little use when comparing sorting algorithms. We will choose as the measure of complexity the amount of paperwork to be done when executing the algorithm by hand, and develop a model of computation whose formal complexity measure reflects this amount of paperwork sufficiently closely.

In the following we will discuss properties of complexity measures, develop an appropriate computation model, which will be SAL over a compound data type of strings, and, in the following chapter, define various classes of algorithmic machines. A well-established algorithmic machine is the Turing machine, developed by Alan M. Turing in the 1930s.

Although this brings some loss in uniformity, complexity results will largely be presented using this Turing machine, because the major part of the relevant literature is cast in the framework of Turing machines.

9.1 Complexity measures

Consider the following sorting problem:

> Given is a list L of strings. The problem is to find a list L_s, which consists of all the strings occurring on L, but presented in ascending order. Strings are ordered by length, and, if the lengths are equal, alphabetically.

The following algorithm solves the problem.

```
input: L, an array[1:n] of strings over the alphabet V
output: Ls, an array[1:n] of strings over the alphabet V
        {Ls is a sorted version of L}

method:
   Ls:= L;
   for i:= n − 1 to 1 do
A:   for j:= 1 to i do
        if Ls(j) > Ls(j + 1) then
           temp:= Ls(j + 1);
           Ls(j + 1):= Ls(j);
           Ls(j):= temp
        fi
B:   od
C:od
```

The inner for-loop scans the array elements 1 up to $i+1$, keeping track of the maximum value encountered so far. When $j = i$, $L_s(i+1)$ is the maximum of the elements $L_s(1)$ up to $L_s(i+1)$. The outer loop repeats this process of bubbling the maximum element towards the end of the array.

Thus the algorithm maintains the truth of the following properties:

- $L_s(i+2)$ up to $L_s(n)$ are sorted; they consist of the largest $n-(i+1)$ elements of L. This property holds at points A and C (for $i=0$).
- $L_s(j) \geqslant L_s(k)$ for all $k \leqslant j$. This property holds at point B.

To characterize the complexity of an algorithm, two complexity measures are frequently used, namely time complexity and space complexity.

The time complexity is the time spent executing the algorithm. Time however, is not a meaningful notion in our computation models; therefore, instead of the time used, we consider the number of operations performed. This is also a better measure than time itself, because the amount of time actually spent in executing the program is dependent on the technology used.

Space is not a meaningful notion in our models either. We use the number of elements involved, in the above example the size of the array L, i.e. n, or we take into account the implicit representation. We have not formally fixed the underlying data type of the above example, but a reasonable measure could be the sum of the lengths of the strings in the array L.

For the time being, we concentrate on time complexity, that is, the number of operations performed. In the above example, we have the following types of operations:

- arithmetical operations
- an operation, $<$, to compare array elements, that is, strings
- (implicit) operations to access array elements.

The comparison operation is the crucial one in this application. The remaining operations, used for manipulating the data and testing boundary conditions, do not reflect properties of the sorting process.

Thus to analyse the complexity of this algorithm and the sorting problem in general, we assume the following compound data type:

$$\mathbf{D} = (V^*, N, \{\text{TRUE}, \text{FALSE}\}; >, +_1, \dot{-}, \ldots, \neq_0),$$

where the predicate $>: V^* \times V^* \to \{\text{TRUE}, \text{FALSE}\}$ is the only operation involving V^*, the Type Of Interest (TOI). The remaining operations are auxiliary operations. Note that we have no implicit representation of the elements of V^* in the above data type. Thus the only reasonable space measure in this case is the number of strings involved in the sorting process.

The time complexity of the algorithm is defined to be the number of times the comparison operator $>$ is used in a terminating computation. In the current example, this comparison operation is executed $\frac{1}{2}n(n-1)$ times, independent of the strings involved. Even if L happens to be already sorted, the algorithm performs these $\frac{1}{2}n(n-1)$ comparisons.

The algorithm can be improved by noting whether misplaced elements have been found during a pass of the inner loop, and if so, at which value of the index j. If the elements happened to be in ascending order, the computation can be terminated after a single sweep. Otherwise, if j_0 was the highest index such that $L_s(j_0) > L_s(j_0+1)$, and $L_s(j_0)$ and $L_s(j_0+1)$ have been swapped in consequence, we set the value of the outer loop variable to $j_0 - 1$. This results in the following algorithm:

```
swapped:= n - 1;
while swapped ≥ 1 do

   i:= swapped;
   swapped:= 1;
   for j:= 1 to i do
      if Ls(j) > Ls(j + 1) then
         swapped:= j;
         temp:= Ls(j + 1);
         Ls(j + 1):= Ls(j);
         Ls(j):= temp
      fi
   od;
   swapped:= swapped - 1
od
```

In the above algorithm, the number of comparisons is still $\frac{1}{2}n(n-1)$ in the worst case, but on the average, the number of comparisons will be less. (The average number of comparisons can be shown to be approximately $\frac{1}{4}n^2$.)

Note how the complexity of the algorithm is expressed as a function $c: N \to N$, determined as follows:

(1) Given an instance x of the problem, first determine the size $s(x)$ of the instance. ($s(x)$ is assumed to be a natural number.) In our running example the size of an instance is the number of elements to be sorted.

(2) Let $s(x) = n$. Two possibilities are of importance:

   (a) *Worst-case complexity*
   $c(n) = c(s(x)) = \max\{\#(y) \mid s(y) = n\}$, where $\#(y)$ is the number of counted operations, in our example comparisons, performed by the algorithm in its computation on input y. Thus we consider the performance of the algorithm for that instance of size n on which the performance is poorest.

(b) *Average-case complexity*
This is the average, over all legal instances of the problem, of the number of operations performed. The average is defined relative to a given probability distribution $p(y)$ on the set of all legal instances of the problem. That is, p is a function from the set of all legal instances to the closed interval $[0, 1]$ of the set of all real numbers, and $\Sigma_y p(y) = 1$. Then $c(n) = \Sigma_{s(y)=n} p(y). \# (y)$.

These measures are both of practical importance. The average-case complexity gives some indication of the cost you must expect when running the algorithm, assuming your probability distribution has been well chosen.

In the theory of algorithms, we are more concerned about the worst-case performance, because this measure captures the complexity of the problem. If many instances of a problem are easy to solve, but some are very hard, then these latter instances embody the complexity of the problem, whereas the former only hide this complexity. Since it is the complexity of problems which interests us rather than the performance of an individual algorithm, we must concentrate on the worst-case complexity.

Similar reasoning leads us to the observation that we must study asymptotic worst-case complexity, that is, the worst-case complexity of very large problem instances.

For every algorithmically solvable problem an algorithm can be developed which is very efficient on small problem instances, by pre-computing the solutions to all problem instances having a size less than some chosen bound m, and storing these solutions in a table which the program can access. Thus, for instances of the problem having size less than m, the complexity of the program is equivalent to that of accessing this table, that is, something between constant (direct access to the elements of the table), and $\log_2(M)$ (searching an ordered list of these small instances).

To avoid such misleading results, we have to consider the complexity of arbitrarily large instances, that is, the asymptotic worst-case complexity. The complexity needs to be determined only up to a constant factor. Thus instead of $\frac{1}{2}n^2 - n$, it suffices to write $c = O(n^2)$, pronounced 'c is of the order at most n^2'. ($g = O(f)$ if and only if $(\exists m)(\exists c)(\forall n)[\text{if } n \geqslant m \text{ then } g(n) \leqslant cf(n)]$, see Section 1.2.)

Ignoring constants of proportionality is permitted because it is possible to speed up any given algorithm by any given constant factor, by using a more powerful underlying data type. In our running example, this would mean assuming a 'comparison' operation, which determines which of three given strings is the least, the largest, and which is the one in the middle. It is not immediately obvious that this linear speed-up is always possible; this is a corollary to Theorem 11.2 (page 274).

Thus to discuss the complexity of a problem and the complexities of algorithms solving this problem:

- choose an underlying data type which fits the application
- consider the asymptotic worst-case complexity
- the calculation of the complexity has to be accurate up to a constant of proportionality only.

9.2 Which data type to choose

In the preceding section, we have chosen a particular data type $\mathbf{D} = (V^*, N, \{\text{TRUE}, \text{FALSE}\}; >, +_1, \dot{-}, \ldots, \neq_0)$, which could be used to assign practically meaningful complexities to sorting algorithms. This data type is of no use whatsoever when discussing, for instance, matrix multiplication algorithms.

There even exist sorting algorithms which cannot be adequately discussed using this data type, e.g. the so-called 'bucket-sort' algorithm. This algorithm is best described in terms of numbers, that is, we have a list of natural numbers, and require a sorted version of it. The essential trick of the algorithm is that elements to be sorted can be, and actually are, used as indices in arrays, i.e. used to access array elements.

```
input: L, an array[1:n] of natural numbers, each no larger than m.
output: Ls, an array[1:n] of natural numbers, Ls is a sorted version of L.
method:
  {B is an array[1:m] of natural numbers, used to indicate how often
  the numbers between 1 and m occur in L}

  for i:= 1 to m do B[i]:= 0 od;
  for i:= 1 to n do B[L[i]]:= B[L[i]] + 1 od;

  j:= 1;
  for i:= 1 to n do
     while B[j] = 0 do j:= j + 1 od;
     Ls[i]:= j;
     B[j]:= B[j] − 1
  od
```

The complexity of this algorithm, that is, the number of array accesses performed, is $O(n+m)$. If $m = O(n)$, the above algorithm is $O(n)$ and outperforms any sorting algorithm based on comparisons, because every algorithm based on comparisons requires at least $O(n \log(n))$ comparisons. (For a proof of this, see for example Knuth (1973).)

Thus the data type we chose to discuss the sorting problem and algorithms solving it cannot be used for all sorting algorithms. Therefore this data type does not encompass the essential ingredients of all algorithmic methods solving the sorting problem. Nevertheless, this data type and the corresponding complexity analysis of algorithms is very useful in practice, as long as the discussion is limited to a particular class of algorithms. It is not sufficient in the theory of algorithms, however, because we want to compare the complexities of problems, and not just complexities of some algorithms solving a particular problem.

The algorithmic complexity of a problem is defined as follows:

Definition 9.1

The *algorithmic complexity of a problem P* is a function $f: N \to N$ such that:

(1) for all algorithms A solving P, we have $f = O(c_A)$. That is, the order of f is at most equal to the order of the complexity c_A of algorithm A; in other words, f is a lower bound to the complexity of algorithms solving P.

(2) there is at least one algorithm B, such that $c_B = O(f)$. That is, f is an upper bound of the complexity of algorithm B, which also solves problem P. Thus algorithm B is optimal, as regards complexity. ■

To compare the complexities of problems, we must choose an underlying data type which can be used in any algorithm. Furthermore, and this is the crucial point, the complexities assigned according to this data type must agree with the complexities observed in practice.

The compound data type **N** (Definition 3.1) adopted for our study of recursion theory cannot be used. It does not lead to complexities which are valid in current computing practice. This is due to the implicit data representation. As discussed in Section 2.5, the data type behaves as if natural numbers were represented as polynomial expressions, because numbers can be constructed from the given constants of the data type and given inputs to a program, only by applying operations available in the data type. For instance, $+_1(+_1(+_1(+_1(+_1(0)))))$ is the shortest polynomial expression to construct the number 5 from the constants and operations available in the data type. Owing to this implicit representation, adding e.g. 17 to x takes 17 steps. Thus the complexity of the addition operation becomes too high for it to be of practical significance.

More powerful data types using numbers as their TOI, e.g.

$$(N, \{\text{TRUE}, \text{FALSE}\};\ \times, /, +, \dot{-}, 0, 1, \textit{even}, \neq_0)$$

cannot be used either. Employing these more powerful data types, we get complexities which are too low according to current computing practice.

Consider the following macro program over the data type

$$(N, \{\text{TRUE}, \text{FALSE}\};\ \times, /, +, \dot{-}, 0, 1, \textit{even}, \neq_0).$$

```
input: m, n∈N, {n ≥ 0, m ≥ 1}
output: p {p = m^n}
method:
  p:= 1; y:= m; x:= n;
  {m^n = p.y^x and x ≥ 0}

  while x ≠ 0 do
    if even(x) then
      y:= y × y;
      x:= x/2
    else
      p:= p × y;
      y:= y × y;
      x:= (x ∸ 1)/2
    fi
  od

  {m^n = p.y^x and x = 0}
  {p = m^n}
```

The algorithm computes m^n using $\log(n)$ multiplications. Assuming that all operations available in our data type take unit time, the algorithm takes $O(\log(n))$ time, whereas it already takes $O(n\log(m))$ time to write down the number m^n in decimal or binary notation. Therefore, the complexity of the exponentiation operation according to the above program is too low for it to be practically valid.

In Chapter 2, we have discussed at length that all computation actually boils down to manipulation of representations of data elements. Thus we have two options.

(1) Choose an abstract data type such as **N** (Definition 3.1). Then the functions computable according to the model SAL(**N**) are functions $N^n \rightarrow N^k$. Mathematics has developed an adequate vocabulary and adequate tools to discuss these functions. The computations, however, are performed on numbers represented using the data representation implicitly defined by the data type. Thus the cost of computation is determined according to this implicit representation. If this implicit representation is not equivalent to the representation used in practice, as is the case in our data type **N**, we obtain complexity results which are not practically valid. As regards the power of algorithmic methods, we can use the full expressiveness of mathematics, because computability is expressed directly in terms of well-understood mathematical objects.

(2) Choose a data type of a low level of abstraction, such as

$$\mathbf{S} \triangleq (BIN, \{\text{TRUE}, \text{FALSE}\}; i, head, tail, put_0, put_1, 0, 1, \neq_0, \neq_1)$$

(see Example 2.4 on page 28). In this case the implicit data representation is equivalent to the representation used in practice, e.g. 5 is represented by the polynomial $put_1(put_0(put_1(0)))$. There are of course other, and much more complicated, polynomial expressions representing the number 5, but the point is that there also exists a 'short' polynomial expression representing 5. The clever algorithm will use these short expressions to represent numbers. The cost of computation according to the SAL(S) model of computation is therefore practically valid.

The inconvenience of the data type is that the computable functions are functions $BIN^n \to BIN^k$. Mathematics has not yet developed a large body of tools to discuss such functions on strings with intellectual economy. Thus we must use codings (see Definition 2.3 on page 29) to represent these computable functions $BIN^n \to BIN^k$ as functions $N^n \to N^k$ on a higher level of abstraction, a level on which we can discuss these functions more easily.

Of the above, option (1) is the less laborious one. This option is chosen to study the power of algorithmic methods, where the question is the *existence* of an algorithmic method, with no concern for the cost of this method.

To develop complexity theory, we have no choice but to adopt option (2). At first glance intuitition does not seem to favour this, since computers have such powerful instructions as, for instance, multiply instructions. Note, however, that these instructions are applicable only to 'small' numbers fitting in a single machine word. Large numbers must be represented using sequences of machine words, whence multiply instructions and the like are no longer directly applicable. Since we have to consider asymptotic worst-case complexity, these very large instances are the ones we are dealing with.

If you are studying algorithms to solve a particular problem, e.g. matrix multiplication, it may very well be justified to ignore the finiteness of the word length, and simply count the number of relevant operations, e.g. scalar multiplications, performed. In such cases, choose a data type which, in that application, separates the better algorithms from the worse in a practically valid way.

Since we must discuss the complexities of problems rather than just the complexities of a specific class of algorithms solving a specific problem, we cannot ignore the finiteness of word length, and assume for instance that multiplication of arbitrarily large numbers can be performed in one single step.

In the above, it has been assumed that the time consumed by the computation induced by some algorithm is proportional to the number of operations performed, and that every operation takes unit time, no matter to what argument values it is applied. As we have seen, this gives misleading complexity results. Instead of abandoning the chosen data type, presumably the best or most natural one for the application, one can charge more time for the execution of an operation. For instance, charge for the execution of $x + y$ an amount $\log(\max\{x, y\})$ units. Using this so-called logarithmic cost, misleading results can be avoided. Since logarithmic cost is natural only when manipulating numbers, and much less so when comparing strings, for instance, we will not use non-uniform cost for the operations provided with the data type.

9.3 Predicates and predicate versions of optimization problems

In discussing the complexity of problems, we will restrict ourselves to membership problems; that is, we consider predicates instead of arbitrary functions.

In preceding chapters, we have divided the class of all predicates into three subclasses, namely

- the recursively decidable predicates
- the recursively partially decidable, or recursively enumerable, predicates
- the remaining predicates, those which are not recursively enumerable.

In Chapters 7 and 8 we have seen that the recursively enumerable predicates are not all equally complex. The reducibility relations $\leqslant_m$ and $\leqslant_T$ give very intricate subdivisions of the class of all recursively enumerable predicates. However, all recursively decidable predicates are mutually equivalent relative to $\equiv_m$ and $\equiv_T$.

In complexity theory we concentrate on the recursively decidable predicates. We want to classify these according to the cost of evaluating these predicates.

If this classification is to be of practical significance, we have to use a data type whose TOI is the set V^* of all strings over some finite alphabet V, as has been argued in the preceding sections. Thus we have to reword Definition 4.3 (page 74), relating functions to predicates, in accordance with the string data type.

Definition 9.2

Let V^* be the set of all strings over the alphabet $V = \{a_1, a_2, \ldots, a_m\}$, and ε the empty string. One of the letters of V plays a special role, and we select the letter a_1 for this.

(1) Let f be any partial function from $(V^*)^n$ to $(V^*)^k$. The predicate P_f corresponding to f is defined as follows:

$$P_f \triangleq \lambda w[\mathit{first\_letter}(\pi_1^k(f(w))) = a_1]$$

where *first_letter*$(\varepsilon)=\varepsilon$, and *first_letter*$(aw)=a$ for any $a\in V$ and $w\in V^*$.

(2) Let P be any predicate on strings having n arguments. The set $L_P\subseteq(V^*)^n$ corresponding to P is defined as follows:

$$L_P \triangleq \{x\in(V^*)^n \mid P(x)=\text{TRUE}\}$$

(3) Let L be any subset of $(V^*)^n$. The function $\chi_L:(V^*)^n\rightarrow V^*$ corresponding to L is defined as follows:

$$\chi_L \triangleq \lambda w[\text{if } w\in L \text{ then } a_1 \text{ else } \varepsilon]$$

χ_L is called the *characteristic function* of L. ■

Many practical problems are optimization problems rather than membership problems. To be able to discuss the complexity of such problems, we have to construct membership versions of optimization problems. To illustrate the procedure, the 'Travelling Salesman Problem' (TSP) will be used.

EXAMPLE 9.1

TSP – optimization version

Given an $n\times n$ matrix D over the non-negative integers, the element d_{ij} is defined to be the distance from city i to city j. The problem is to find a tour $i_1(=1), i_2,\ldots,i_{n+1}=i_1(=1)$ which visits all the cities exactly once and minimizes the total distance travelled.

Since we ask for a tour of minimal cost, that is to say of minimal distance, the above problem is an optimization problem. A predicate, or membership version of an optimization problem can be obtained by asking whether or not there exists a solution having a cost less than some given value, instead of asking for the solution of optimal cost.

EXAMPLE 9.2

TSP – predicate version

Given an $n\times n$ matrix over the non-negative integers and a bound b, the problem is to decide whether or not there exists a tour with a total distance at most b.

If the cost function is relatively easy to evaluate, the predicate version of a problem cannot be much harder than the optimization version, because once you can solve the optimization version of the problem you can also solve the predicate version of it at little extra cost. In the TSP you could solve the predicate version as follows:

(1) solve the optimization version;

(2) evaluate the cost of the optimal tour found;

(3) answer = if (the cost is at most b) then 'yes' else 'no'.

Under the realistic assumption that the time required for solving the optimization problem is not of a lower order of magnitude than the time needed to calculate the cost of a given solution, the predicate version of the problem is no harder than the optimization version.

There is no general method to solve the optimization version of a problem given some way to solve the predicate version, although in some cases – and TSP happens to be one of them – it can indeed be shown that the optimization version is no harder than the predicate version.

Finally, we have to write instances of the problem as strings over an arbitrary, but fixed finite alphabet. The representation must be concise; no extraneous information is allowed in the representation of problem instances. In particular, numbers must be represented in decimal, binary, or something like that, but *not unary*. Using unary representation of numbers and adding extraneous symbols to the description of a problem instance amounts to padding this description with meaningless symbols. This in turn results in deceptive time complexity functions, because the size assigned to the instance becomes too large.

Once we have agreed to restrict outselves to predicates, and have also agreed to represent instances of problems as strings over some finite alphabet, we can identify problems with membership problems of sets of strings.

EXAMPLE 9.3

TSP – set version

Assume that the alphabet consists of the numerals 0 up to 9, brackets] and [to be used for grouping, a separation symbol ¢ used for tuple formation, and a string termination symbol $.

The predicate version of the TSP in the representation as described above is equivalent to the membership problem of the set

$TSP \triangleq \{w_1 ¢ w_2 ¢ \cdots ¢ w_m \$ \mid (\exists n \geqslant 1)[n^2+1=m]$ and for all j, $1 \leqslant j \leqslant m$, w_j is either the numeral 0 or a string of one or more numerals of which the first is nonzero, and the distance matrix defined by letting w_2 to w_m represent the elements $d_{11}, \ldots, d_{1n}, d_{21}, \ldots, d_{2n}, \ldots, d_{nn}$ of D, allows a tour with total distance at most $w_1\}$

In the example above the brackets are not used, but for representing more complicated structures such as graphs they come in handy.

If we represent problems as sets of strings over a finite alphabet, the set of all strings over this alphabet is divided into three subsets:

(1) the set of all valid encodings of instances of the problem for which the answer is 'yes'

(2) the set of all valid encodings of instances of the problem for which the answer is 'no'

(3) the set of all strings over the alphabet which are not valid encodings of instances of the problem (in the above example the strings 1$1, 5¢20¢1443 etc. belong to this subset of invalid strings).

The first subset is identified with the problem itself, and is denoted by some mnemonic such as *TSP*. The second subset is called the *complementary* problem, and corresponds to the negated predicate. In our current example this subset is denoted by *co-TSP*. Note that in set-theoretic terms *co-TSP* is not the complement of *TSP*, because there are strings which are not valid encodings of problem instances and therefore belong neither to *TSP* nor to *co-TSP*.

9.4 Non-deterministic algorithms

There are many practical problems whose complexity is unknown up to this day. In those cases there is a large gap between the performance of the best algorithm known, and the lower bound we are able to prove. The Travelling Salesman Problem discussed in the preceding section is an example of this. The best algorithm known has an exponential worst-case asymptotic time complexity, but we are unable to show that more than polynomial time is necessary.

In these cases the predicates contain existential quantifiers, and the only solution method known uses exhaustive search, that is, it consists of trying all possibilities. Verifying whether or not a given candidate indeed constitutes a solution is generally fairly easy. However, we are unable to show that testing whether a candidate is a solution is strictly easier than solving the original problem, that is, whether or not exhaustively searching the set of all candidates is an essential ingredient of any algorithm solving this original problem.

To study the relation between the complexity of a problem and the complexity of the corresponding candidate verification problem, a new class of algorithms is introduced – the *non-deterministic algorithms*. The general idea is that a non-deterministic algorithm does not need to find a satisfactory candidate by searching, but instead can pick any candidate, and then only has to check that its choice has been satisfactory.

For instance, the Travelling Salesman Problem can be solved by the following non-deterministic algorithm:

input: *an instance of TSP as defined in Example 9.3, that is, a string* $w_1 ¢ w_2 ¢ \cdots ¢ w_m \$$

output: *if a tour of the cities having travel distance at most* w_1 *exists, then there must exist a terminating computation giving* y *for an answer. If such a tour does not exist, then either a computation does not terminate, or otherwise the answer produced is unequal to* y.

method:

(1) *If the input is ill-formed, then enter some non-terminating computation; otherwise compute in such a way that* $n^2+1=m$. *(The input is ill-formed if there is no such* n, *or for some* j w_j *is neither the numeral* 0 *nor a string of one or more digits of which the first is nonzero.)*

(2) *pick any sequence* $1, c_2, \ldots, c_{n-1}, 1$ *of length* n *consisting of natural numbers between* 1 *and* $n-1$ *inclusive.*

(3) *check that this sequence is a tour visiting every city but the first exactly once, and has travel distance at most* w_1. *That is,*

(a) *check that the only number occurring more than once in the sequence is the number* 1

(b) *check that every number between* 1 *and* $n-1$ *inclusive occurs in the sequence*

(c) *compute the distance travelled as determined by the distance matrix with entries* w_2 *up to* w_m

(d) *check that the distance found is less than or equal to* w_1

If the result is positive, that is, a satisfactory candidate has been chosen, terminate the computation with answer y; *otherwise, enter some non-terminating computation.*

Note that if the choice proved to be unsatisfactory, the above algorithm does not reconsider the choice, but enters a non-terminating computation instead. This does not violate the specified input/output behaviour. If there is a satisfactory tour, there must *exist* a terminating computation; a proper choice, leading to a satisfactory terminating computation, must be *possible*. This is the *only* requirement. You *do not* have to worry about how to find this satisfactory terminating computation. Not having to worry about this is precisely the reason for introducing non-deterministic programs.

In a programming language, e.g. SAL, non-determinism is introduced by means of an added control construct, the *either statement*.

⟨*statement*⟩::= **either** ⟨*sequence*⟩ **or** ⟨*sequence*⟩ **od**

The intention of **either** S_1 **or** S_2 **od** is that exactly one of the sequences S_1 and S_2 is executed. Which one, however, is left undetermined.

SAL extended with the either statement as defined above will be denoted by ND-SAL. In this section we will discuss ND-SAL(**N**), i.e. ND-SAL over our standard data type **N**. We will precisely define the meaning of non-deterministic programs, that is, the set of all computations described by the program. Furthermore, we will show that non-deterministic programs are equivalent to deterministic programs as far as computability is concerned. This then leaves the question of whether non-deterministic programs always are more efficient that deterministic ones. First consider the following example.

EXAMPLE 9.4

The following ND-SAL(**N**) program has a terminating computation on input x if and only if x is a composite number. The output resulting from a terminating computation is 1.

```
input: x∈N
output: z∈N
method:

  d:= 0; enough:= 1;
  while enough ≠ 0 do
    either enough:= 0
    or d:= d + 1
    od
  od;
  if (x ≤ 1 or (1 < d < x and d divides x)) then
    z:= 1
  else
    while true do d:= d + 1 od
  fi
```

If x is not composite, then x is a prime number. Therefore, $x \geq 2$ and x has no proper divisor. Thus the condition of the if statement is not satisfiable and only the else part can be executed. Hence if x is not a composite number, there is no terminating computation.

If x is a composite number then the possibilities are:

(1) The first while loop does not terminate because the sequence 'enough:= 0' is never executed.

(2) The first while loop does terminate and leaves some value d. Now we have three possibilities:

(a) $x \leqslant 1$. The then part will be executed no matter what the value of d is. The computation terminates and z will have value 1.

(b) $x \geqslant 2$ but a useless value for d has been computed. Either d is out of range or d is not a divisor of x. In this case the else part is executed and the computation does not terminate.

(c) $x \geqslant 2$ and a useful value has been computed for d, i.e. a proper divisor of x. In this case the then part is executed, the computation terminates and z gets the value 1.

Hence, if x is a composite number, then, among many non-terminating computations, there is at least one terminating computation, and there may be many terminating computations. All terminating computations, however, produce the same result, namely, z becomes equal to 1.

The preceding non-deterministic algorithm works by picking a natural number and checking that it is a divisor; a deterministic algorithm must also describe *how* to pick this number, which seems to be less easy.

Note that although it is very easy to write an ND-SAL program which terminates exactly on the non-primes, it is a far from trivial task to write an ND-SAL program which terminates exactly on the primes. (There does exist an ND-SAL program which terminates exactly on the prime numbers and runs in polynomial time; see Pratt (1975).)

The semantics of a program, and also of a non-determinstic program, is the set of all computations described by the program. These computations will be defined as sequences of single steps, in the same way as was done for SAL(**N**) programs in Section 3.4.

Write a given ND-SAL(**N**) program (n, k, p, P) on successive lines, one entity per line; i.e. a line contains:

- an assignment statement
- a while-head **while** $x_i \neq 0$ **do**
- the keyword **either**
- the keyword **or**, or
- the keyword **od**

Number the lines from 1 upwards, place behind each while-head the line number of the corresponding while-tail **od**, and behind each while-tail the line number of the corresponding while-head. Place behind the keyword **either** the line number of the corresponding keyword **or**, behind this keyword **or** the line number of the corresponding keyword **od**, and finally behind this keyword **od** the number 0, to distinguish it from a while-tail.

EXAMPLE 9.5

```
1:  while x2 ≠ 0 do      :8
2:      either           :4
3:          x1 := x1 + 1
4:      or               :7
5:          x1 := x1 + 1;
6:          x2 := x2 ∸ 1
7:      od               :0
8:  od                   :1
```

Definition 9.3

Let $Q = (n, k, p, P)$ be an ND-SAL(**N**) program, and let its lines be numbered from 1 to f. A *state vector* of Q is a tuple (s, x), $s \in \{1, 2, 3, \ldots, f, f + 1\}$, $x \in N^p$. The state vector is called *initial* if $s = 1$ and x_{n+1} to x_p are 0, and called *final* if $s = f + 1$. ■

In vector (s, x) the first component s determines which statement must be executed next. It acts as a pointer in the program text. The second component gives the current value of the program variables.

Definition 9.4

Let $Q = (n, k, p, P)$ be an ND-SAL(**N**) program, having its lines numbered from 1 to f. Let V_Q be the set of all state vectors of P.

State vector $v_1 = (s, x) \in V_Q$ is transformed into state vector $v_2 = (t, y) \in V_Q$ in a single computation step (notation $v_1 \vdash_Q v_2$), if and only if:

(1) s: ⟨*assignment*⟩ occurs in the labelled version of Q, $t = s + 1$ and $y_r = x_r$ for all r such that $1 \leqslant r \leqslant p$ and $r \neq i$, and the value of y_i depends on the assignment statement as specified in Table 9.1.

(2) Otherwise, the line with number s contains a keyword. In that case $y = x$, and the value of t is determined as follows:

  (a) if s: **while** $x_i \neq 0$ **do** $:s'$ occurs in the labelled version of Q, then $t = s + 1$ if $\neq_0(x_i)$ and $t = s' + 1$ otherwise

  (b) if s: **od** $:s'$, with $s' \neq 0$, and s': **while** $x_i \neq 0$ **do** $:s$ occur in

Table 9.1 Semantics of assignment statements.

⟨*assignment*⟩	Value of y_i
$x_i := 0$	$y_i = 0$
$x_i := x_j$	$y_i = x_j$
$x_i := x_j + 1$	$y_i = +_1(x_j)$
$x_i := x_j \dot{-} 1$	$y_i = \dot{-}_1(x_j)$

the labelled version of P, then $t = s' + 1$ if $\neq_0(x_i)$ and $t = s + 1$ otherwise

(c) if s: **either** :s' occurs in the labelled version of P, then

 (i) $t = s + 1$ or
 (ii) $t = s' + 1$

(d) if s: **or** :s' occurs in the labelled version of P, then $t = s' + 1$

(e) if s: **od** :0 occurs in the labelled version of P, then $t = s + 1$ ■

This definition specifies the single steps in the computation determined by the program. A computation then, is a sequence of single steps $u \vdash_Q u_1 \vdash_Q u_2 \vdash_Q \cdots \vdash_Q v$. The computation *terminates* with state vector v if it can proceed no further, i.e. if there is no state vector w such that $v \vdash_Q w$. This is the case if and only if v is a final state vector.

As in Section 3.4 the notation $u \vdash_Q^+ v$ means that u can be transformed into v by a computation consisting of $k \geqslant 1$ steps, and the notation $u \vdash_Q^* v$ means that a computation of $k \geqslant 0$ steps exists which transforms u into v.

Note that the relation $\vdash_Q$ is not necessarily single-valued. A state vector may have zero, one, or two successors. A SAL(**N**) program defines a single computation on any given input. The computation may or may not terminate. It can be represented as a finite or infinite linear list. An ND-SAL(**N**) program may define many computations on a given input. These computations can be arranged in a tree, as in Figure 9.1.

Because the relation $\vdash_Q$ is not a partial function, ND-SAL(**N**) programs do not compute partial functions, but (partial) relations.

Definition 9.5

Let $Q = (n, k, p, P)$ be an ND-SAL(**N**) program having its lines numbered from 1 to f. The meaning of the program is the following relation $M_Q \subseteq N^p \times N^p$:

$$M_Q \triangleq \{(a, b) \mid \text{there is a terminating computation } (1, a) \vdash_Q^* (f + 1, b)\}$$

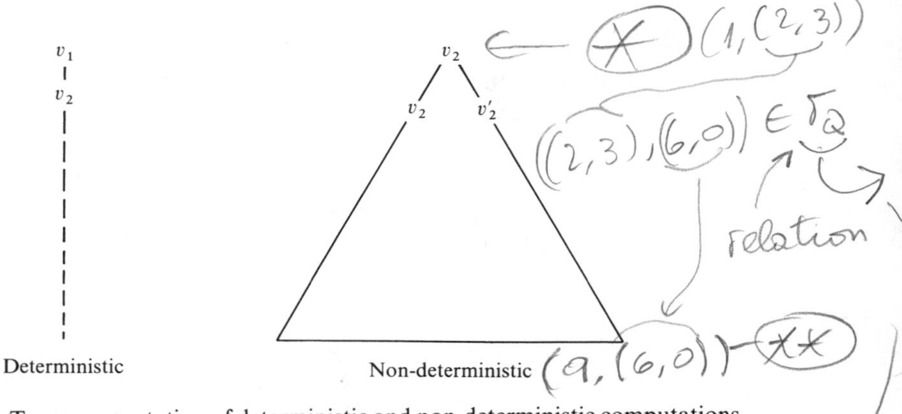

Figure 9.1 Tree representation of deterministic and non-deterministic computations.

The relation $r_Q \subseteq N^n \times N^k$ computed by the program is determined as follows:

$$r_Q \triangleq \{(a, \pi^p_{1\ldots k}(b)) | (a, 0, \ldots, 0, b) \in M_Q\}$$

where $\pi^p_{1\ldots k}$ is the function $\lambda x_1 x_2 \cdots x_p[(x_1, x_2, \ldots, x_k)]$, projecting a p-dimensional vector on its first k components. ■

In the following we will write $\vdash$ instead of $\vdash_Q$ when it is clear which program Q is involved.

EXAMPLE 9.6

Consider the ND-SAL(**N**) program $(2, 2, 2, P)$, where P is the sequence given in Example 9.5. The program computes the relation

$$\{(x_1, x_2, z, 0) | x_1, x_2 \geqslant 0, \text{if } x_2 = 0 \text{ then } z = x_1 \text{ else } z \geqslant x_1 + x_2\}$$

A computation on input $(2, 3)$ may run as follows:

$$\begin{aligned}
&(1,(2,3)) \vdash (2,(2,3)) \vdash (5,(2,3)) \vdash (6,(3,3)) \vdash (7,(3,2)) \vdash (8,(3,2)) \\
&\quad \vdash (2,(3,2)) \vdash (3,(3,2)) \vdash (4,(4,2)) \vdash (8,(4,2)) \\
&\quad \vdash (2,(4,2)) \vdash (5,(4,2)) \vdash (6,(5,2)) \vdash (7,(5,1)) \vdash (8,(5,2)) \\
&\quad \vdash (2,(5,1)) \vdash (5,(5,1)) \vdash (6,(6,1)) \vdash (7,(6,0)) \vdash (8,(6,0)) \\
&\quad \vdash (9,(6,0))
\end{aligned}$$

We will now proceed to show that the computations of ND-SAL(**N**) programs are effective, that is to say that these computations can be simulated by computations of SAL(**N**) programs. To be precise, we show that a subset $W \subseteq N^n \times N^k$ is an ND-SAL(**N**) computable relation if and only if W is a recursively enumerable set.

Lemma 9.1

Let (n, k, p, P) be an ND-SAL(**N**) program. Then the relation $M_{(n,k,p,P)} \subseteq N^n \times N^k$ is a recursively enumerable set.

Proof

We derive a deterministic program whose range is equal to $M_{(n,k,p,P)}$. Thus this relation is recursively enumerable as a set.

Let P_d be the SAL(**N**) sequence obtained from the ND-SAL(**N**) sequence P by replacing all either statements '**either** S_1 **or** S_2 **od**' by macro statements '**if** (s is even) **then** $s := s/2; S_1$ **else** $s := (s-1)/2; S_2$ **fi**, where s is a variable not occurring in P.

Let $(n+1, n+k, p+1+n, Q)$ be the following SAL(**N**) program:

input: $x_1, \ldots, x_n, s$
output: $z_1, \ldots, z_n, x_1, \ldots, x_k$
method:
  $z_1 := x_1; \ldots; z_n := x_n;$
  $\{z_1$ to z_n are new variables only used to copy the input values to the output$\}$
  P_d

The variable s in the above program specifies for each either statement encountered in the execution of the ND-SAL(**N**) sequence P which alternative will be selected. For each terminating computation of the ND-SAL(**N**) program, there is a value of s which specifies exactly the sequence of selections corresponding to this specific terminating computation.

If $(x, y) \in M_{(n,k,p,P)}$, then there is a terminating computation of P on input x with result y and therefore a value s such that $f_Q(x, s) = (x, y)$. On the other hand, if $f_Q(x, s) = (x, y)$ then there is a terminating computation of P_d and thus a terminating computation of P on input x with result y, and therefore $(x, y) \in M_{(n,k,p,P)}$. Therefore $M_{(n,k,p,P)} = range(f_Q)$, whence the relation computed by the ND-SAL(**N**) program (n, k, p, P) is recursively enumerable as a set.

∎

The proof of the lemma above corroborates our intuitive description of non-deterministic programs as programs consisting of a guessing stage followed by a deterministic verification stage. In terms of the above lemma, the guessing stage picks a value of s, whereas the verification stage checks that this value of s determines a terminating compuation with a satisfactory outcome.

Lemma 9.2

For any recursively enumerable set $W \subseteq N^n \times N^k$ there exists an ND-SAL(**N**) program P such that $W = M_P$.

Proof

Since W is recursively enumerable there exists a SAL(**N**) program $(n+k, 1, p, P_W)$ such that $W = \{w \mid f(w)\downarrow\}$, where f is the function computed by the program $(n+k, 1, p, P_W)$. Consider the following ND-SAL(**N**) program Q:

input: $x_1, \ldots, x_n$
output: $y_1, \ldots, y_k$
method:
  $z_1 :\in N; \ldots; z_k \in N;$
  $y_1 := z_1; \ldots; y_k := z_k;$
  $f(x_1, \ldots, x_n, z_1, \ldots, z_k)$

In this program '$z_i :\in N$' is a macro statement which has the following expansion.

$z_i := 0$; *enough* := 1;
while *enough* $\neq 0$ **do**
  either $z_i := z_i + 1$
  or *enough* := 0
  od
od

Obviously $(x, y) \in W$ if and only if $(x, y) \in M_Q$. ■

Thus, the relations computed by ND-SAL(**N**) programs are all and only those which are recursively enumerable as sets.

It remains to associate predicates with relations as computed by ND-SAL(**N**) programs.

Definition 9.6

Let $r \subseteq N^n \times N$ be any relation. The predicate P_r corresponding to the relation r is defined as follows:

$$P_r \triangleq \lambda x[(\exists y > 0)[(x, y) \in r]]$$ ■

The predicate P_r corresponding to a relation r according to the above definition is identical to the predicate corresponding to r by Definition 4.3 (page 74) if the relation r happens to be a (partial) function.

From Lemmas 9.1 and 9.2 it follows that a predicate is ND-SAL(**N**) computable if and only if the predicate is SAL(**N**) computable. This leaves the question of whether a predicate can be computed by a non-deterministic program more efficiently than by a deterministic one. A specific instance of this question is the question of whether the set **P** of all predicates computable in polynomial time by deterministic programs is equal to the set **NP** of all predicates computable in polynomial time by non-deterministic programs. The answer to this question is still unknown although most people believe that $\mathbf{P} \neq \mathbf{NP}$.

9.5 The data type V*

In Section 9.2 it has been argued that one must take a data type having the set of all strings over some finite alphabet as TOI in order to obtain results which are valid with respect to current computing practice. Defining such a data type is the purpose of this section.

We must find string operations which are both powerful enough, and sufficiently simple, so that the assumption that they are single-step operations is valid. We choose operations which manipulate the front end of a string.

Think of a string as the value of a pushdown stack, with the leftmost symbol of the string as the top of the stack. For instance *abacb*, has top-symbol *a*, and the current length of the stack is 5. The operations which can be performed on strings are the operations which can be performed on stacks:

- pop the stack, remove the first symbol of the string,
- push a symbol on the stack, add a symbol to the left end of the string,
- read the top symbol of the stack, determine the leftmost symbol of the string.

Definition 9.7

Let $V \triangleq \{a_1, a_2, \ldots, a_m\}$ be a non-empty alphabet of symbols, and V^* the set of all strings over this alphabet.

The data type $\mathbf{V}^* \triangleq (V^*, \{\text{TRUE}, \text{FALSE}\}; pop, push, top, \varepsilon, a_1, \ldots, a_m, \neq_\varepsilon, \equiv)$ consists of:

(1) two sorts, namely the TOI V^*, and the sort $\{\text{TRUE}, \text{FALSE}\}$ consisting of the Boolean values.

(2) and $m + 6$ operators:

(a) *pop*: $V^* \to V^*$, defined as follows:

$$pop \triangleq \lambda w[\text{if } (\exists a \in V)(\exists v \in V^*)[w = av] \text{ then } v \text{ else } \varepsilon]$$

(b) *push*: $V^* \times V^* \to V^*$, defined as follows:

$$push \triangleq \lambda aw[\text{if } a \in V \text{ then } aw \text{ else } w]$$

(c) *top*: $V^* \to V^*$, defined as follows:

$$top \triangleq \lambda w[\text{if } (\exists a \in V)(\exists v \in V^*)[w = av] \text{ then } a \text{ else } \varepsilon]$$

(d) $m+1$ nullary operators, in other words, available constants, namely ε and a_1 to a_m inclusive

(e) two predicates, namely $\neq_\varepsilon$ and $\equiv$, which are defined as follows:

(i) $\neq_\varepsilon \triangleq \lambda w[\text{if } (w \neq \varepsilon) \text{ then TRUE else FALSE}]$
(ii) $\equiv \triangleq \lambda xy[\text{if } (\exists a \in V)(\exists u \in V^*)(\exists v \in V^*)[(x = au) \text{ and } (y = av)] \text{ then TRUE else FALSE}]$ ■

It would be more natural to take three sorts, V, V^*, and {TRUE, FALSE}. We could then define the operations in a way which is intuitively more acceptable, for example *push*: $V \times V^* \to V^*$, instead of *push*: $V^* \times V^* \to V^*$. Also, we could define $\equiv$ as an equality predicate on V, instead of an equivalence predicate on V^*. But then we would need typed variables in our programming languages. Furthermore, we would need partial operations, because *top* should be a function from V^* to V, and *top*(ε) should be undefined.

To avoid these complications, we restrict ourselves to two sorts, and define e.g. *push*: $V^* \times V^* \to V^*$ in such a way that it behaves as expected for the intended values, and does something harmless when called with unsuitable argument values, e.g. *push*(*ab*, *abc*) = *abc*.

Note that extensions such as *push*(*ab*, *abc*) = *ababc*, are not allowable, because in actual practice string concatenation cannot be a single-step operation, and therefore should not be supplied as such in the data type.

9.6 Syntax, semantics, and examples of ND-SAL(V*) programs

The definitions of syntax and semantics to be given closely follow those for ND-SAL(**N**) (Sections 3.2 and 9.4). Separate definitions for the deterministic case are not necessary, because a deterministic program is just a non-deterministic program which happens to contain no either statements.

Definition 9.8

V is assumed to be a non-empty, finite alphabet of symbols.

⟨*character*⟩ ::= V.
⟨*identifier*⟩ ::= $\{x_i \mid i = 1, 2, 3, \ldots\}$.

$\langle \textit{assignment} \rangle ::= \langle \textit{identifier} \rangle := \varepsilon \,|$
$\quad \langle \textit{identifier} \rangle := \langle \textit{character} \rangle \,|$
$\quad \langle \textit{identifier} \rangle := \langle \textit{identifier} \rangle \,|$
$\quad \langle \textit{identifier} \rangle := \textit{pop}(\langle \textit{identifier} \rangle) \,|$
$\quad \langle \textit{identifier} \rangle := \textit{push}(\langle \textit{identifier} \rangle, \langle \textit{identifier} \rangle) \,|$
$\quad \langle \textit{identifier} \rangle := \textit{top}(\langle \textit{identifier} \rangle).$

$\langle \textit{statement} \rangle ::= \langle \textit{assignment} \rangle \,|$
$\quad \textbf{while}\ \langle \textit{identifier} \rangle \neq \varepsilon\ \textbf{do}\ \langle \textit{sequence} \rangle\ \textbf{od} \,|$
$\quad \textbf{while}\ \langle \textit{identifier} \rangle \equiv \langle \textit{character} \rangle\ \textbf{do}\ \langle \textit{sequence} \rangle\ \textbf{od} \,|$
$\quad \textbf{while}\ \langle \textit{identifier} \rangle \equiv \langle \textit{identifier} \rangle\ \textbf{do}\ \langle \textit{sequence} \rangle\ \textbf{od} \,|$
$\quad \textbf{either}\ \langle \textit{sequence} \rangle\ \textbf{or}\ \langle \textit{sequence} \rangle\ \textbf{od}.$

$\langle \textit{sequence} \rangle ::= \langle \textit{statement} \rangle \,|$
$\quad \langle \textit{sequence} \rangle ; \langle \textit{statement} \rangle.$

$\langle \textit{program} \rangle ::= \{(n, k, p, \langle \textit{sequence} \rangle) \mid n \geqslant 0, k \geqslant 1 \text{ and } n, k \leqslant p\}.$ ■

The definition of the semantics of ND-SAL(**V***) below closely follows Section 9.4.

Write a given ND-SAL(**V***) program (n, k, p, P) on successive lines, one entity per line; i.e. a line contains:

- an assignment statement
- a while-head **while** $x_i \neq \varepsilon$ **do**
- a while-head **while** $x_i \equiv x_j$ **do**
- the keyword **either**
- the keyword **or**, or
- the keyword **od**

Number the lines from 1 upwards, place behind each while-head the line number of the corresponding while-tail **od**, and behind each while-tail the line number of the corresponding while-head. Place behind the keyword **either** the line number of the corresponding keyword **or**, behind this keyword **or** the line number of the corresponding keyword **od**, and finally behind this keyword **od** the number 0, to distinguish it from a while-tail.

EXAMPLE 9.7

```
1:  while x2 ≠ 0 do          :8
2:     either                :4
3:        x1 := push(a, x1)
4:     or                    :7
5:        x1 := push(a, x1);
6:        x2 := pop(x2)
```

```
7:     od                  :0
8:  od                     :1
```

Definition 9.9

Let $Q=(n,k,p,P)$ be an ND-SAL($\mathbf{V}^*$) program, and let its lines be numbered from 1 to f. A *state vector* of Q is a tuple (s,x), $s\in\{1,2,3,\ldots,f,f+1\}$, $x\in(V^*)^p$. The state vector is called *initial* if $s=1$ and x_{n+1} to x_p are empty, i.e. ε, and called *final* if $s=f+1$. ■

In vector (s,x) the first component s determines which statement must be executed next. The component acts as a pointer in the program text. The second component gives the current value of the program variables.

Definition 9.10

Let $Q=(n,k,p,P)$ be an ND-SAL($\mathbf{V}^*$) program, having its lines numbered from 1 to f. Let V_Q be the set of all state vectors of Q.

State vector $v_1=(s,x)\in V_Q$ is transformed into state vector $v_2=(t,y)\in V_Q$ in a single computation step (notation $v_1 \vdash_Q v_2$) if and only if:

(1) if s: ⟨*assignment*⟩ occurs in the labelled version of Q, then $t=s+1$ and $y_r=x_r$ for all r, $1\leqslant r\leqslant p$, $r\neq i$, and the value of y_i depends on the assignment as specified in Table 9.2. Note the difference between the two occurrences of *push*. In the left-hand column, *push* occurs as a syntactical object; in the right-hand column it denotes an operation belonging to a particular data type, namely that of Definition 9.7. The same holds for the two occurrences of *pop* and *top*.

(2) Otherwise, the line with number s contains a keyword. In that case $y=x$ and t is determined as follows:

(a) if s: **while** ⟨*condition*⟩ **do** :s', where ⟨*condition*⟩

Table 9.2 Semantics of assignment statements.

⟨*assignment*⟩	Value of y_i
$x_i := \varepsilon$	$y_i=\varepsilon$
$x_i := a$, $(a\in V)$	$y_i=a$
$x_i := x_j$	$y_i=x_j$
$x_i := push(x_j, x_m)$	$y_i=push(x_j,x_m)$
$x_i := pop(x_j)$	$y_i=pop(x_j)$
$x_i := top(x_j)$	$y_i=top(x_j)$

is $\langle identifier \rangle \neq \varepsilon$, or $\langle identifier \rangle \equiv \langle character \rangle$, or $\langle identifier \rangle \equiv \langle identifier \rangle$, occurs in the labelled version of P, then $t = s + 1$ if $\langle condition \rangle$ evaluates to TRUE, and $t = s' + 1$ otherwise.

(b) if s: **od** :s', with $s' \neq 0$, and s': **while** $\langle condition \rangle$ **do** :s occur in the labelled version of P, then $t = s' + 1$ if $\langle condition \rangle$ evaluates to TRUE, and $t = s + 1$ otherwise.

(c) if s: **either** :s' occurs in the labelled version of P, then

(i) $t = s + 1$ or
(ii) $t = s' + 1$

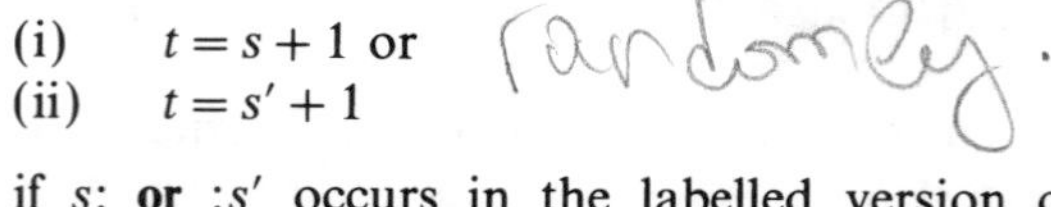

(d) if s: **or** :s' occurs in the labelled version of P, then $t = s' + 1$

(e) if s: **od** :0 occurs in the labelled version of P, then $t = s + 1$ ■

This definition specifies the single steps in the computation determined by the program. A computation, then, is a sequence of single steps $u \vdash_Q u_1 \vdash_Q u_2 \vdash_Q \cdots \vdash_Q v$. The computation *terminates* with state vector v if it can proceed no further, i.e. if there is no state vector w such that $v \vdash_Q w$. This is the case if and only if v is a final state vector.

Because the relation $\vdash_Q$ is not a partial function, ND-SAL(**V***) programs do not compute partial functions but (partial) relations.

Definition 9.11

Let $Q = (n, k, p, P)$ be an ND-SAL(**V***) program having its lines numbered from 1 to f. The meaning of the program is the following relation $M_Q \subseteq (V^*)^p \times (V^*)^p$:

$$M_Q \triangleq \{(a, b) \mid \text{there is a terminating computation } (1, a) \vdash_Q^* (f + 1, b)\}$$

The relation $r_Q \subseteq (V^*)^n \times (V^*)^k$ computed by the program, is determined as follows:

$$r_Q \triangleq \{(a, \pi^p_{1\ldots k}(b)) \mid (a, \varepsilon, \ldots, \varepsilon, b) \in M_Q\}$$

where $\pi^p_{1\ldots k}$ is the function $\lambda x_1 x_2 \cdots x_p[(x_1, x_2, \ldots, x_k)]$, projecting a p-dimensional vector on its first k components. ■

As a special case, consider deterministic programs, that is, ND-SAL(**V***) programs which contain no either statements. Then the meaning M_Q of program Q and the relation r_Q computed by the program are functions. In those cases we will write $M_Q(x) = y$ instead of $(x, y) \in M_Q$ and $r_Q(x) = y$ instead of $(x, y) \in r_Q$.

Owing to the primitiveness of the available operators and control constructs, writing ND-SAL(**V***) programs is as tedious as writing SAL(**N**) programs. Therefore we will use macros in ND-SAL(**V***) programs, as we did in SAL(**N**) programs.

EXAMPLE 9.8

In the examples below, a is an arbitrary but fixed symbol from the alphabet V.

(1) An equality test on the elements of V^*, returning a if x and y are equal; otherwise ε.

```
input: x, y
output: z
method:

  z:= a;

  while x ≡ y do
     x:= pop(x);
     y:= pop(y)
  od;

  while x ≠ ε do
     x:= ε;
     z:= ε
  od;

  while y ≠ ε do
     y:= ε;
     z:= ε
  od
```

Note that $x \equiv y$ holds true if and only if x and y are both non-empty, and their leftmost symbols are the same.

(2) Expansion of the control macro **if** $x \neq \varepsilon$ **then** S_1 **else** S_2 **fi**.

```
yes:= x; no:= a;

while yes ≠ ε do
   yes:= ε; no:= ε;
   S1
od;

while no ≠ ε do
   no:= ε;
   S2
od
```

(3) Evaluation of Boolean expressions, where $x \in V^*$ represents FALSE if x is the empty string ε and TRUE otherwise.

Let $P(x)$ and $Q(y)$ be macro calls, evaluating the predicates P and Q.

$z := \neg P(x)$: **if** $P(x) \neq \varepsilon$ **then** $z := \varepsilon$ **else** $z := a$ **fi**

$z := P(x) \vee Q(y)$: $z := push(top(Q(y)), P(x))$

$z := P(x) \wedge Q(y)$: **if** $P(x) \neq \varepsilon$ **then** $z := Q(y)$ **else** $z := \varepsilon$ **fi**

(4) Reversing a string from V^*.

```
input: x
output: y
method:

  while x ≠ ε do
    y:= push(top(x), y);
    x:= pop(x)
  od
```

(5) String concatenation.

```
input: x, y
output: z{z = xy}
method:

  u:= reverse(x);
  z:= y;
  while u ≠ ε do
    z:= push(top(u), z);
    u:= pop(u)
  od
```

(6) Addition of numbers in binary representation. Assume that $\{0,1\} \subseteq V$, and let $x_1 x_2 \cdots x_m$ represent the number

$$\sum_{i=1}^{m} x_i . 2^{i-1}$$

i.e. the leftmost symbol of x is interpreted as the least significant digit in the binary representation of the number.

```
input: x, y
output: z
method:
  c:= 0;
  {initialization of carry}
  while x ≠ ε ∨ y ≠ ε ∨ c ≡ 1 do
```

```
        if top(x) ≠ ε then p:= top(x) else p:= 0 fi;
        if top(y) ≠ ε then q:= top(y) else q:= 0 fi;
        if p ≡ 0 ∧ q ≡ 0 ∧ c ≡ 0 then s:= 0; c:= 0 fi;
        ⋮
        if p ≡ 1 ∧ q ≡ 1 ∧ c ≡ 1 then s:= 1; c:= 1 fi
        z:= push(s, z)
    od;
    z:= reverse(z)
```

(7) The following is an ND-SAL(**V***) program which computes the *partition* predicate. Given is a set S of natural numbers, and the problem is to decide whether there is a $J \subseteq S$ such that

$$\sum_{x \in J} x = \sum_{x \in (S-J)} x$$

We assume that $\{0, 1, a\} \subseteq V$, and that the instance $(x_1, x_2, \ldots, x_m) \in N^m$ is represented by the string $w_1 a w_2 a \cdots a w_m \in V^*$, where w_i is the binary representation, without leading zeros, of x_i, the most significant digit being the leftmost symbol of w_i.

```
input: x
output: y
method:

  in:= ε; out:= ε;
  while x ≠ ε do

    if top(x) ∉ {0, 1} then
      while 0 ≠ ε do g:= pop(g) od
    fi;

    g:= ε;
    while top(x) ∈ {0, 1} do
      g:= push(top(x), g);
      x:= pop(x)
    od;

    if top(x) ≡ a then x:= pop(x) fi;
    either in:= in + g
    or out:= out + g
    od
  od;

  if in ≠ out then
    while 0 ≠ ε do g:= pop(g) od
  fi;
  y:= a
```

Let the size of the instance under consideration be n, i.e. the initial value of x is a string of length n. Then there are no more than n numbers involved. Furthermore, no program will ever have a value with more than n letters in a terminating computation. Since each addition requires no more than $O(n)$ steps, and there are no more than n additions to perform, terminating computations require no more than $O(n^2)$ steps. Thus the above is a non-deterministic algorithm computing the partition predicate, and running in polynomial time. No deterministic algorithm is known which runs in polynomial time and computes the partition predicate.

9.7 SAL(N) and SAL(V*) are equivalent

SAL(**N**) computable functions are functions $N^n \to N^k$, whereas SAL(**V***) computable functions are functions $(V^*)^n \to (V^*)^k$. Thus we must compare SAL(**N**) and SAL(**V***) relative to a coding, as defined in Section 2.2 (see Definition 2.3 on page 29). The proofs have no special interest, but they are presented for completeness. Find suitable codings yourself, before reading the proofs below.

Lemma 9.3

F(SAL(**V***)) incorporates F(SAL(**N**)) relative to the coding $c \triangleq \lambda n[a^n]$, where a is an arbitrary but fixed letter of V, and a^n denotes a sequence of n as. (Thus unary number representation is used in this coding.)

Proof

Let $A = (n, k, p, P)$ be a SAL(**N**) program. Translate this program into a SAL(**V***) program $B = (n, k, p, Q)$ by substituting:

(1) $x_i := pop(x_j)$ for $x_i := x_j \dot{-} 1$,

(2) $x_i := push(a, x_j)$ for $x_i := x_j + 1$,

(3) **while** $x_i \neq \varepsilon$ **do** for **while** $x_i \neq 0$ **do**.

Let V_A be the set of all state vectors of A, and V_B the set of all state vectors of B. If $(s, x) \in V_A$, then $(s, c(x)) \in V_B$. Also, if $(s, x) \vdash_{\overline{A}} (t, y)$, then $(s, c(x)) \vdash_{\overline{B}} (t, c(y))$. By induction on m, it follows that if $(s, x) \vdash_{\overline{A}} (s_1, x_1) \vdash_{\overline{A}} \cdots \vdash_{\overline{A}} (s_m, x_m)$ then $(s, c(x)) \vdash_{\overline{B}} (s_1, c(x_1)) \vdash_{\overline{B}} \cdots \vdash_{\overline{B}} (s_m, c(x_m))$. Therefore $\{(c(x), c(y)) | y = M_A(x)\} \subseteq \{(x, y) | y = M_B(x)\}$, and consequently F(SAL(**V***)) incorporates F(SAL(**N**)). ■

In Definition 2.3 (page 29), we stated that 'the function $f: N^n \to N^k$ is incorporated in the function $g: W^n \to W^k$ relative to the coding c if and only if $\{(c(x), c(y)) | y = f(x)\} \subseteq \{(u, v) | v = g(u)\}$. Clearly, this definition is readily extended to relations. Relation $s \subseteq W^n \times W^k$ incorporates relation $r \subseteq N^n \times N^k$ relative to the coding c if and only if $\{(c(x), c(y)) | (x, y) \in r\} \subseteq s$.

Thus we can compare computation models employing non-deterministic control, in particular, ND-SAL($\mathbf{V}^*$) and ND-SAL($\mathbf{N}$). The proof of the above lemma shows that R(ND-SAL($\mathbf{V}^*$)), the set of all relations computed by ND-SAL($\mathbf{V}^*$) programs, incorporates R(ND-SAL($\mathbf{N}$)), the set of all relations computed by ND-SAL($\mathbf{N}$) programs, relative to the coding c.

Lemma 9.4

Let the symbols of the alphabet be numbered, $V = \{a_1, a_2, \ldots, a_m\}$. Then F(SAL($\mathbf{N}$)) incorporates F(SAL($\mathbf{V}^*$)) relative to the coding c defined as follows:

$$c(\varepsilon) = 0$$

$$c(a_{i_1} a_{i_2} \cdots a_{i_s}) = \langle i_1, i_2, \ldots, i_s \rangle \dot{-} 1$$

where $\langle\,\rangle$ is our standard Gödel numbering, given in Definition 5.1 (page 97).

Proof

The proof is similar to the proof of the preceding lemma, but more laborious. Let $A = (n, k, p, P)$ be a SAL($\mathbf{V}^*$) program. Translate this into a SAL($\mathbf{N}$) program $B = (n, k, q, Q)$ as follows:

(1) *pop* and *push* statements are considered SAL($\mathbf{N}$) macro statements, and are expanded accordingly. We have already used these macros in our universal program; see Section 5.3.

(2) $x_i := top(x_j)$ is replaced by the SAL($\mathbf{N}$) macro $x_i := \tau(x_j, 1)$, and then expanded accordingly (see Section 5.2).

(3) $x_i \neq \varepsilon$ is replaced by $x_i \neq 0$.

(4) $x_i \equiv x_j$ is considered a SAL($\mathbf{N}$) macro, and expanded according to the definition below:

input: x, y
output: z
method:

if $x_i \neq 0 \wedge x_j \neq 0 \wedge \tau(x_i, 1) = \tau(x_j, 1)$ **then** $z := 1$
else $z := 0$
fi

$x_i \equiv a_j$ is also considered a SAL($\mathbf{N}$) macro, its expansion is similar to the one above.

Let V_A denote the set of all state vectors of the SAL($\mathbf{V}^*$) program A, and V_B the set of all state vectors of the SAL($\mathbf{N}$) program B.

The variables x_1 to x_p of the resulting SAL($\mathbf{N}$) program B correspond to the variables x_1 to x_p of the SAL($\mathbf{V}^*$) program A, the remaining variables x_{p+1} to x_q are auxiliary, and are used in the expansion of the macros. To notationally distinguish between these two classes, we write $(s,(v_1,v_2))\in V_B$, meaning that v_1 represents the values of the variables x_1 to x_p, and v_2 represents the values of the remaining variables.

If $(s,x)\in V_A$, then $(s,(c(x),v_2))\in V_B$ for some $v_2\in N^{q-p}$. Also, if $(s,x)\vdash_A (t,y)$, then for any v_2 there is a computation

$$(s,(c(x),v_2))\vdash_B (s_1,(c(x),v_{2,1}))\vdash_B \cdots$$
$$\vdash_B (s_r,(c(x),v_{2,r}))\vdash_B (t,(c(y),v_{2,r+1}))$$

for some $v_{2,r+1}$.

By induction on the length of the computation, we have if $(s,x)\vdash_A^* (t,y)$, then for any v_2 there is a computation $(s,(c(x),v_2))\vdash_B^* (t,(c(y),v_2'))$, for some $v_2'\in N^{q-p}$.

Consequently,

$$\{(c(x),c(y))\,|\,y=f_A(x)\}\subseteq\{(x,y)\,|\,y=f_B(x)\}$$

and therefore F(SAL($\mathbf{N}$)) incorporates F(SAL($\mathbf{V}^*$)). ■

In summary, SAL($\mathbf{N}$) and SAL($\mathbf{V}^*$) are equivalent models of computation, and ND-SAL($\mathbf{N}$) and ND-SAL($\mathbf{V}^*$) are also equivalent models of computation. Although these models are equivalent with respect to computing power, and both determine the same, up to a coding, class of computable functions, the models are *not equivalent* with respect to complexity. In SAL($\mathbf{V}^*$) you *can* use efficient representations of data, whereas in SAL($\mathbf{N}$) you are restricted to the use of unary data representation.

9.8 Exercises

9.1 Write a non-deterministic algorithm which has a terminating computation on an input string of the form $10^{i_1}10^{i_2}\cdots 10^{i_k}$ if there is some set

$$I\subseteq\{1,2,\ldots,k\} \text{ for which } \sum_{j\in I} i_j=\sum_{j\notin I} i_j.$$

9.2 Write an ND-SAL($\mathbf{V}^*$) program with a polynomial time complexity that solves the Hamilton circuit problem.

9.3 Write an ND-SAL(**V***) program computing the function λx[if (x is prime) then 1 else 0].

9.4 Prove that for every non-deterministic algorithm P computing a *partial function* f_P, there exists a deterministic algorithm Q such that $f_P = f_Q$.

9.5 Design a non-deterministic sorting algorithm with worst-case complexity $O(n)$ ($n =$ the number of elements to be sorted).

9.6 Write a non-deterministic SAL(**V***) program which determines whether a given Boolean expression is satisfiable; this means that there is an assignment to the variables occurring in the expression, for which the expression evaluates to TRUE. Design the algorithm such that the time complexity is linear in the length of the expression.

9.7 Prove that for every ND-SAL(**V***) program P computing a *partial function* f_P, there exists a SAL(**V***) program Q such that $f_P = f_Q$ and $T_Q(n) = O(2^{T_P(n)})$.

9.8 Prove that every SAL(**V***) enumerable relation is ND-SAL(**V***) computable and vice versa.

9.9 References

Analyses of time and space requirements of various algorithms and a discussion of how to perform such analyses can be found in many books on the design of algorithms; see for example Horowitz and Sahni (1978).

R(ND-SAL(N)) = R(ND-SAL(V*))

F(SAL(N)) = F(SAL(V*))

tue 18 jul 85

Chapter 10
Algorithmic Machines

In this chapter we will discuss machines manipulating data using the operations provided by a given data type.

The difference between programs and machines is largely psychological. From a mathematical point of view, both programs and machines are finite mathematical objects, having infinite sets of computations associated with them. Nevertheless, machines allow a natural physical interpretation, whereas programs intuitively need a machine to be executed on.

One of the machines to be discussed is the Turing machine, invented by Alan M. Turing in the 1930s. Quite a sizable part of the literature on the complexity of computations is formulated in terms of Turing machines. It is therefore useful to thoroughly acquaint yourself with this type of machine. You should be able to transform intuitive arguments to formal arguments using Turing machines, as easily as you can transform them using programs.

10.1 Introduction

Let **A** be a compound data type. A SAL(**A**) program (n, k, p, P) is a finite mathematical object. Our definitions associate with this object an infinite set of computations. Using this infinite set the function $f_{(n,k,p,P)}$, 'computed by the program', is defined. The expression 'computed by the program', is an abuse of terminology. The program exists as a mathematical object, but does not do anything. We also do not have a physical interpretation of a program as an object which does something, such as manipulating numbers.

Instead of programs over a data type **A**, we can consider algorithmic machines over the data type **A**. These machines are intended to be single-purpose machines; a machine is designed to compute a particular function, just as a program is designed to specify how to compute a particular function. A machine is a finite mathematical object specified by a finite text, just as a program is specified by a finite text.

Mathematically speaking, there is not much of a difference between programs and machines. Both are fixed, finite 'inanimate', mathematical objects, to which a meaning can be assigned at will, although this assignment of meaning is normally done so as to conform to intuition. Nevertheless, a machine has a natural physical interpretation whereas a program does not.

In the following we will consider SAM(**A**), a class of Simple Algorithmic Machines over the data type **A**, consisting of algorithmic machines defined using the following concepts;

(1) *Registers.* Every machine has a memory which consists of a finite, machine-dependent number of registers $x_1, \ldots, x_p$. The registers $x_1, \ldots, x_n$ and $x_1, \ldots, x_k$, $n \geq 0$ and $k \geq 1$, are the input and output registers. Each register may hold an arbitrary data element of **A**. Registers in a machine are the equivalents of variables in a program.

(2) *States.* The current state of a machine determines what action the machine will perform next. For every machine there is a finite, machine-dependent set of possible states, and the machine is always in one of those states. A state is comparable to a line number of a SAL program having its lines numbered from 1 to f, as required in the definition of the semantics of the program.

(3) *Instructions.* Two types of instructions are available – data manipulation instructions and control instructions.

(a) The data manipulation instructions are register assignments. There is an assignment for each operation of the underlying data type. The general format is

$$q_1: \quad x_i := o(x_{j_1}, x_{j_2}, \ldots, x_{j_m}) \quad :q_2$$

where q_1 and q_2 are machine states.

The intention is that if the machine is in state q_1, it performs the register assignment and then brings itself into state q_2. It is assumed that the hardware of the machine does all this in a single step.

(b) Control instructions perform a test on the contents of the registers, and change the state of the machine accordingly. For each predicate supplied by the underlying data type there is a corresponding control instruction. The general format is

$$q_1: \ \textbf{if } p(x_{j_1}, x_{j_2}, \ldots, x_{j_m}) \textbf{ then } q_2 \textbf{ else } q_3$$

where q_1, q_2 and q_3 are machine states.

There is no need for an unconditional change of the machine state, because this can be accomplished using dummy assignments, q_1: $x_1 := x_1 : q_2$.

Finally, a machine specification is a finite set of action items, where an action item is either a data manipulation instruction, or a control instruction.

The following specifies a SAM(**N**) machine, a Simple Algorithmic Machine over the data type **N**, which doubles its input.

1: $x_2 := x_1$:2
2: **if** $\neq_0(x_2)$ **then** 3 **else** 5
3: $x_1 := +_1(x_1)$:4
4: $x_2 := \dot{-}_1(x_2)$:2

The meaning of a machine is the set of all computations it induces. The computations performed by a machine are defined using state vectors; these definitions are almost identical to those defining the computations induced by a SAL program. The only difference is that there is no need to introduce line numbers, since action items already include the required information about the state of the machine.

A Simple Algorithmic Machine has a physical interpretation as depicted in Figure 10.1.

We will take classes of machines of the type described above as our principal models of mechanical computing devices, because the machines are very close to the programs we have used to formalize computability, and because the difference between SAM(**V***) machines and Turing machines is only very small.

Computers as they exist today are not adequately modelled by the special-purpose machines described above. Computers are general-purpose machines, which can execute an arbitrary program, and while executing a program, behave as the corresponding special-purpose machines.

J. C. Shepherdson and H. E. Sturgis have developed a model, the Unlimited Register Machine (URM), a universal machine which can be loaded with a finite set of action items, that is, loaded with a machine specification, or equivalently, with a program, and which can then execute this specification. The machine is sketched in Figure 10.2.

The URM has an infinite data memory, consisting of addressable registers, in which can be stored arbitrary integers. Separate from the data memory, there is an infinite program memory, which can store the finite set of action items to be executed.

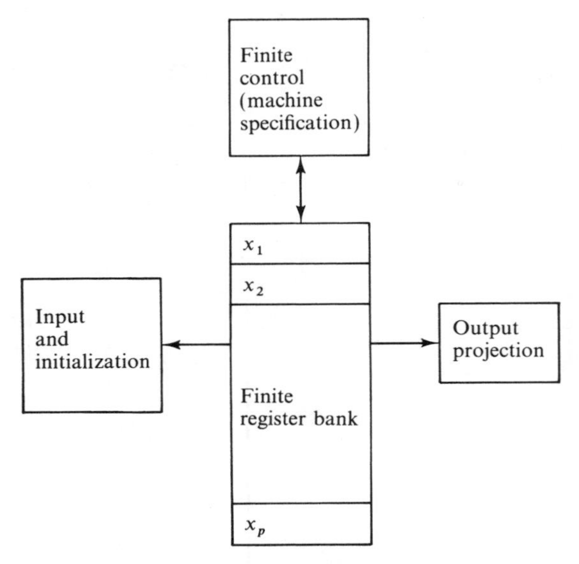

Figure 10.1 A SAM(**A**) machine.

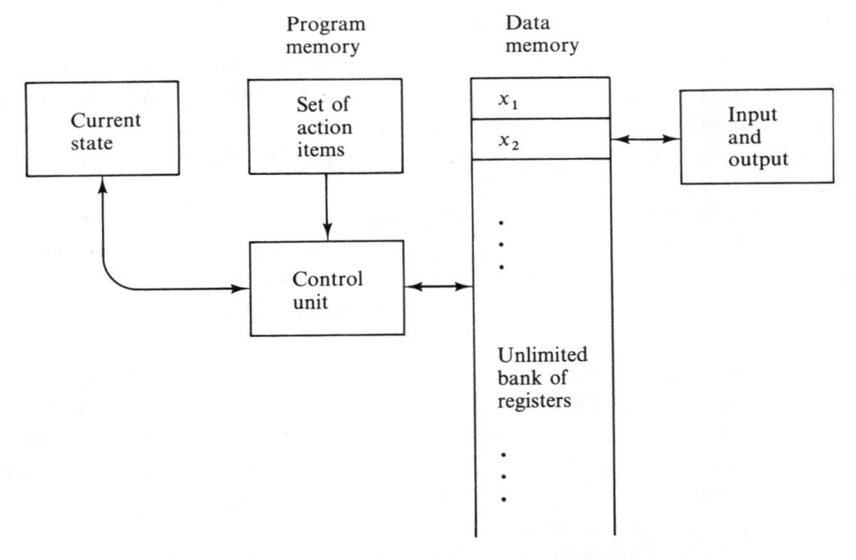

Figure 10.2 The Unlimited Register Machine, URM.

The control unit fetches an action item according to the current state, executes the specified action, thereby appropriately updating the data memory, and then makes the next state of the executed action item the new current state of the URM. This cycle is then repeated.

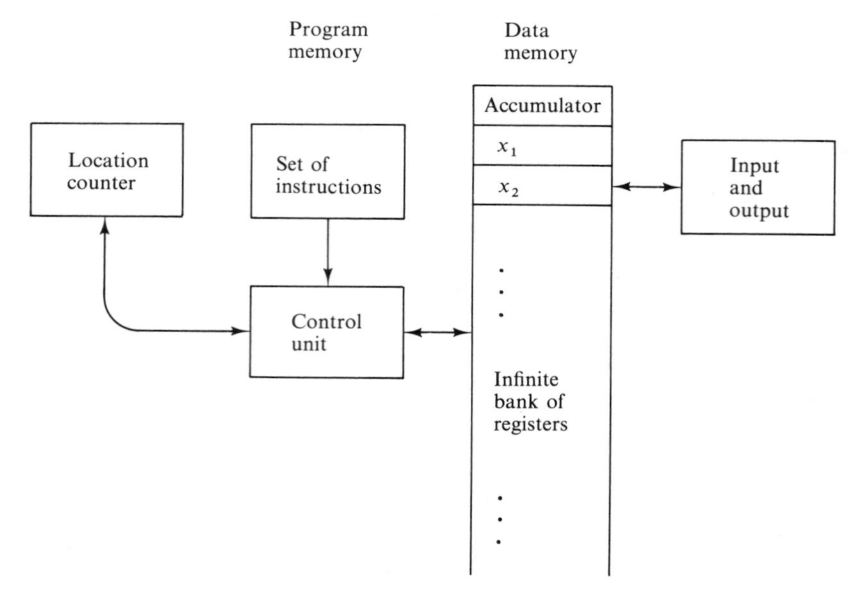

Figure 10.3 A Random Access Machine, RAM.

A somewhat more realistic model of existing computers is the Random Access Machine (RAM). Just like the URM, the RAM has an unbounded program memory and an unbounded data memory. The instruction set of the RAM resembles that of existing computers. The RAM is sketched in Figure 10.3.

A machine cycle consists of fetching the instruction as determined by the location counter and then executing this instruction, thereby appropriately updating the data memory and the location counter.

The RAM has the following types of instructions:

(1) *Data manipulation instructions.* These are determined by the data type assumed. It is customary to assume the data type $(N, \{\text{TRUE}, \text{FALSE}\}; \times, /, +, \dot{-}, even, \neq_0)$ with $\times$ as multiplication, $/$ as division with remainder, $+$ as addition, and $\dot{-}$ as subtraction of natural numbers, and predicates *even* and $\neq_0$ respectively to determine whether a natural number is even or unequal to zero.

All arithmetic is performed using a special register, the accumulator. There are instructions to add a number x to the contents of the accumulator, subtract x from it, multiply the contents of the accumulator by x, and so forth. The number x, called the operand, is specified in the instruction as follows:

(a) directly, as a number, e.g. $ADD = 5$ adds 5 to the contents of the accumulator.

(b) by giving the index, that is, the address, of a register, e.g. *ADD* 5 adds the contents of register x_5 to the contents of the accumulator.

(c) indirectly, by giving the address of a register whose contents is the address of the register which contains the value to be used, e.g. *ADD* *5 adds to the accumulator the contents of the register whose address is the contents of register number 5.

Note that the power of the RAM depends on the choice of these arithmetic instructions. For instance, incorporating the multiplication instruction makes the RAM too powerful to be a practically valid machine model, as we have seen in Section 9.2.

(2) *Control instructions.* The instructions of a program for the RAM are assumed to be stored in some addressable program memory, one instruction per memory location. The instructions are stored in successive locations, so that the next instruction can normally be found at the next highest address of the program memory. Thus, sequential execution is implemented by increasing the contents of the location counter by 1, after performing e.g. an *ADD* instruction. Special instructions are available to alter this sequential flow of control:

(a) *HALT*, which terminates the computation.

(b) *JMP* x (jump to x) which stores the operand x in the location counter. The instruction at this place in memory will be executed next.

(c) *JNZ* x (jump to x if accumulator is nonzero).

(3) *Input/output instructions.* The input for the RAM, a sequence of say k numbers, is written on a so-called input tape. The machine reads numbers from the input tape. After a number has been read, the read head is shifted so that the next number on the tape can be read. The number read can be stored in some arbitrary data register, e.g. *READ* x reads a number from the input tape and stores it in the register with index x.

The output is written on an output tape. We assume that after a number has been written on the tape, the write head is shifted, so that another number can written. For example *WRITE* x writes the operand x on the output tape.

An instruction set for a RAM is given in Table 10.1. In describing the meaning of the instructions, the following notation is used:

(1) $c(i)$ denotes the contents of register i.

(2) $v(a)$ denotes the value of operand a, that is,

(a) $v(=i)=i$,

(b) $v(i)=c(i)$, and

(c) $v(*i)=c(c(i))$.

Table 10.1 Instruction set of RAM.

Instruction	Meaning
LOAD a	$AC := v(a)$
STORE i	$c(i) := AC$
STORE $*i$	$c(c(i)) := AC$
ADD a	$AC := AC + v(a)$
SUB a	$AC := AC \dot{-} v(a)$
MULT a	$AC := AC \times v(a)$
DIV a	$AC :=$ integer part of $AC/v(a)$
READ i	$c(i) :=$ current input number, shift read head
READ $*i$	$c(c(i)) :=$ current input number, shift read head
WRITE a	print $v(a)$ on output tape, shift write head
JMP a	$LC := v(a)$
JNZ a	if $AC \neq 0$ then $LC := v(a)$ else $LC := LC + 1$
HALT	machine halts

(3) AC denotes the contents of the accumulator.

(4) LC denotes the contents of the location counter.

RAMs are faster than URMs and SAMs having the same underlying data type, because RAMs can use indirect addressing. This speed-up is realistic as long as the addresses used are sufficiently small, that is, are within the limits of some finite address space. It is therefore not always justified to use this facility of arbitrary indirect addressing when calculating the asymptotic complexity of problems.

Let P be a program requiring time $t(n)$ to run on a RAM. The computations induced by this program can be simulated by a SAM which uses our standard Gödel numbering scheme to code the contents of the RAM memory used by P. Because the Gödel access functions (see Section 5.2) require polynomial time on a SAM($\mathbf{V}^*$), the computation induced by P can be simulated by a SAM($\mathbf{V}^*$) in time $p(t(n))$, where p is some polynomial. Thus, problems solvable in polynomial time on a RAM are also solvable in polynomial time using SAM($\mathbf{V}^*$)s.

We will not discuss the relative merits of URMs and RAMs, but proceed instead with our study of the complexity of problems on the basis of SAM($\mathbf{V}^*$), and Turing machines, bearing in mind that a general-purpose machine executing a particular program can be identified with a special-purpose machine designed to perform the same calculations.

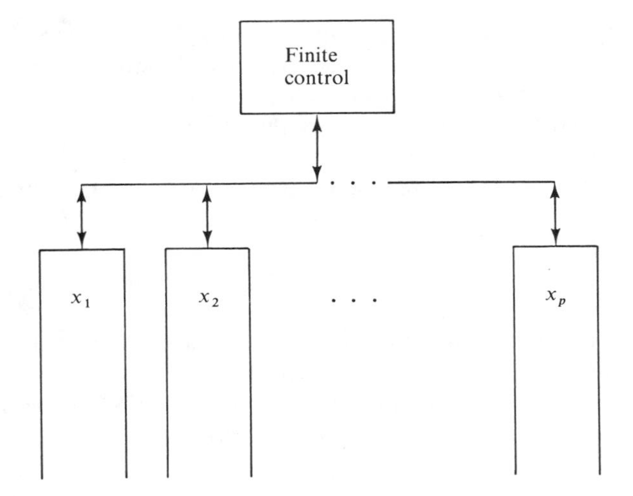

bank of pushdown stacks

Figure 10.4 A SAM(**V***) machine.

10.2 SAM(V*) machines and SAL(V*) programs

In Section 9.5 the data type **V*** was introduced. The data elements of the data type **V*** are finite strings over the finite alphabet V. The available operations concern the front end of the string. A string is thought of as the value of a pushdown stack, with the leftmost symbol of the string as the top of the stack. The available operations are

(1) $x := pop(y)$, remove y's leftmost symbol and assign the result to x

(2) $x := push(y, z)$, add $y \in V$ to the left-hand end of string z, and assign the result to x

(3) $x := top(y)$, assign the leftmost symbol of y to x

(4) and the predicates

  (a) $x \neq \varepsilon$, test whether the string x is the empty string

  (b) $x \equiv y$, test whether the leftmost symbols of x and y, if any, are equal.

In these terms, the data memory of a SAM(**V***) machine consists of a finite, machine-dependent number of pushdown stores (see Figure 10.4).

The boxes 'input and initialization' and 'output projection', given in Figure 10.1, are not present in Figure 10.4. The functions represented by these boxes are not incorporated in the machine model.

As you will see from the definition of the computations induced by SAMs, there is neither a first step in which registers are initialized nor a final step projecting register contents on output values. The same applies to SAL programs: we assume that the variables are properly initialized and that the output values are obtained from the values of the variables, but neither of these is by an explicit action induced by the program.

Initializing and projecting are taken care of simply by definition; we only consider computations starting on initial state vectors, thus assuming proper initialization as a starting condition, and define the result of the computation as if obtained by projection.

We did not show initialization and output projection in programs, and we will no longer make these explicit in our machines.

We will now define the classes of deterministic and of non-deterministic SAM($\mathbf{V}^*$) machines, and show that these machines are computationally equivalent to deterministic and non-deterministic SAL($\mathbf{V}^*$) programs. (The definition of ND-SAL($\mathbf{V}^*$), and example programs, can be found in Section 9.6.)

Definition 10.1

The universal set $U(\mathbf{V}^*)$ of action items is defined as follows:

$$
\begin{aligned}
U(\mathbf{V}^*) \triangleq\ & \{(s\colon x_i := \varepsilon : t) \mid i \in N, s, t \in N\} \cup \\
& \{(s\colon x_i := a : t) \mid i \in N, a \in V, s, t \in N\} \cup \\
& \{(s\colon x_i := x_j : t) \mid i, j \in N, s, t \in N\} \cup \\
& \{(s\colon x_i := push(x_j, x_m) : t) \mid i, j, m \in N, s, t \in N\} \cup \\
& \{(s\colon x_i := pop(x_j) : t) \mid i, j \in N, s, t \in N\} \cup \\
& \{(s\colon x_i := top(x_j) : t) \mid i, j \in N, s, t \in N\} \cup \\
& \{(s\colon \textbf{if } x_i \neq \varepsilon \textbf{ then } t_1 \textbf{ else } t_2) \mid i \in N, s, t_1, t_2 \in N\} \cup \\
& \{(s\colon \textbf{if } x_i \equiv a \textbf{ then } t_1 \textbf{ else } t_2) \mid i \in N, a \in V, s, t_1, t_2 \in N\} \cup \\
& \{(s\colon \textbf{if } x_i \equiv x_j \textbf{ then } t_1 \textbf{ else } t_2) \mid i, j \in N, s, t_1, t_2 \in N\}.
\end{aligned}
$$

The elements of $U(\mathbf{V}^*)$ are called action items. The numbers occurring in the action items are called *states*; those denoted by s are called *current states*, those denoted by t, t_1, and t_2 are called *next states*. The objects denoted by x_i and x_j are called *register identifiers*.

An ND-SAM($\mathbf{V}^*$) machine is a 4-tuple (n, k, p, M), where $n \geqslant 0$, $k \geqslant 1$, $n, k \leqslant p$, and M is a finite subset of $U(\mathbf{V}^*)$, such that all register identifiers occurring in M have an index at most p. The machine is deterministic if no two different action items have the same current state; otherwise the machine is non-deterministic. The set of deterministic SAM($\mathbf{V}^*$) machines is denoted by D-SAM($\mathbf{V}^*$). ■

The meaning of an ND-SAM($\mathbf{V}^*$) machine is the set of all computations it

Table 10.2 Semantics of ND-SAM($\mathbf{V}^*$) items.

$\langle action \rangle$	Value of y_i
$x_i := \varepsilon$	$y_i = \varepsilon$
$x_i := a$, $(a \in V)$	$y_i = a$
$x_i := x_j$	$y_i = x_j$
$x_i := push(x_j, x_m)$	$y_i = push(x_j, x_m)$
$x_i := pop(x_j)$	$y_i = pop(x_j)$
$x_i := top(x_j$	$y_i = top(x_j)$

induces. These computations and the relation computed by the machine are defined using state vectors, in the same way as was done for programs.

Definition 10.2
Let $Q = (n, k, p, M)$ be an ND-SAM($\mathbf{V}^*$) machine, and let f be such that all states occurring in M either as current states or as next states are at most equal to f. A state vector of Q is a $(p+1)$-tuple

$$(s, x_1, x_2, \ldots, x_p) \in \{1, 2, \ldots, f+1\} \times (V^*)^p$$

The first component gives the current state of the machine, the remaining components give the current contents of the registers.

$(s, x_1, x_2, \ldots, x_p)$ is an initial state vector if and only if $s = 1$ and $x_{n+1} = x_{n+2} = \cdots = x_p = \varepsilon$; it is a final state vector if and only if there is no action item in M with current state s.

State vector (s, x) is transformed into state vector (t, y) in a single step (notation $(s, x) \vdash_Q (t, y)$) if and only if

(1) an action item (s: $\langle action \rangle$:t) occurs in M, and $y_r = x_r$ for all r such that $1 \leqslant r \leqslant p$ and $r \neq i$, and the value of y_i depends on the $\langle action \rangle$ as specified in Table 10.2. Note the difference between the two occurrences of *push*, *pop*, and *top*. The leftmost are occurrences as syntactical objects in a machine specification, the rightmost occurrences denote operations supplied by the underlying data type $\mathbf{V}^*$.

(2) an action item (s: **if** $\langle condition \rangle$ **then** t_1 **else** t_2), where $\langle condition \rangle$ is $x_i \neq \varepsilon$, $x_i \equiv a$, or $x_i \equiv x_j$, occurs in M, $y = x$, and $t = t_1$ if $\langle condition \rangle$ evaluates to TRUE, and $t = t_2$ otherwise. ■

A computation performed by the machine is a sequence $u \vdash_Q u_1 \vdash_Q \cdots \vdash_Q u_m$. The computation terminates with state vector u_m, if it can proceed no further, that is, if u_m is a final state vector.

Table 10.3 Construction of a set of action items from a program.

Statement of program	Corresponding action item(s) of machine
s: $\langle$*assignment*$\rangle$	(s: $\langle$*assignment*$\rangle$ $:s+1$)
s: **while** $x_i \neq \varepsilon$ **do** $:t$	(s: **if** $x_i \neq \varepsilon$ **then** $s+1$ **else** $t+1$)
s: **while** $x_i \equiv a$ **do** $:t$	(s: **if** $x_i \equiv a$ **then** $s+1$ **else** $t+1$)
s: **while** $x_i \equiv x_j$ **do** $:t$	(s: **if** $x_i \equiv x_j$ **then** $s+1$ **else** $t+1$)
s: **od** $:t$ $(t \neq 0)$	(s: $x_1 := x_1$ $:t$)
s: **either** $:t$	(s: $x_1 := x_1$ $:s+1$) (s: $x_1 := x_1$ $:t+1$)
s: **or** $:t$	(s: $x_1 := x_1$ $:t+1$)
s: **od** $:0$	(s: $x_1 := x_1$ $:s+1$)

The notations $u \vdash^{+}_{Q} v$ and $u \vdash^{*}_{Q} v$ mean that the state vector u can be transformed into state vector v by a computation of k steps, where k is respectively $\geqslant 1$ or $\geqslant 0$.

Definition 10.3

The meaning of the ND-SAM($\mathbf{V^*}$) machine $Q=(n,k,p,M)$ is the following relation $m_Q \subseteq (V^*)^p \times (V^*)^p$

$$m_Q \triangleq \{(a,b) \mid \text{there is a computation } (1,a) \vdash_Q (s,x) \vdash_Q \cdots \vdash_Q (f,b) \text{ where } (f,b) \text{ is a final state vector}\}$$

The relation $r_Q \subseteq (V^*)^n \times (V^*)^k$ computed by the machine, is determined as follows:

$$r_Q \triangleq \{(a, \pi^p_{1\ldots k}(b)) \mid (a,\varepsilon,\ldots,\varepsilon,b) \in m_Q, \text{ and } a \in (V^*)^n, \text{ and } b \in (V^*)^p\}$$

■

To show that ND-SAM($\mathbf{V^*}$) and ND-SAL($\mathbf{V^*}$) are equivalent as models of computability, we construct an ND-SAM($\mathbf{V^*}$) machine which computes the same relation as a given ND-SAL($\mathbf{V^*}$) program, and vice versa.

Let (n,k,p,P) be an ND-SAL($\mathbf{V^*}$) program having its lines numbered from 1 to f. Construct an ND-SAM($\mathbf{V^*}$) machine (n,k,p,M) to simulate the computations of the given program. The set M of action items is constructed according to Table 10.3.

Obviously, the ND-SAM($\mathbf{V^*}$) machine constructed according to Table 10.3 computes the same relation as the original ND-SAL($\mathbf{V^*}$) program, and therefore R(ND-SAL($\mathbf{V^*}$)) $\subseteq$ R(ND-SAM($\mathbf{V^*}$)). Note that the

computations induced by the machine are slightly longer than those induced by the program, due to the inefficient transcription of a program line containing an **od** statement.

Constructing a SAL program to simulate the computations performed by a SAM machine is more complicated, because the programming language SAL does not contain a goto statement.

Let (n, k, p, M) be an ND-SAM(**V***) machine, and a some element of V. We construct an ND-SAL(**V***) program (n, k, q, P) of the following general structure.

```
continue:= a;
state:= initial_state;
while continue ≠ ε do
   continue:= ε;
   if state = s_1 then
      continue:= a;
      {ND-SAL(V*) statements replacing the action items in M with
      current state s_1}
   fi;
   if state = s_2 then
      continue:= a;
      {ND-SAL(V*) statements replacing the action items in M with
      current state s_2}
   fi;
   ⋮
   fi
od
```

In the above ND-SAL(**V***) program, the variable *state* codes the current state of the SAM(**V***) machine being simulated. Simulating the execution of a single action item requires one iteration of the while loop to find the appropriate sequence of ND-SAL(**V***) statements. If this sequence has been found and executed, the variable *state* is updated appropriately, and another iteration is carried out. If no ND-SAL(**V***) sequence has been found, the execution ceases. Thus, the running time of the ND-SAL(**V***) program is a machine-dependent constant times the running time of the ND-SAM(**V***) machine.

The ND-SAL(**V***) sequence replacing the action items in M with current state s is constructed as follows:

(1) If there is only a single action item with current state s, then this item is replaced by an ND-SAL(**V***) sequence as follows:

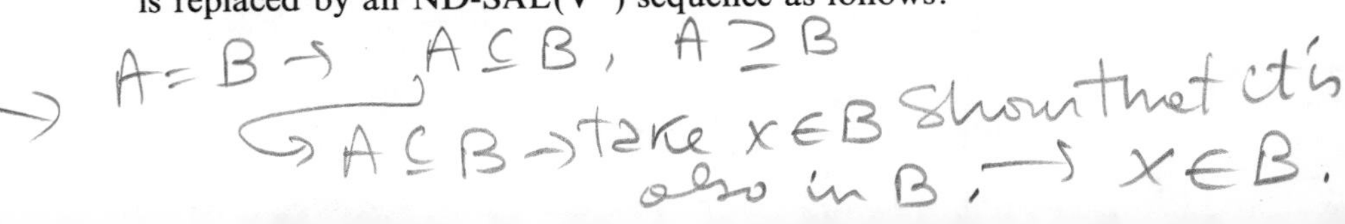

Item	*Sequence*
(s: ⟨*assignment*⟩ :t)	⟨*assignment*⟩; *state*:= t_r;
(s: **if** ⟨*condition*⟩ **then** t_1 **else** t_2)	**if** ⟨*condition*⟩ **then** *state*: = t_{1r} **else** *state*:= t_{2r} **fi**

In an ND-SAM(**V***) machine, the states are natural numbers, which have to be represented as strings in the ND-SAL(**V***) program. In the replacement scheme above, the representation of the natural number t is denoted by t_r.

The details of this representation are irrelevant, because there is but a finite, fixed, machine-dependent number of states to represent.

(2) If there are several action items with the same current state, these are grouped using the either statement, and then replaced individually as described above. For example, (s: *action*$_1$:t_1), (s: *action*$_2$:t_2), and (s: *action*$_3$:t_3) are replaced by

> **either** ⟨*sequence replacing action*$_1$⟩; *state*:= t_1
>
> **or either** ⟨*sequence replacing action*$_2$⟩; *state*:= t_2
>
>   **or** ⟨*sequence replacing action*$_3$⟩; *state*:= t_3
>
>   **od**
>
> **od**

The result of expanding all the macros in the program constructed as described above is an ND-SAL(**V***) program which computes the relation computed by the given ND-SAM(**V***) machine. This is summarized in the following theorem.

Theorem 10.1

ND-SAM(**V***) machines are computationally equivalent to ND-SAL(**V***) programs. Furthermore, there are constants, c_1 and c_2, such that a relation which is ND-SAL(**V***) computable in time $t(n)$ is SAM(**V***) computable in time $c_1t(n)$, and a relation ND-SAM(**V***) computable in time $t(n)$, is ND-SAL(**V***) computable in time $c_2t(n)$. ■

10.3 Turing machines

In 1936 Alan M. Turing published a paper, in which he defined a class of abstract machines, now called Turing machines, and defended the following proposition.

> Any process which could naturally be called an effective procedure can be realized by a Turing machine.

In his 1936 paper, Turing gives a sequence of arguments supporting this thesis. Below is a quotation from this paper, in which the essential features of the machine are deduced from the way in which a person performs computations. In the quotation, 'the computer' refers to this person, and not to a mechanical device of some sort.

> Computing is normally done by writing certain symbols on paper. We may suppose this paper is divided into squares like a child's arithmetic book. In elementary arithmetic the two-dimensional character of the paper is sometimes used. But such a use is always avoidable, and I think that it will be agreed that the two-dimensional character of paper is no essential of computation. I assume then that the computation is carried out on one-dimensional paper, i.e. on a tape divided into squares. I shall also suppose that the number of symbols that may be printed is finite. If we were to allow an infinity of symbols, then there would be symbols differing to an arbitrarily small extent. The effect of this restriction of the number of symbols is not very serious. It is always possible to use sequences of symbols in the place of single symbols. Thus an Arabic numeral such as 17 or 999999999999999 is normally treated as a single symbol. Similarly in any European language words are treated as single symbols (Chinese, however, attempts to have an enumerable infinity of symbols). The difference from our point of view between the single and compound symbols is that the compound symbols, if they are too lengthy, cannot be observed at one glance. This is in accordance with experience. We cannot tell at a glance whether 9999999999999999 and 999999999999999 are the same.
>
> The behaviour of the computer at any moment is determined by the symbols he is observing, and his 'state of mind' at that moment. We may suppose that there is a bound *B* to the number of symbols or squares which the computer can observed at one moment. If he wishes to observe more, he must use successive observations. We will also suppose that the number of states of mind which need be taken into account is finite. The reasons for this are of the same character as those which restrict the number of symbols. If we admitted an infinity of states of mind, some of them will be 'arbitrarily close' and will be confused. Again the restriction is not one which seriously affects computation, since the use of more complicated states of mind can be avoided by writing more symbols on the tape.
>
> Let us imagine the operations performed by the computer to be split up into 'simple operations' which are so elementary that it is not easy to imagine them further divided. Every such operation consists of some change of the physical system consisting of the computer and

his tape. We know the state of the system if we know the sequence of symbols on the tape, which of these are observed by the computer (possibly with a special order), and the state of mind of the computer. We may suppose that in a simple operation not more than one symbol is altered. Any other changes can be split up into simple operations of this kind. The situation in regard to the squares whose symbols may be altered in this way is the same as in regard to the observed squares. We may, therefore, without loss of generality, assume that the squares whose symbols are changed are always 'observed' squares.

Beside these changes of symbols, the simple operations must include changes of distribution of observed squares. The new observed squares must be immediately recognisable by the computer. I think it is reasonable to suppose that they can only be squares whose distance from the closest of the immediately previously observed squares does not exceed a certain fixed amount. Let us say that each of the new observed squares is within L squares of an immediately previously observed square.

In the above quotation, Turing derives essential features of machines which can perform computations:

(1) The machine has a finite set of states representing the finite set of 'states of mind'.

(2) The machine has an unbounded memory, composed of infinitely many cells of limited capacity. We can think of the memory as an infinite tape divided into squares, these are the memory cells. Each cell has a limited capacity, it can store a single symbol belonging to a finite alphabet of possible symbols. Furthermore, the memory access functions are extremely simple. Individual squares cannot be addressed directly, the machine can only move from the currently observed square to a 'physically close' square.

(3) The machine can perform only the simplest operations, namely

   (a) reading the symbol written on the currently observed square,
   (b) writing a symbol, belonging to the given finite set of possible symbols, on the currently observed square,
   (c) and moving to another square of the tape.

We imagine that the tape is equipped with a 'read/write head'. The square underneath this read/write head is the currently observed square. The symbol on this square can be read and rewritten. The read/write head can remain in position, or be shifted one square to the left, or one square to the right. From the quotation above, we conclude that the number of observed squares is finite, but not necessarily one. Thus we could have a finite number of read/write heads.

(4) The machine has a finite control unit, which specifies the operation of the machine in terms of the simple operations discussed above.

The Turing machines we are about to examine are equipped with a finite number p of tapes, instead of a single tape. This is an advantage in designing machines, because data can be stored on individual tapes in accordance with semantic content, which makes it easier to understand how a particular machine works. These extra tapes do not increase computing power, but do increase computing speed as we shall see.

Note that the tapes are infinite, extending towards infinity both to the left and to the right. All these infinitely many cells contain symbols belonging to the given finite alphabet of possible symbols. We assume that all the squares which have never been underneath the read/write head are blank, that is, contain a special symbol of the alphabet called the blank symbol.

A particular Turing machine is given by a machine specification, a finite set of action items, similar to the specification of a SAM(**V***) machine. The action items have the following general format.

$$(q_1, u, v, d, q_2)$$

q_1 is the current state, u is the symbol on the currently observed square underneath the read/write head, v is the symbol written on this currently observed square, d represents the direction in which the read/write head is to be moved, S for keeping it stationary, R for moving one square to the right, and L for moving one square to the left, and finally, q_2 is the next state.

If the machine has p tapes, then u is a p-dimensional vector $(u_1, u_2, \ldots, u_p)$, where u_i is the symbol read by the read/write head on tape i. Also, v and d are p-dimensional vectors, the ith components of which determine which symbol is to be written on the currently observed square of tape i, and whether, and if so, how, the read/write head on tape i will move.

The contents of a tape of a Turing machine (see Figure 10.5) can be represented using two pushdown stacks. One stack represents the contents of the tape to the left of the read/write head, the other pushdown stack represents the contents of the square underneath the read/write head and the contents of the tape to the right of the read/write head. The contents can be represented in this way, because the pushdown stacks need only store a finite part of the tape contents, such that this part includes all squares marked with a non-blank symbol. Thus, a Turing machine with p tapes, using an alphabet V of symbols (including the blank symbol) can be simulated by

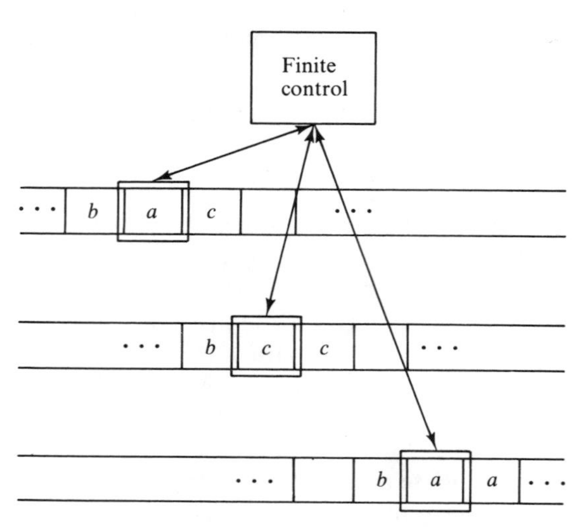

Figure 10.5 A Turing machine with p tapes.

a SAM(V*) machine using $2p$ registers. This equivalence will be made precise in the next section.

We will now formally define Turing machines and computations induced by Turing machines.

Definition 10.4

A non-deterministic, p-tape Turing machine over the (finite) alphabet V is a tuple $(n, k, p, V, B, Q, q_0, M)$, where

(1) n, k, and p are natural numbers, $n \geqslant 0$, $k \geqslant 1$, $p \geqslant n$, and $p \geqslant k$. These numbers specify that the machine has p tapes, of which tape 1 to tape n are used for input, and tape 1 to tape k are used for output.

(2) V is a non-empty, finite alphabet; it specifies the set of tape symbols and V is called the *tape alphabet*.

(3) $B \in V$ is the symbol used to denote blank.

(4) Q is a non-empty, finite set. Its elements are called the states of the machine.

(5) $q_0 \in Q$ is called the initial state, it is used in defining the relation computed by the Turing machine.

(6) $M \subseteq Q \times V^p \times V^p \times D^p \times Q$, where $D = \{S, L, R\}$, is called the machine specification. Its elements are called action items. In an action item (p, u, v, d, q)

(a) p is the current state
(b) q is the next state

(c) u_i is the symbol read from tape i
(d) v_i is the symbol written on tape i
(e) d_i is the movement of the read/write head of tape i.

The machine is deterministic, if for each $q \in Q$ and $u \in V^p$ there is at most one action item $(q, u, v, d, q_n) \in M$, in other words, if M is a partial function from $Q \times V^p$ to $V^p \times D^p \times Q$.

The set of all non-deterministic Turing machines over the alphabet V is denoted by NDTM($\mathbf{V}^*$), the set of all deterministic Turing machines over the alphabet V is denoted by DTM($\mathbf{V}^*$). ■

The definition above differs in two respects from the definition of ND-SAM($\mathbf{v}^*$) machines in Definition 10.1 (page 238).

(1) The alphabet V of tape symbols is carried along in the notation. This is customary, but also a notational advantage, because we will use Turing machines over many different alphabets, choosing the alphabet as is convenient for the application at hand.
(2) The set Q of machine states is an arbitrary set instead of a finite subset of N. This makes the formalism more verbose, but is done nevertheless, to avoid undue deviation from what is customary in the relevant literature.

We turn now to the computations induced by Turing machines. These computations and the relation computed by the machine are defined using *instantaneous descriptions*.

Let $T = (n, k, p, V, B, Q, q_0, M)$ be an NDTM($\mathbf{V}^*$). An instantaneous description, or *ID* for short, is used to specify the current state of the machine, the current contents of the tapes, and the current positions of the read/write heads. Because the tapes of the machine are infinite, *ID*s are infinite objects. Since only finite portions of the tapes contain non-blanks, we can finitely represent an *ID* by only specifying the contents of finite parts of the tapes, which encompass all squares of the tapes written with non-blanks. We use the following notation for *ID*s:

$$((\alpha_1, \alpha_2, \ldots, \alpha_p), q, (\beta_1, \beta_2, \ldots, \beta_p))$$

This *ID* indicates that the Turing machine is currently in state q, and that the content of tape i is $B^\infty \alpha_i \beta_i B^\infty$, that is, an infinite sequence of blanks followed by $\alpha_i \beta_i$ followed by an infinite sequence of blanks; it also specifies that the read/write head of tape i is currently positioned above the square containing the leftmost symbol of $\beta_i B^\infty$, $1 \leqslant i \leqslant p$.

The notation for *ID*s is not unique, Bs may be added to the left of the α_is and to the right of the β_is at will. To make the notation unique, we

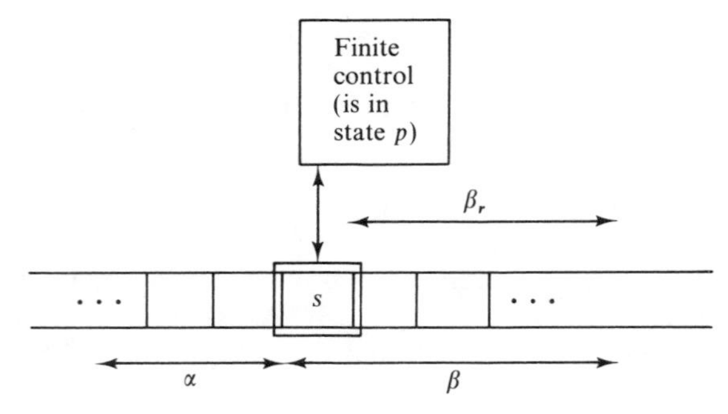

Figure 10.6 Before a single step of a Turing machine.

will henceforth require that the representation is succinct, that is, does not include superfluous Bs.

A computation induced by a non-deterministic Turing machine is defined to be a sequence of steps, where each step transforms an *ID* into another *ID*, in accordance with the machine specification, as described below.

Let (α, p, β) be an *ID* of a single-tape Turing machine (Figure 10.6). Suppose that (p, s, t, R, q) is an action item belonging to the specification M of our Turing machine. According to this action item, the above *ID* (α, p, β) is transformed into the *ID* (α_n, q, β_n) (Figure 10.7).

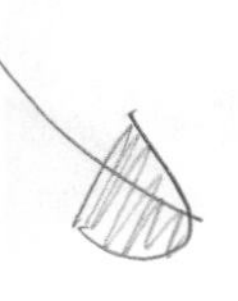

Definition 10.5

Let $T=(n, k, p, V, B, Q, q_0, M)$ be a non-deterministic, p-tape Turing machine over the alphabet V.

(1) An element $((\alpha_1, \ldots, \alpha_p), q, (\beta_1, \ldots, \beta_p))$ of $(V^*)^p \times Q \times (V^*)^p$ is an *instantaneous description* (*ID*) of T if for all i, $1 \leqslant i \leqslant p$,

(a) $\alpha_i = \varepsilon$, or the leftmost symbol of α_i is unequal to B, and
(b) $\beta_i = \varepsilon$, or the rightmost symbol of β_i is unequal to B.

(2) The *ID* $((\alpha_1, \ldots, \alpha_p), q, (\beta_1, \ldots, \beta_p))$ is *initial* if $q = q_0$ and $\alpha_1, \ldots, \alpha_p$ and $\beta_{n+1}, \ldots, \beta_p$ are ε.

(3) Let

$$c_1 = ((\alpha_1, \alpha_2, \ldots, \alpha_p), q_1, (\beta_1, \beta_2, \ldots, \beta_p))$$

and

$$c_2 = ((\gamma_1, \gamma_2, \ldots, \gamma_p), q_2, (\delta_1, \delta_2, \ldots, \delta_p))$$

be *ID*s of the Turing machine T. Then *ID* c_1 is transformed into *ID* c_2 in a single-step notation $c_1 \vdash_T c_2$, if there is an action item $(q_1, (u_1, u_2, \ldots, u_p), (v_1, v_2, \ldots, v_p), (d_1, d_2, \ldots, d_p), q_2)$ in M, and for all i, $1 \leqslant i \leqslant p$, γ_i and δ_i are determined as follows.

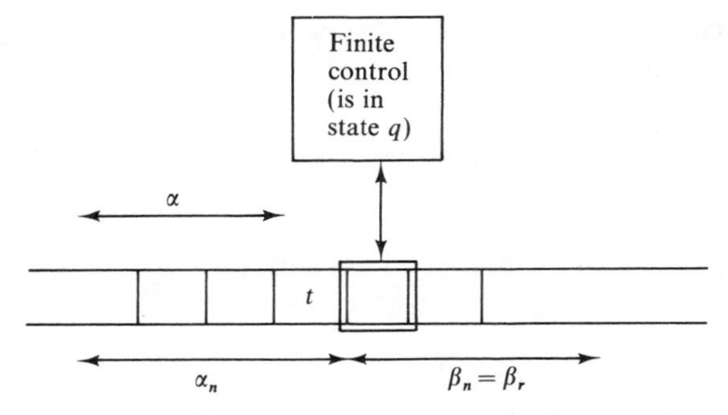

Figure 10.7 After a single step of a Turing machine.

Table 10.4 The contents of tape i dependent on head movement.

$d_i = S$	$d_i = R$	$d_i = L$
$\gamma_i = \alpha_i$	$\gamma_i = \alpha_i v_i$	$\gamma_i = \eta_i$
$\delta_i = v_i \omega_i$	$\delta_i = \omega_i$	$\delta_i = a_i v_i \omega_i$

Let $a_i \in V$ and $\eta_i \in V^*$ be such that $\alpha_i = \eta_i a_i$ if $\alpha_i \neq \varepsilon$, and $a_i = B$, $\eta_i = \varepsilon$ otherwise. Similarly, let $b_i \in V$ and $\omega_i \in V^*$ be such that $\beta_i = b_i \omega_i$ if $\beta_i \neq \varepsilon$, and $b_i = B$, $\omega_i = \varepsilon$ otherwise.

Then b_i must be equal to u_i for all i, $1 \leqslant i \leqslant p$. The values of γ_i and δ_i depend on the movement of the read/write head as specified in Table 10.4. If $c_1 \vdash_T c_2$, then c_2 is called a successor *ID* of c_1.

(4) An *ID* of the Turing machine T is *final* if it has no successor *ID*. ■

A computation performed by a non-deterministic Turing machine is a finite sequence $c_1 \vdash_T c_2 \vdash_T \cdots \vdash_T c_t$. The computation terminates with *ID* c_t, if it can proceed no further, that is, if c_t has no successor configuration and is therefore a final *ID*. The notations $u \vdash_T^+ v$ and $u \vdash_T^* v$ mean that u can be transformed into v by a sequence of k steps, where k is respectively $\geqslant 1$ and $\geqslant 0$.

Definition 10.6

The meaning of a non-deterministic Turing machine $T = (n, k, p, V, B, Q, q_0, M)$ is the following relation $m_T \subseteq (V^*)^p \times (V^*)^p$.

$$m_T \triangleq \{(\alpha, \beta\gamma) \mid \text{there is a computation } (\varepsilon^p, q_0, \alpha) \vdash_T^* (\beta, q, \gamma), \text{ where } (\beta, q, \gamma) \text{ is a final } ID \text{ of } T, \text{ and } \alpha, \beta, \gamma \in (V^*)^p\}$$

The relation $r_T \subseteq (V^*)^n \times (V^*)^k$ computed by the Turing machine T is determined as follows:

$$r_T \triangleq \{(\alpha, \pi^p_{1\dots k}(\beta)) | ((\alpha, \varepsilon, \dots, \varepsilon), \beta) \in m_T, \text{ and } \alpha \in (V^*)^n \text{ and } \beta \in (V^*)^p\}$$

where $\pi^p_{1\dots k}$ is the function $\lambda x_1 x_2 \cdots x_p[(x_1, x_2, \dots, x_k)]$, projecting a p-dimensional vector on its first k components. ■

The time and space complexity of a non-deterministic Turing machine are defined in the same way as for **SAM(V*)** machines and **SAL(V*)** programs.

(1) *Time complexity.* As time is not a meaningful concept in relation to the computations induced by Turing machines, we take for time complexity the length, i.e. the number of single steps, of a terminating computation.

(2) *Space complexity.* Space used is a much more natural concept in relation to Turing machines than it is in any of the other models discussed. Space used is simply the number of tape squares used. To be precise

(a) the space used by the *ID* $((\alpha_1, \alpha_2, \dots, \alpha_p), q, (\beta_1, \beta_2, \dots, \beta_p))$ is the maximum of $\{length\_of(\alpha_i \beta_i) | 1 \leq i \leq p\}$.

(b) the space used by a computation $c_1 \vdash_T c_2 \vdash_T \cdots \vdash_T c_t$ is the maximum of $\{space\_used\_by(c_i) | 1 \leq i \leq t\}$.

According to this definition, we count the space needed to represent the input as space used by the Turing machine. To avoid this, we will assume that tapes 1 to n, the input tapes, are read-only. The space used by the Turing machine is then the space used on tapes $n+1$ to p, in the sense defined above. If we wish to stress that the space required to represent the input is ignored, we refer to the *workspace* used by the machine.

In the following, we will use $\vdash$ instead of $\vdash_T$ when the machine involved is clear from the context.

EXAMPLE 10.1

We construct a Turing machine to compare two strings over an alphabet V. If the strings are equal, the result must be a, where a is an arbitrary but fixed symbol belonging to V. If the strings are not equal, the output must be the empty string ε. (Compare this machine with the first program of Example 9.8).

The machine will have two tapes. The alphabet of tape symbols is $V \cup \{B\}$, where $B \notin V$ denotes the blank symbol. The machine will have three states q_0, q_1 and q_2, of which q_0 is the initial state. The set of action items is constructed as follows.

(1) The machine reads symbols from tape 1 and from tape 2, replaces them by blanks and moves the read/write heads to the right. As long as the symbols read remain equal and non-blank, this same action is repeated.

This behaviour is achieved by adding to the machine description M an action item $(q_0,(x,x),(B,B),(R,R),q_0)$ for every $x \in V$.

(2) If the strings are equal, the machine ends up reading blanks from both tapes. In this case the machine prints the required answer, a symbol a, and then the computation ceases. This behaviour is achieved by adding to the machine description M the action item $(q_0,(B,B),(a,B),(S,S),q_2)$.

(3) If the machine detects a difference between the strings, it must clear the tapes. Detection of a difference between the strings on tape 1 and tape 2 is achieved by adding to the machine description M an action item $(q_0,(x,y),(B,B),(R,R),q_1)$ for every pair $x, y \in (V \cup \{B\})$ such that $x \neq y$.

(4) If the machine is in state q_1, a difference has been detected, and the tapes must now be cleared. This behaviour is achieved by adding to the machine description M an action item $(q_1,(x,z),(B,B),(R,R),q_1)$ for every pair $x, z \in (V \cup \{B\})$ such that x and z are not both equal to B.

An example of a computation is

$((\varepsilon,\varepsilon),q_0,(ab,ab)) \vdash$

[$((B,b),q_0,(b,b))$ is succinctly represented by]

$((\varepsilon,\varepsilon),q_0,(b,b)) \vdash$

$((\varepsilon,\varepsilon),q_0,(\varepsilon,\varepsilon)) \vdash$

[now the machine reads (B,B) from the tapes]

$((\varepsilon,\varepsilon),q_2,(a,\varepsilon))$

[this is a final ID, there is no action item with current state q_2.]

A pictorial representation of this computation is given in Figure 10.8.

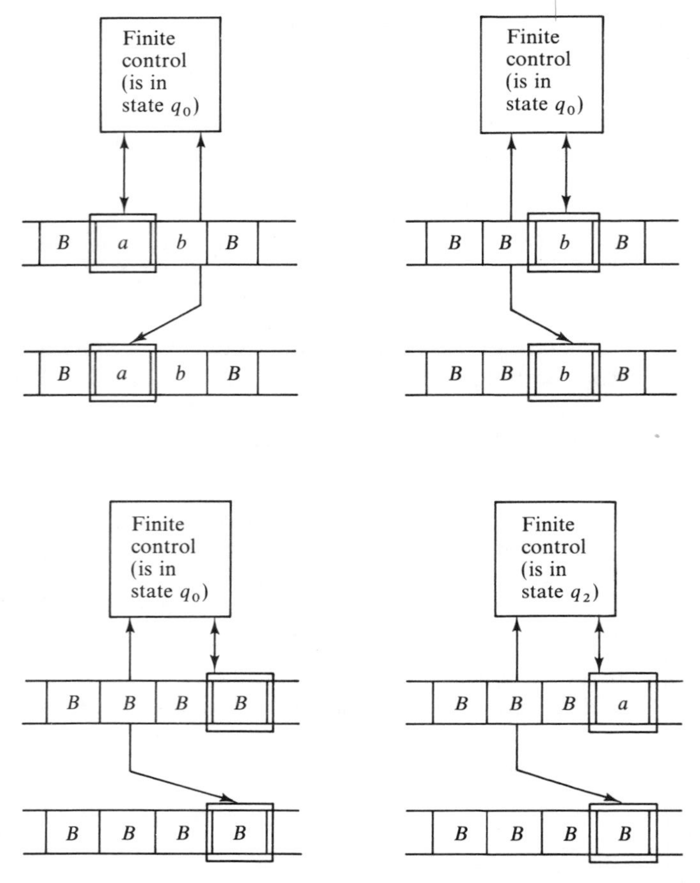

Figure 10.8 An example of a computation on input (ab, ab).

EXAMPLE 10.2

We construct a Turing machine to add numbers in binary representation. (Compare this machine with program (6) of Example 9.8.) Assume that $V = \{0, 1, B\}$, and that B denotes the blank symbol. Let $x_1 x_2 \cdots x_m$ represent the number $\sum_{i=1}^{m} x_i . 2^{i-1}$, i.e. the leftmost symbol of x is interpreted as the least significant digit in the binary representation of the number. The answer produced on tape 1 must use the same representation.

The machine operates by adding the two given numbers digit by digit in the obvious way. The sum of two digits is written on tape 1 and a possible carry is coded in the state of the machine, that is, the machine is in state q_0 if there is no carry, and in state q_1 otherwise. The initial state is state q_0, of course.

The machine specification M consists of the following action items:

(1) $(q_0, (0,0), (0,0), (R,R), q_0)$
(2) $(q_0, (0,1), (1,1), (R,R), q_0)$
(3) $(q_0, (0,B), (0,B), (R,R), q_0)$

(4) $(q_0, (1,0), (1,0), (R,R), q_0)$
(5) $(q_0, (1,1), (0,1), (R,R), q_1)$
(6) $(q_0, (1,B), (1,B), (R,R), q_0)$

(7) $(q_0, (B,0), (0,0), (R,R), q_0)$
(8) $(q_0, (B,1), (1,1), (R,R), q_0)$

(9) $(q_1, (0,0), (1,0), (R,R), q_0)$
(10) $(q_1, (0,1), (0,1), (R,R), q_1)$
(11) $(q_1, (0,B), (1,B), (R,R), q_0)$

(12) $(q_1, (1,0), (0,0), (R,R), q_1)$
(13) $(q_1, (1,1), (1,1), (R,R), q_1)$
(14) $(q_1, (1,B), (0,B), (R,R), q_1)$

(15) $(q_1, (B,0), (1,0), (R,R), q_0)$
(16) $(q_1, (B,1), (0,1), (R,R), q_1)$
(17) $(q_1, (B,B), (1,B), (R,R), q_0)$

Because the representations of the two numbers need not be equally long, it is necessary to include the action items having one of the symbols read from the tapes equal to B.

The addition is finished if two Bs are read from the tapes, and there is no remaining carry, i.e. the machine is in state q_0.

Table 10.5 gives a computation of the Turing machine specified above.

According to Definition 10.6 the output of the Turing machine does not depend on the position of the read/write heads, thus, it is not necessary to reposition the read/write head of tape 1.

In the current example we have one output tape, tape 1. Therefore, the contents of tape 2 in a final *ID* is immaterial. For the Turing machine specified above, we have chosen to leave the contents of tape 2 undisturbed. Any other choice is equally valid.

Our definition of what constitutes a non-deterministic Turing machine differs from what is customary. The customary definition is more compatible with the idea that a Turing machine computes a predicate than with the idea that a Turing machine computes a relation, or a function. Complexity theory is mainly concerned with predicates (see Section 9.3), thus the customary

Table 10.5 Example computation of a Turing machine adder.

ID	Calculation	Result		Action item used
		sum	carry	
$((\varepsilon, \varepsilon), q_0, (101, 11))$	$1+1+0$	0	1	5
$((0, 1), q_1, (01, 1))$	$0+1+1$	0	1	10
$((00, 11), q_1, (1, \varepsilon))$	$1+B+1$	0	1	14
$((000, 11B), q_1, (\varepsilon, \varepsilon))$	$B+B+1$	1	0	17
$((0001, 11BB), q_0, (\varepsilon, \varepsilon))$	terminal *ID* of Turing machine			

definition is often more convenient. It differs in the following way from our earlier Definition 10.4 (page 246).

(1) A predicate is assumed to have a single argument, which is a string over a given alphabet Σ. The tape alphabet V of the Turing machine is a superset of Σ. The set $V - \Sigma$ contains auxiliary symbols used by the Turing machine.

(2) The machine is not required to produce output. Instead, a subset $A \subseteq Q$ is specified. The predicate computed by the machine is TRUE for the given argument, if there is a computation which brings the machine in a state belonging to A, and FALSE otherwise.

(3) Because the input is a single string belonging to Σ^*, it is also very natural to say that the machine *accepts a language*. Instead of saying that the predicate on this input evaluates to TRUE or to FALSE, we say that the machine *accepts* the input, or that the machine *rejects* the input. The language accepted by the Turing machine is the set of all inputs accepted by the machine. This latter terminology corresponds to the set version of the predicate.

Definition 10.7

A non-deterministic Turing *acceptor* is an 8-tuple $T = (p, V, \Sigma, B, Q, q_0, A, M)$ where

(1) $p \geqslant 1$ is the number of tapes of the machine.

(2) V is a finite non-empty alphabet, consisting of all the symbols that can be printed on the tape squares.

(3) Σ is a non-empty subset of V.

(4) B is blank symbol of the Turing machine.

(5) Q is a finite non-empty set, the elements are the states of the machine.

(6) $q_0 \in Q$ is the initial state.

The states represent the information we wish to store.

(7) A is a subset of Q, its elements are called *accepting states.*
(8) M is the machine specification.

Computations of the machine are defined using instantaneous descriptions as in Definition 10.5. The machine *accepts* the following subset $L(T)$ of Σ^*.

$$L(T) \triangleq \{\omega \in \Sigma^* \mid \text{there is a computation } ((\varepsilon,\ldots,\varepsilon), q_0, (\omega, \varepsilon,\ldots,\varepsilon)) \vdash_T^* ((\alpha_1, \alpha_2,\ldots,\alpha_p), q, (\beta_1, \beta_2,\ldots,\beta_p)) \text{ such that } q \in A\}$$

Note that it is not required that the *ID* $((\alpha_1, \alpha_2,\ldots,\alpha_p), q, (\beta_1, \beta_2,\ldots,\beta_p))$ is final. It must be possible to reach an accepting state, but the computation does not necessarily terminate at that, or any later point.

The set $L(T)$ is also called the *language accepted by T.* (T itself is also called a non-deterministic Turing machine.) ■

EXAMPLE 10.3

We will discuss a Turing machine which accepts the language $\{a^n b^n \mid n \geqslant 1\}$. The machine has two tapes, of which tape 1 is the input tape, and it operates as follows:

(1) as are copied from tape 1 to tape 2 until the first b is encountered.

(2) Then the as on tape 2 are compared with the bs on tape 1. If they extend over an equal number of squares, the machine enters an accepting state.

Formally, $T = (2, \{a, b, B\}, \{a, b\}, B, \{q_0, q_1, q_2\}, q_0, \{q_2\}, M)$, where M consists of the following action items:

$(q_0, (a, B), (a, a), (R, R), q_0)$

$(q_0, (b, B), (b, B), (S, L), q_1)$

$(q_1, (b, a), (b, B), (R, L), q_1)$

$(q_1, (B, B), (B, B), (S, S), q_2)$

$(q_2, (B, B), (B, B), (S, S), q_2)$

Note that T rejects all words belonging to $\{a, b\}^* - \{a^n b^n \mid n \geqslant 1\}$ via terminating computations, whereas accepting computations proceed indefinitely, using the action item $(q_2, (B, B), (B, B), (S, S), q_2)$. Thus they proceed indefinitely, although nothing actually changes.

In the following we will mainly use Turing machines in the sense of the definition above. Our earlier Definition 10.4, however, is better suited to convey the similarity between programs, simple algorithmic machines in general and Turing machines in particular.

10.4 Computational equivalence between NDTM(V*) and SAM(V*)

In this section we will show that non-deterministic Turing machines are computationally equivalent to non-deterministic simple algorithmic machines.

We can construct a SAM(**V***) machine which simulates the computations induced by a given non-deterministic Turing machine T, and uses, up to a constant factor, the same amount of time and space.

The converse is more difficult. We can construct a non-deterministic Turing machine which simulates the computations induced by a given SAM(**V***) machine, and uses, up to a constant factor, the same amount of space, but needs significantly more time than the SAM(**V***) machine. The extra time needed by the Turing machine is consumed transporting data from one tape to some other tape, as required to implement register assignments (s: $x_i := x_j$:t).

The use of general assignments, such as $x := push(a, y)$, which imply copying the value of one register into the next, as opposed to assignments of the more limited form $x := push(a, x)$, which require no copying, is not entirely justified in practice. After all, copying an unbounded amount of information in a single step is not possible in practice.

Thus the time needed by Turing machines is a better estimate of time requirements than is the time needed by simple algorithmic machines, or equivalently SAL(**V***) programs. (All examples of programs and machines discussed so far have only made use of the more limited form of assignment.)

Let (n, k, p, M) be an ND-SAM(**V***) machine, B a symbol not belonging to V, and $W \triangleq V \cup \{B\}$. We construct an equivalent non-deterministic Turing machine $(n, k, p, W, B, Q, q_0, M_T)$ as follows.

The set Q of states of the Turing machine consists of all natural numbers which occur in the specification M of the SAM(**V***) machine, either as current state, or as next state. The initial state of the Turing machine is '1', i.e. the initial state of the SAM(**V***) machine. Q also contains states denoted by '$s.n$', where s is a state occurring in the specification of the SAM(**V***) machine as current or next state, and n is some number. These states are used in action items of the Turing machine intended to simulate an action item (s: $\langle action \rangle$:t) of the SAM(**V***) machine.

The current value of register x_i of the SAM(**V***) machine is stored as the contents of the tape i of the Turing machine. The read/write head will normally be on the square containing the leftmost symbol of x_i, but will move around when necessary for data transport etc.

The set M_T of action items of the Turing machine is obtained by adding to M_T action items necessary to simulate the behaviour induced by the action items of the SAM(**V***) machine as discussed below.

Consider the register assignment (s: x_i:= $push(x_m, x_j)$:t). The Turing machine simulates this register assignment as follows:

(1) Tape i is cleared by overwriting non-blanks with blanks. Simultaneously, the read/write head on tape j is moved to the rightmost symbol on tape j. This is achieved by the first two groups of action items, having current state s, or s.1.

(2) Then the contents of tape j are copied to tape i. When copying is finished, the read/write head of tape j is on the leftmost non-blank symbol, if this exists, and the read/write head of tape i is on the first blank to the left of any non-blank portion of tape i. This behaviour is achieved using the action items with current state s.2.

(3) Then the machine checks whether the read/write head of tape m is reading a non-blank. If so, the machine writes this symbol on tape i and then checks that tape m contains no other non-blanks; if not, the read/write head is shifted to the right, which brings the head into the correct position. This behaviour is achieved using the action items with current state s.3 or $s.x$, $x \in V$.

To perform the tasks discussed above, the action items shown in Table 10.6 are added to M_T.

Other assignment action items of the SAM(**V***) machine are similarly translated into sets of action items for the Turing machine.

Consider the control action item (s: **if** $x_i \equiv x_j$ **then** t.1 **else** t.2). To simulate this action item, add the action items in Table 10.7 to the specification M_T of the Turing machine.

The other control action items of the SAM(**V***) machine are similarly translated into sets of action items for the Turing machine.

Simulating a register assignment, such as x_i:= x_j, requires a maximum of $\{length(x_i), length(x_j)\}$ steps. In a computation of the SAM(**V***) machine of length t, all register values have length at most t. The Turing machine constructed above can therefore simulate this computation in no more than t^2 steps. In other words, the time complexity of the Turing machine is quadratic in the time complexity of the SAM(**V***) machine.

Let us now construct an ND-SAM(**V***) machine, which simulates the computations of a given Turing machine.

The idea is to use two pushdown stacks, i.e. two registers of the SAM(**V***) machine to store the contents of a single tape of the Turing machine, as sketched in Figure 10.9.

Table 10.6 Part of specification of a Turing machine to simulate a SAM(**V***) machine.

Action item	u_i	v_i	d_i	u_j	v_j	d_j	u_r	v_r	d_r
$(s, u, v, d, s.1)$	x	B	R	y	y	R	z	z	S
$(s, u, v, d, s.1)$	B	B	S	y	y	R	z	z	S
$(s, u, v, d, s.1)$	x	B	R	B	B	S	z	z	S
$(s, u, v, d, s.3)$	B	B	S	B	B	S	z	z	S
$(s.1, u, v, d, s.1)$	x	B	R	y	y	R	z	z	S
$(s.1, u, v, d, s.1)$	B	B	S	y	y	R	z	z	S
$(s.1, u, v, d, s.1)$	x	B	R	B	B	S	z	z	S
$(s.1, u, v, d, s.2)$	B	B	S	B	B	L	z	z	S
$(s.2, u, v, d, s.2)$	B	y	L	y	y	L	z	z	S
$(s.2, u, v, d, s.3)$	B	B	S	B	B	R	z	z	S

for all r such that $1 \leqslant r \leqslant p$, $r \neq i$, $r \neq j$, and all $x, y \in V$ and $z \in W$

				u_m	v_m	d_m			
$(s.3, u, v, d, s.x)$	B	x	S	x	x	S	z	z	S
$(s.3, u, v, d, t)$	B	B	R	B	B	S	z	z	S
$(s.x, u, v, d, t)$	B	x	S	B	B	L	z	z	S
$(s.x, u, v, d, t)$	B	B	R	z	z	L	z	z	S

for all r such that $1 \leqslant r \leqslant p$, $r \neq i$, $r \neq m$, and all $x \in V$ and $z \in W$

Table 10.7 Part of a specification of a Turing machine to simulate a SAM(**V***) machine.

Action item	u_i	v_i	d_i	u_j	v_j	d_j	u_r	v_r	d_r
$(s, u, v, d, t.1)$	x	x	S	x	x	S	y	y	S
$(s, u, v, d, t.2)$	B	B	S	B	B	S	y	y	S
$(s, u, v, d, t.2)$	z_1	z_1	S	z_2	z_2	S	y	y	S

for all r such that $1 \leqslant r \leqslant p$, $r \neq i$, $r \neq j$, and all $x \in V$ and $y, z_1, z_2 \in W$ such that $z_1 \neq z_2$

The specification of the SAM(**V***) machine is obtained by translating every action item of the Turing machine into a set of action items for the simple algorithmic machine.

Consider the action item (s, u, v, d, t) of the Turing machine. It must be checked first whether the action item of the Turing machine is applicable.

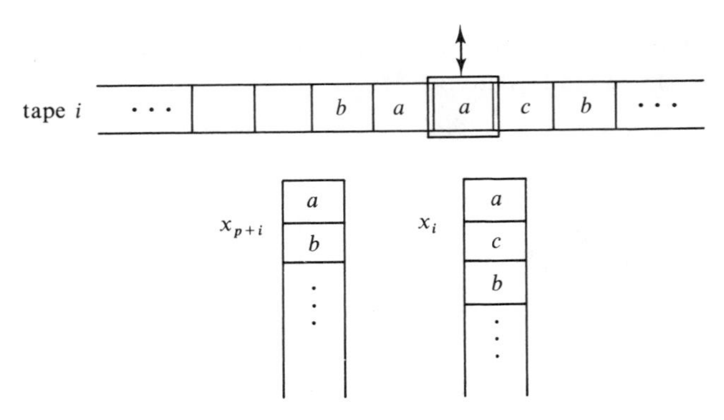

Figure 10.9 Representing the contents of a Turing machine tape in SAM(**V***) machine registers.

If so, registers x_1 and x_{p+1} are updated first, then x_2 and x_{p+2} are updated, and so on.

Checking whether an action item is applicable amounts to checking whether or not $u_r = top(x_r)$ for all r, $1 \leqslant r \leqslant p$. This is achieved, using the action items below. (We use one extra register, r, to temporarily store the top symbol of the register currently under consideration.)

(s: $r := top(x_1)$:$s.1$)

($s.1$: **if** $r \neq \varepsilon$ **then** $s.1.2$ **else** $s.1.1$)

($s.1.1$: $r := B$:$s.1.2$)
($s.1.2$: **if** $r \equiv u_1$ **then** $s.1.3$ **else** s_n)
($s.1.3$: $r := top(x_2)$:$s.2$)

($s.2$: **if** $r \neq \varepsilon$ **then** $s.2.2$ **else** $s.2.1$)

($s.2.1$: $\cdots$
($s.2.2$: $\cdots$
($s.2.3$: $\cdots$
⋮
($s.p$: **if** $r \neq \varepsilon$ **then** $s.p.2$ **else** $s.p.1$)

($s.p.1$: $r := B$:$s.p.2$)
($s.p.2$: *if* $r \equiv u_p$ **then** $t.1$ **else** s_n)

In the above, s_n stands for the current state of the next action item of the Turing machine to be tried.

Having established that the action item (s, u, v, d, t) is indeed applicable, the register values must be updated. Updating registers x_1 and x_{p+1} is achieved by the following action items.

($t.1$: $x_1 := pop(x_1)$:$t.1.1$)

The remaining action items intended to update x_1 and x_{p+1} depend on the movement of the read/write head.

(1) $d_1 = S$

$(t.1.1\colon x_1 := push(v_1, x_1) :t.2)$

(2) $d_1 = R$

$(t.1.1\colon x_{p+1} := push(v_1, x_{p+1}) :t.2)$

(3) $d_1 = L$

$(t.1.1\colon x_1 := push(v_1, x_1)\colon t.1.2)$

$(t.1.2\colon r := top(x_{p+1}) :t.1.3)$
$(t.1.3\colon \textbf{if } r \neq \varepsilon \textbf{ then } t.1.5 \textbf{ else } t.1.4)$
$(t.1.4\colon r := B :t.1.5)$

$(t.1.5\colon x_1 := push(r, x_1) :t.1.6)$
$(t.1.6\colon x_{p+1} := pop(x_{p+1}) :t.2)$

A similar sequence of action items is used to update the contents of the other registers $x_2, \ldots, x_p$ and $x_{p+2}, \ldots, x_{p+p}$.

The action items introduced to update x_p and x_{p+p} also make the SAM($\mathbf{V}^*$) machine enter the next state t corresponding to the action item (s, u, v, d, t) of the Turing machine. This is achieved by putting t as next state, where we have $t.2$ in the set of action items above.

Thus a single Turing machine action item requires quite a lot of SAM($\mathbf{V}^*$) action items. Nevertheless, this number of action items used to simulate a single Turing machine action item is bounded; it does not depend on the contents of the tapes. Therefore, the time complexity of the SAM($\mathbf{V}^*$) machine is linear in the time complexity of the Turing machine.

In summary: if a relation is computable by an ND-SAM($\mathbf{V}^*$) machine in time $t(n)$ and space $s(n)$, then this relation can also be computed by an NDTM($\mathbf{W}^*$) in time $O(t^2(n))$ and space $O(s(n))$, where $W = V \cup \{B\}$, and $B \notin V$ the blank symbol. Conversely, if a relation is computable by an NDTM($\mathbf{V}^*$) in time $t(n)$ and space $s(n)$, then this relation can also be computed by an ND-SAM($\mathbf{V}^*$) machine in time $O(t(n))$ and space $O(s(n))$.

10.5 Other string-oriented computation systems

In this section, we will describe two models of computation, which use strings over some given finite alphabet as fundamental objects, namely *storage modification machines* and *rewriting systems*.

Storage modification machines use a memory structure which is quite different from the memory structure of the machines discussed so far. At any moment, the memory consists of a finite number of cells, and each cell has a fixed, machine-dependent number of pointers to cells of the memory. Furthermore, there is an unlimited supply of cells, which can be added to the memory during the computation. It is the structure imposed by the pointers between the cells which is used to store information; there are no explicit data elements stored in a cell.

Rewriting systems are representations of non-deterministic algorithms. There are no variables, no assignments, and no control statements. A rewriting system specifies a finite set of rules $\alpha \to \beta$, stating that the string α may be replaced by string β. The system operates on a single object, a string over a given finite alphabet, by repeatedly applying some rule. An application of a rule consists of selecting an occurrence of the left-hand side of the rule, and then replacing this occurrence by the right-hand side of the rule. It is not easy to write algorithms as rewriting systems, due to the lack of control facilities. Rewriting systems are of central importance to the theory of *formal languages*; the formal grammars used to define programming languages, for instance, are rewriting systems.

As these systems are not used in what follows, this section can be skipped.

10.5.1 Storage modification machines

The memory of a Turing machine consists of one or more tapes, divided into squares. The structure of the memory is invariant, the values to be stored are represented by printing symbols on the squares of the tapes.

Storage modification machines, developed by A. Schönhage in the 1970s, do it the other way round. The memory cells contain a fixed, finite, machine-dependent number of pointers, indexed by a finite, machine-dependent alphabet Δ, to other memory cells; the cells contain no other information. The memory is assumed to be finite, but indefinitely extendible during the computation. Extension occurs by requesting a new cell and incorporating this cell into the already existing memory structure.

Thus, the memory can be identified with an edge-labelled directed graph as depicted in Figure 10.10. There is a single node which is directly accessible, the remaining nodes being accessible only via the pointers. The directly accessible node is called the *origin*.

Formally, a Δ-store is a triple $(X, 0, p)$, where X is the finite set of nodes, $0 \in X$ is the origin, and p is a family p_a, $a \in \Delta$, of functions $p_a : X \to X$. The notation $p_a(x) = y$ means that the pointer with label a, originating from node x goes to node y.

With every Δ-store $(X, 0, p)$ is associated a map $p^* : \Delta^* \to X$, defined as follows

$$p^*(\varepsilon) = 0, \text{ and}$$
$$p^*(wa) = p_a(p^*(w)) \text{ for all } a \in \Delta \text{ and } w \in \Delta^*.$$

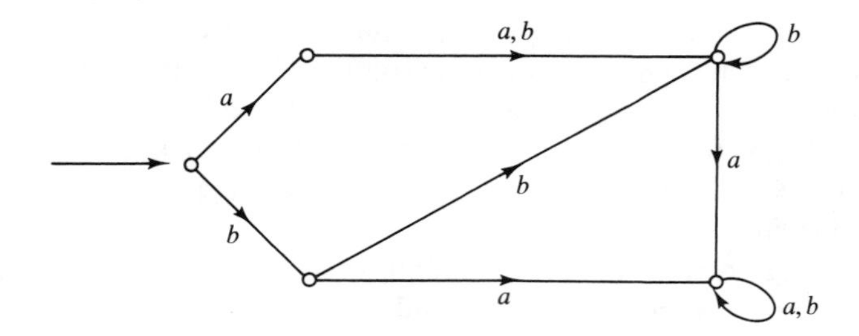

Figure 10.10 Contents of a Δ-store of a storage modification machine.

Thus $p^*(w)$ is the node reached by starting at the origin 0, and following the path along the edges labelled as specified by the word $w \in \Delta^*$. We call w a *name* of node $p^*(w)$.

A storage modification machine consists of a Δ-store and a finite control, which is represented as a program. A program is a sequence of labels and instructions, where a label is an identifier followed by a colon, and an instruction is as described below.

input $L_{a_1}, L_{a_2}, \ldots, L_{a_n}$;

The format depends on the input alphabet, which is assumed to be $\{a_1, a_2, \ldots, a_n\}$ in the above. The instruction reads an input symbol from an input string, and then transfers control to the appropriate label L_{a_i}. If the input string is exhausted, the input instruction has no effect, and the next instruction from the program is executed.

output s;

The symbol s belonging to the output alphabet is appended to the output string.

goto L;

Control is transferred to label L.

halt;

Computation ceases, implicitly this also happens when control passes the end of the program.

new w;

This instruction requests a new memory cell. The new cell is embedded in the existing Δ-store as sketched in Figure 10.11. After embedding, w is a name for the added node.

set w **to** v;

This instruction changes a pointer such that w becomes a name for the node with name v. The pointer change is depicted in Figure 10.12.

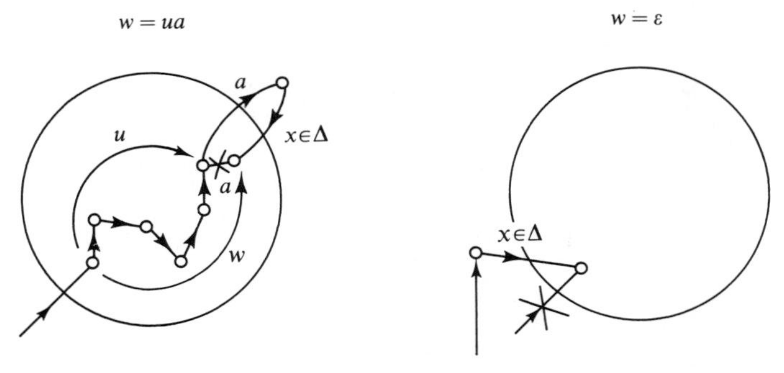

Figure 10.11 The instruction **new** w.

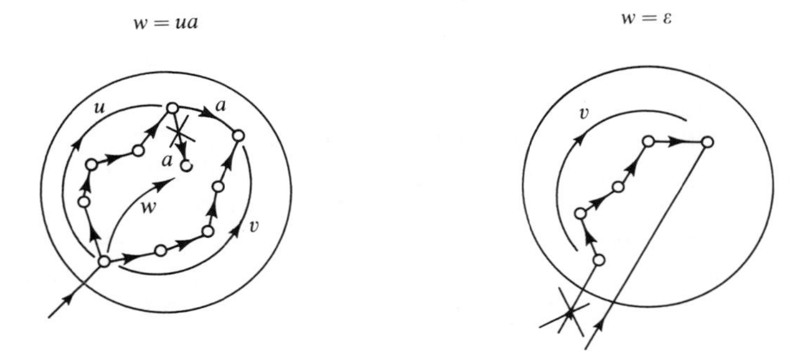

Figure 10.12 The instruction **set** w **to** v.

if $u = v$ **then** ⟨*instruction*⟩;

Here ⟨*instruction*⟩ refers to one of the unconditional instructions discussed above. The ⟨*instruction*⟩ is executed if u and v both name the same node in the current Δ-store. Otherwise control is transferred immediately to the next instruction of the program.

if $u \neq v$ **then** ⟨*instruction*⟩

Similar to the preceding command, but in this case ⟨*instruction*⟩ is executed if u and v do not name the same node in the Δ-store.

The computation induced by a program is a sequence of steps, where each step corresponds to an instruction of the program. The computation starts with the first instruction of the program and an initial Δ-store $(\{0\}, 0, p)$, where $p_a(0) = 0$, for all $a \in \Delta$.

EXAMPLE 10.4

Let $\Delta = \{a, b\}$. Consider the following program.

new *aa*;
set *bb* to *a*;
halt;

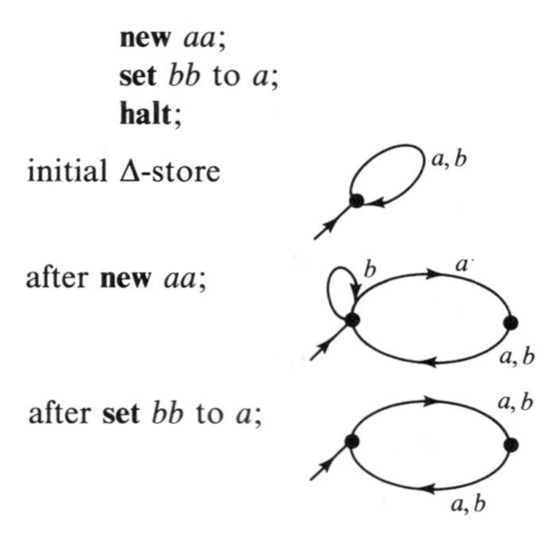

Storage modification machines have the same computational power as Turing machines, unlimited register machines, SAL(**N**) programs, etc. Below, we show how to simulate a SAM(**N**) program by a storage modification machine.

We have a finite number of registers containing natural numbers. These registers and their values can be represented in a Δ-store, with $\Delta = \{s, p, v\}$ as shown below for six registers r_0 to r_5.

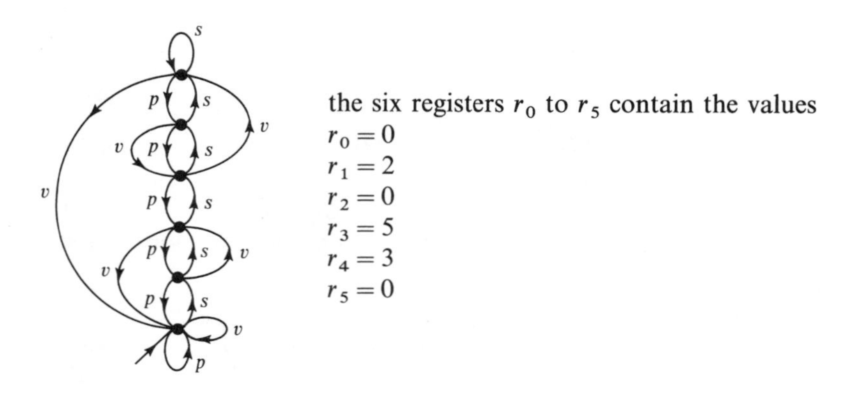

Nodes in the Δ-store serve two purposes.

(1) A node represents a natural number; the origin represents 0. The *s* pointers give the successor of a number, and the *p* pointers give the predecessor of a number. Because the Δ-store is finite, there is a largest represented number, which is its own successor in the current Δ-store. If a larger number is needed during the computation, a new node must be added to the Δ-store, as in the example below.

(2) A node of the Δ-store also represents a register. The v pointer gives the current value of this register. The index of this register is of course the natural number which is also represented by this node.

Using this representation, we can simulate register assignments.

The following storage modification machine program simulates the register assignment $r_3 := r_2 + 1$.

```
⋮
if ssv = ssvs then goto create;
continue: set sssv to ssvs;
⋮
create: new ssvs;
        set ssvsv to ε;
        set ssvss to ssvs;
        goto continue;
```

We assume that registers two and three have been properly initialized. Thus we do not have to check that *ss* and *sss* name different nodes. In the if statement, it is determined whether or not *ssv* and *ssvs* name the same node; in other words, whether the number represented by *ssv* is the largest currently represented number. If it happens to be the largest currently represented number, a new node is obtained and embedded in the Δ-store as a representation of the successor of number *ssv*, and also as a register with current value 0.

Subtractions are even simpler. If a node representing the number n exists, then a node representing the predecessor of n also exists.

Thus $r_3 := r_2 \dot{-} 1$ is implemented by the statement

```
set sssv to ssvp;
```

Storage modification machines are computationally equivalent to the other models of computation, SAL(**N**), SAL(**V***), Turing machines, and so forth. Storage modification machines are fast. A. Schönhage has shown how to compute the product of two n-bit numbers in linear time. The fastest-known Turing machine implementation of multiplication requires $O(n \log(n) \log(\log(n)))$.

10.5.2 Rewriting systems

Let V be some finite, non-empty alphabet, and V^* the set of all strings over this alphabet. A *rewriting system* is a pair (V, R), where R is a finite subset of $V^* \times V^*$. Instead of $(\alpha, \beta) \in R$, we write $\alpha \to \beta$. These objects $\alpha \to \beta$ are called rewrite rules, or production rules. The intention is to apply a rule $\alpha \to \beta$ to a string $\gamma \in V^*$, by replacing some occurrence of α in γ by β, resulting in a string δ. We write $\gamma \Rightarrow \delta$ to mean that δ can be obtained from γ by applying some rewrite rule in R.

Formally,

$\gamma \Rightarrow \delta$ if and only if there is a rewrite rule $\alpha \to \beta$ in R, and strings $\eta_1, \eta_2 \in V^*$, such that $\gamma = \eta_1 \alpha \eta_2$ and $\delta = \eta_1 \beta \eta_2$.

A rewriting system determines, or computes a relation $\Rrightarrow$ from V^* to V^*, defined as follows:

$\gamma \Rrightarrow \delta$ if and only if there are strings $\gamma_0, \gamma_1, \ldots, \gamma_n$, such that

- $\gamma_0 = \gamma$ and $\gamma_n = \delta$
- $\gamma_i \Rightarrow \gamma_{i+1}$, for all $i, 0 \leqslant i < n$
- $\delta = \gamma_n$, and δ cannot be rewritten any further, that is, for no γ_{n+1} do we have $\gamma_n \Rightarrow \gamma_{n+1}$.

The relation $\Rrightarrow$ is in general not a function, because the rewrite relation $\Rightarrow$ is not a function since there is a choice as to what rule to apply, and as to which occurrence of the left-hand string of this rule to rewrite. Thus a rewriting system specifies a *non-deterministic* system.

Normal Markov algorithms constitute a class of deterministic rewriting systems. A normal Markov algorithm is a rewriting system (V, R), where the set R of production rules is ordered. The production rules form a list, there is a first rule, a second rule, etc. Furthermore R contains at least one production rule, which is designated as *final*. The rewriting process terminates immediately when a final production rule has been applied. Final production rules are denoted by $\alpha \twoheadrightarrow \beta$. The rewriting process is made deterministic, and consequently the rewrite relation $\Rightarrow$ becomes a function, by adding two requirements, namely that the first applicable rule must be used, and that this rule must be used to rewrite the leftmost occurrence of its left-hand side. The yield relation $\Rrightarrow$ is defined in terms of $\Rightarrow$, exactly as for rewriting systems in general.

The following are examples of normal Markov algorithms:

(1) Let $V = \{1\}$. Strings over V are thus unary represented natural numbers, that is the number n is represented by a string of n consecutive 1s, denoted by 1^n. Let $R = \{1 \twoheadrightarrow 11, \varepsilon \twoheadrightarrow 1\}$. Then $1^n \Rrightarrow 1^m$ if and only if $m = n + 1$. That is, this normal Markov algorithm computes the successor function $+_1$.

(2) The normal Markov algorithm (V, R) with $V = \{1\}$ and $R = \{111 \to 11, 11 \twoheadrightarrow 1\}$, computes the function $sg = \lambda w[\text{if } w = \varepsilon \text{ then } \varepsilon \text{ else } 1]$.

It can be shown that both normal Markov algorithms and general rewriting systems over an alphabet V are computationally equivalent to SAL(**N**), relative to a suitably chosen coding.

10.6 Exercises

10.1 Prove that every recursive function is computable on a URM.

10.2 Prove that every recursive function is computable on a RAM.

10.3 Prove that every computation on a URM can be simulated on a RAM and vice versa. (This proves that URM and RAM have the same computing power.)

10.4 Prove that every computation on a URM can be simulated by a SAL(**N**) program. (Combined with the result obtained in Exercise 10.1 and the fact that

$$REC = \{f \mid f \text{ is SAL}(\mathbf{N}) \text{ computable}\}$$

this proves that the programming language SAL(**N**) and the URM have the same computing power.)

10.5 Prove that every computation on a single-tape Turing machine can be simulated by a URM.

10.6 Prove that every computation on a single-tape Turing machine can be simulated by a URM with only two registers.

10.7 Construct a k-tape Turing machine that accepts all palindromes over a given input alphabet. (A palindrome is a string such as 'madamimadam', which reads backwards the same as forwards.)

10.8 Informally describe non-deterministic Turing machines or a non-deterministic SAL(**V***) programs that will accept the following languages.

(a) The set of all strings $10^{i_1}10^{i_2}\cdots 10^{i_k}$ such that $i_r = i_s$ for some r and s such that $1 \leqslant r \leqslant s \leqslant k$.

(b) The set of all strings $xcy \in \{a, b, c\}$ such that x is a subword of y and $x, y \in \{a, b\}^*$.

10.9 Let the language $L \subseteq \Sigma^*$ be accepted by a single-tape Turing machine with state set Q and tape alphabet Γ which uses no more than $S(n)$ cells on any input of length n. Let $card(Q) = s$ and $card(\Gamma) = r$, where $card(X)$ denotes the number of elements of the set X. Prove that every $w \in L$ of length n can be accepted within at most $s.S(n).r^{S(n)}$ steps.

10.10 Prove that every computation of a non-deterministic Turing machine can be simulated by a deterministic Turing machine and vice versa.

10.7 References

The algorithmic machines as described in this chapter are generalizations of various computation models in the literature. The idea of defining effective computability by a mathematical model of a machine was introduced by Turing (1936) and Post (1936). Turing (1936) proved the equivalence of his model to the computation model of λ-definable functions. The importance of Turing machines in the development of computability theory and complexity theory is illustrated by the fact that almost all papers about sequential complexity theory use the Turing machine as a computation model.

Variants of Turing machines, like the programmable Turing machine, were introduced by Wang (1957). Other machine models like the URM were introduced by Shepherdson and Sturgis (1963). Markov (1962) introduced a formalism based on rewriting as a model of computation.

Eilenberg (1974) introduced a general framework within which other important kinds of machines such as linear-bounded automata, push-down automata, finite automata, etc. can be studied.

Chapter 11
A Hierarchy of Complexity Classes

In the study of complexity, we restrict ourselves to predicates, or equivalently, membership problems. Furthermore, we *always assume a string data type*, that is, we assume an underlying data type having the set of all strings over some finite alphabet as type of interest (TOI). We can use the data type $\mathbf{V}^*$ (Definition 9.7) and ND-SAL($\mathbf{V}^*$) (Definition 9.8), or, equivalently, use Turing machines (Definition 10.7), as the underlying model of computation.

As is customary, results will be phrased in terms of Turing machines. However, most of the time only the intended behaviour of a machine will be described. Thus the presentation is not very sensitive with respect to the details of the underlying model.

11.1 Complexity classes

Below we discuss the way a complexity is assigned to a subset $L \subseteq \Sigma^*$, also called a *language* over the *alphabet* Σ. This is a particular case of the way a complexity is assigned to a predicate in general as discussed in Sections 9.1 and 9.2.

Let $L \subseteq \Sigma^*$. This set represents the predicate $\lambda x[x \in L]$. Thus, the problem is to determine for an arbitrary $w \in \Sigma^*$ whether or not w belongs to L. Therefore, the word w determines an *instance* of the problem. The *size* of the instance w is the *length* of the string w. The complexity assigned to the language L is the complexity assigned to the optimal Turing machine accepting L. (See Section 10.3 for details about what is to be understood by the length of, and the space used by, a computation of a Turing machine.)

Definition 11.1

Let $TM = (p, V, \Sigma, B, Q, q_0, F, M)$ be a non-deterministic Turing machine accepting L in the sense of Definition 10.7.

(1) The *time complexity* of TM is a function $N \to N$ denoted by T_{TM} and determined as follows:

(a) $T_{TM}(n) = \uparrow$ if there is no word of length n which can be accepted by the Turing machine TM,

(b) $T_{TM}(n) =$ the least m such that all words of length n which can be accepted by TM can be accepted by computations of length at most m.

If for some function $h: N \to N$ we have

$$(\forall n)[\text{if } T_{TM}(n)\downarrow \text{ then } h(n)\downarrow \text{ and } T_{TM}(n) \leqslant h(n)]$$

then TM is said to be *h time bounded.*

(2) The *space complexity* of TM is a function $N \to N$ denoted by S_{TM} and determined as follows:

(a) $S_{TM}(n) = \uparrow$ if there is no word of length n which can be accepted by the Turing machine TM,

(b) $S_{TM}(n) =$ the least m such that all words of length n which can be accepted by TM can be accepted via computations using *workspace* at most m.

Note that we only count *workspace*; we assume that tape 1, containing the input, is a read-only tape, and we only count the workspace used, that is, the cells of the tapes 2 to p.

If for some function $h: N \to N$ we have

$$(\forall n)[\text{if } S_{TM}(n)\downarrow \text{ then } h(n) \text{ and } S_{TM}(n) \leqslant h(n)],$$

then TM is said to be *h-space-bounded.* ■

There are Turing machines such that their time complexity functions are not partial recursive functions. Similarly, there are Turing machines whose space complexity functions are not partial recursive functions. For further details, see Section 11.5.

We turn now to the complexity of *languages*.

Definition 11.2

(1) The *time complexity* of a language L is a function $N \to N$ denoted by T_L and determined as follows:

$T_L = T_A$ if A is a Turing machine accepting L and for all Turing machines B accepting L there are constants c and m such that

$$(\forall n \geqslant m)[\text{if } T_A(n)\downarrow \text{ then } T_A(n) \leqslant c \cdot T_B(n)]$$

T_L is undefined if there is no such Turing machine A.

(2) The *space complexity* of the language L is a function $N \to N$ denoted by S_L and determined as follows:

$S_L = S_A$ if A is a Turing machine accepting L and for all Turing machines B accepting L there are constants c and m such that

$$(\forall n \geqslant m)[\text{if } S_A(n)\downarrow \text{ then } S_A(n) \leqslant c \cdot S_B(n)]$$

S_L is undefined if there is no such Turing machine A. ■

Note that there might be languages whose complexities are undefined, because there is no machine of least complexity. In Section 11.8 it will be shown that there are indeed problems which do not have an 'optimal' algorithm. In particular, a language will be exhibited in such a way that any Turing machine accepting this language can be speeded up by an arbitrary given factor.

Our intention is to partition the class of all recursively decidable problems, that is, all recursively decidable subsets of Σ^*, into *complexity classes*, and study the resulting structure.

Definition 11.3

Let $f: N \to N$ be any total function.

(1) $DTIME(f) \triangleq \{L \subseteq \Sigma^* \mid \Sigma$ is a finite alphabet, and there is a deterministic Turing machine $TM = (p, V, \Sigma, B, Q, q_0, F, M)$ accepting L such that $(\forall n)[\text{if } T_{TM}(n)\downarrow \text{ then } T_{TM}(n) \leqslant f(n)]\}$.

(2) $NDTIME(f) \triangleq \{L \subseteq \Sigma^* \mid \Sigma$ is a finite alphabet, and there is a non-deterministic Turing machine $TM = (p, V, \Sigma, B, Q, q_0, F, M)$ accepting L such that $(\forall n)[\text{if } T_{TM}(n)\downarrow \text{ then } T_{TM}(n) \leqslant f(n)]\}$.

(3) $DSPACE(f) \triangleq \{L \subseteq \Sigma^* \mid \Sigma$ is a finite alphabet, and there is a deterministic Turing machine $TM =$

$(p, V, \Sigma, B, Q, q_0, F, M)$ accepting L such that $(\forall n)[\text{if } S_{TM}(n)\downarrow \text{ then } S_{TM}(n) \leqslant f(n)]\}$.

(4) $NDSPACE(f) \triangleq \{L \subseteq \Sigma^* \mid \Sigma$ is a finite alphabet, and there is a non-deterministic Turing machine $TM = (p, V, \Sigma, B, Q, q_0, F, M)$ accepting L such that $(\forall n)[\text{if } S_{TM}(n)\downarrow$ then $S_{TM}(n) \leqslant f(n)]\}$. ■

We are mainly interested in predicates occupying the lower complexity classes, for example, those in the class $DTIME(\lambda x[x^3])$. Such problems, which are not only recursively decidable, but have a solution which is also feasible in practice, are the subject of the next three chapters. In the current chapter we will consider properties of the complexity classes in general.

11.2 On the size of the tape alphabet

As we have seen in Section 9.2, the complexity assigned to a computation crucially depends on the power of the operations provided by the underlying data type. In string data types, such as $\mathbf{V}^*$, the size of the alphabet V is a measure of the power of these operations. The larger the size of the alphabet, the more powerful the operations. This observation is made precise in the theorems below, which state that a linear decrease in complexity can be obtained by increasing the size of the alphabet.

Firstly, we will show that a linear compression of the required workspace can be obtained by increasing the size of the alphabet.

Theorem 11.1

If $L \subseteq \Sigma^*$ is accepted by an S-space-bounded deterministic Turing machine, then for every $c > 0$ there is a deterministic Turing machine which accepts L and is $c \cdot S \triangleq \lambda n[\max\{1, c \cdot S(n)\}]$ space-bounded. Thus $DSPACE(S) = DSPACE(c \cdot S)$ for any $c > 0$.

Proof

Let $TM_1 = (p, V_1, \Sigma, B_1, Q_1, q_1, F_1, M_1)$ be an S-space-bounded deterministic Turing machine accepting L, and let $c > 0$. If $c > 1$, there is nothing to prove; assume therefore that $c < 1$. Let r be any natural number such that $|S(n)/r| \leqslant c \cdot S(n)$.

We construct a Turing machine $TM_2 = (p, V_2, \Sigma, B_2, Q_2, q_2, F_2, M_2)$ which is to simulate the machine TM_1. Machine TM_2 has tape alphabet $V_2 \triangleq V_1 \cup V_1^r$ and state set $Q_2 \triangleq Q_1 \times \{1, 2, \ldots, r\}^{p-1}$. Thus TM_2 can represent in a single tape cell the contents of r tape cells of TM_1. Also, it can keep track of the positions of the read/write heads of TM_1 in a block of r consecutive tape cells, using the numeric components of its states.

The machine specification M_2 is constructed such that TM_2 simulates the computation steps of TM_1. For example, let $p=3$, $r=4$, and suppose that M_1 contains the following action item:

$$(s,(\sigma,a,c),(\sigma,b,d),(S,R,L),t)$$

Then the machine specification M_2 contains an action item

$$((s,i,j),(\sigma,x_1,y_1),(\sigma,x_2,y_2),(S,D_R,D_L),(t,k,m))$$

where

$$x_1=(u_1,u_2,u_3,u_4) \text{ and } u_i=a,$$
$$x_2=(\bar{u}_1,\bar{u}_2,\bar{u}_3,\bar{u}_4) \text{ and } \bar{u}_i=b, \text{ and } \bar{u}_r=u_r \text{ for } r\neq i$$

$$y_1=(v_1,v_2,v_3,v_4) \text{ and } v_j=c,$$
$$y_2=(\bar{v}_1,\bar{v}_2,\bar{v}_3,\bar{v}_4) \text{ and } \bar{v}_j=d, \text{ and } \bar{v}_r=v_r \text{ for } r\neq j$$

$$k=\text{if } i+1\leqslant 4 \text{ then } i+1 \text{ else } 1$$
$$m=\text{if } j-1\geqslant 1 \text{ then } j-1 \text{ else } 4$$

$$D_R=\text{if } i+1\leqslant 4 \text{ then } S \text{ else } R$$
$$D_L=\text{if } j-1\geqslant 1 \text{ then } S \text{ else } L,$$

for every i,j such that $1\leqslant i,j\leqslant 4$,
and every $u_r,\bar{u}_r,v_r,\bar{v}_r\in\Sigma$, $1\leqslant r\leqslant 4$.

In this way the action items comprising the machine specification M_2 are derived from those of the machine specification M_1 of TM_1. Consequently, $DSPACE(S)\subseteq DSPACE(c\cdot S)$ for any $c>0$. Then $DSPACE(c\cdot S)\subseteq DSPACE(r\cdot c\cdot S)$ for any $r>0$. Choosing $r=1/c$ gives $DSPACE(c\cdot S)\subseteq DSPACE(S)$, whence $DSPACE(S)=DSPACE(c\cdot S)$. ■

The assumption that TM_1 is deterministic is of no consequence in the above proof. Therefore we have the following corollary.

Corollary 11.1
$NDSPACE(S)=NDSPACE(c\cdot S)$ for every $c>0$. ■

The next theorem states that it is also possible to obtain any required linear speed-up by increasing the size of the tape alphabet. The argument is similar to that in the above proof, but more laborious. Also, the time complexity of the machine to be sped up must not be too small.

Theorem 11.2

Let T be some function from N to N such that for all k there is an m such that

$$(\forall n)[\text{if } n \geqslant m \text{ then } T(n) \geqslant kn].$$

In other words, T increases more than linearly. If $L \subseteq \Sigma^*$ is accepted by a T-time-bounded deterministic Turing machine, then for every $c > 0$ there is a deterministic Turing machine which accepts L and is

$$c \cdot T \triangleq \lambda n[\max\{n+1, cT(n)\}]$$

time-bounded. Thus, $DTIME(T) = DTIME(c \cdot T)$ for every $c > 0$, and every such function T.

Proof

If $c > 1$, there is nothing to prove. Assume therefore that $c < 1$.

Let $TM_1 = (p, V_1, \Sigma, B_1, Q_1, q_1, F_1, M_1)$ be a T-time-bounded deterministic Turing machine accepting L. We will construct a deterministic Turing machine $TM_2 = (p+1, V_2, \Sigma, B_2, Q_2, q_2, F_2, M_2)$ which simulates the computations of TM_1. The idea is to compress the tapes as in the proof of the theorem above, and then dispose of moves which leave the read/write heads stationary, or has them oscillating between adjacent squares. TM_2 has $p+1$ tapes, the contents of tape i of TM_1 are represented in compressed form on tape $i+1$ of TM_2.

Choose $V_2 = V_1 \cup V_1^r$, thus combining r cells of TM_1 in a single cell of TM_2. The value of r must be chosen sufficiently large in order to obtain the required speed-up. It will be determined later.

The simulation of TM_1 by TM_2 proceeds as follows.

(1) First a single sweep is made along tape 1 to copy the input from tape 1 in compressed format to tape 2. Then the read/write head is positioned on the leftmost non-blank cell of tape 2. All of this requires $n + \lceil n/r \rceil$ steps.

(2) To simulate the steps of TM_1, TM_2 moves the read/write heads of tapes 2 to $p+1$ one place to the left, then two to the right, and finally one to the left, thus returning to the original position, as sketched in Figure 11.1.

The contents of the visited cells can be stored as a

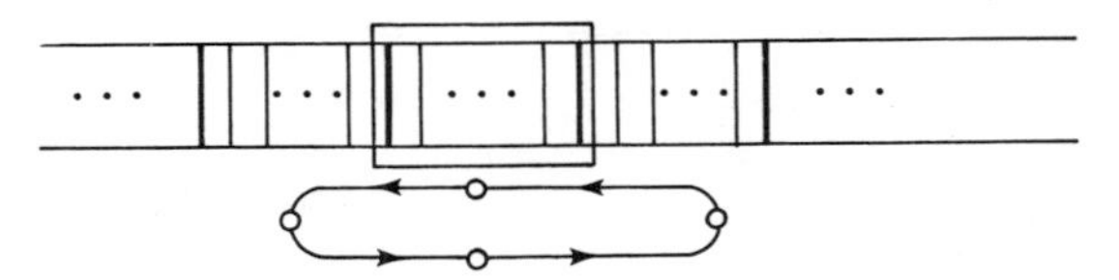

Figure 11.1 Movement of the head of the Turing machine TM_2.

component of the states in the same way the head positions were stored in the preceding proof. That is, let

$$Q_2 = Q_1 \times V_1^r \times V_1^r \times V_1^r \times \{\text{other components}\}.$$

Then the component $V_1^r \times V_1^r \times V_1^r$ can store the contents of $3r$ consecutive cells of TM_1 as represented in 3 cells of TM_2.

Having determined the contents of $3r$ consecutive tape cells of TM_1, TM_2 determines the contents of these $3r$ cells at the time when at least one of the read/write heads of TM_1 leaves this region of $3r$ cells. This takes no time at all; the required information is supplied in the machine specification M_2. There are three possibilities:

(a) TM_1 does not leave this region of $3r$ cells and accepts its input. Then TM_2 enters a dedicated accepting state.

(b) TM_1 does not leave this region and does not accept its input. Then TM_2 enters a dedicated, non-accepting state.

(c) TM_1 does at some moment of time leave this region. The $3r$ cells have some particular contents at that moment of time. TM_2 now re-visits the neighbouring cells, updating their contents as necessary. The updating process is sketched in Figure 11.2, assuming that TM_1 leaves the tape region in the direction R.

This also takes 4 steps. Since TM_1 needs at least r steps to leave the region of $3r$ cells, it takes TM_2 only 8 steps to simulate at least r steps of TM_1.

Thus a computation of TM_1 of $T(n)$ steps requires no more than

$$n + \left\lceil \frac{n}{r} \right\rceil + 8\left\lceil \frac{T(n)}{r} \right\rceil \leqslant n + \frac{n}{r} + 8\frac{T(n)}{r} + 2$$

steps of TM_2. It remains to show that this number can be bounded by $c \cdot T(n)$.

By our assumption on T, for any k there is an m such that for all $n \geqslant m$ we have $T(n) \geqslant kn$, and therefore $n \leqslant T(n)/k$. We can choose $m \geqslant 2$, so that $2 \leqslant m \leqslant n \leqslant T(n)/k$.

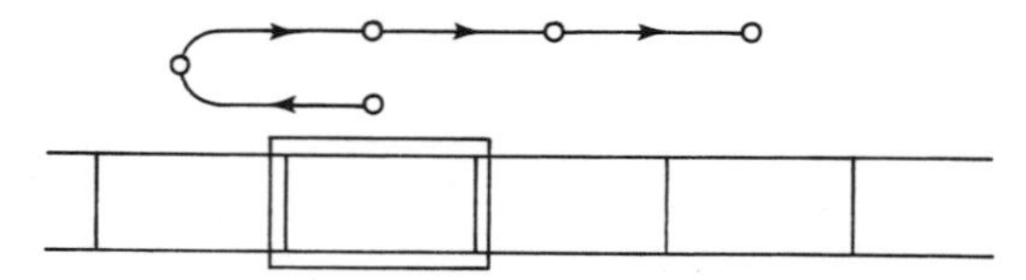

Figure 11.2 Movement of the head of the Turing machine TM_2.

For all $n \geqslant m$ we can bound the number of TM_2-steps as follows:

$$\begin{aligned}\text{number of } TM_2\text{-steps} &\leqslant n + n/r + 8T(n)/r + 2 \\ &\leqslant T(n)/k + T(n)/rk + 8T(n)/r + T(n)/k \\ &= T(n)(2/k + 8/r + 1/rk)\end{aligned}$$

It remains to choose r and k such that $(2/k + 8/r + 1/rk) \leqslant c$. Taking $r \geqslant 16/c$ makes $8/r \leqslant c/2$. Taking $k \geqslant 6/c$ makes $2/k + 1/rk \leqslant 3/k \leqslant c/2$ and we are done.

There are only finitely many words of length less than m. These words are tested during the initial scan of tape 1 and require no extra time. Those words which are to be accepted are stored as a component of the states of TM_2. Thus TM_2 is a $c \cdot T$-time-bounded deterministic Turing machine which accepts L, and consequently, $DTIME(T) = DTIME(c \cdot T)$. ∎

The assumption that TM_1 is a deterministic Turing machine is of little consequence. If TM_1 is non-deterministic, the construction as sketched above results in a non-deterministic Turing machine which accepts L and is $c \cdot T$ time bounded. Thus we have the following corollary.

Corollary 11.2

$NDTIME(T) = NDTIME(c \cdot T)$ for every $c > 0$ and every function $T{:}N \to N$ such that $(\forall k)(\exists m)(\forall n)[\text{if } n \geqslant m \text{ then } T(n) \geqslant kn]$. ∎

The preceding two theorems show that a linear decrease in time and space complexity can be obtained by increasing the size of the tape alphabet. *Decreasing* the size of the alphabet used on the work tapes results in a linear increase of complexity.

Theorem 11.3

If $L \subseteq \Sigma^*$ is accepted by an S-space-bounded, T-time-bounded deterministic Turing machine, then there is a deterministic Turing machine accepting L which uses a two-letter alphabet on its work tapes and is $c \cdot S$-space-bounded, and $c \cdot T$-time-bounded.

Proof

Let $TM_1 = (p, V, \Sigma, B, Q_1, q_1, F_1, M_1)$ be a deterministic S-space-

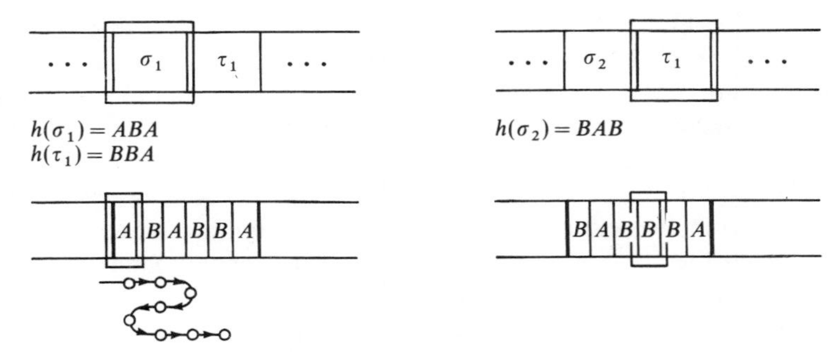

Figure 11.3 Movement of the head of the Turing machine TM_2.

bounded, T-time-bounded Turing machine accepting L. We construct a deterministic Turing machine TM_2 which accepts L and uses only the two-letter alphabet $\{A, B\} \subseteq V$ on its work tapes, the tapes 2 to p.

Let $r = \lceil \log_2 |V| \rceil$ and let $h: V \rightarrow \{A, B\}^r$ be a coding of the symbols of V as strings over the alphabet $\{A, B\}$. TM_2 simulates TM_1 using this coding to represent the symbols of V (see Figure 11.3). To simulate a single step of TM_1, TM_2 proceeds as follows.

(1) Move the read/write heads $r-1$ steps to the right, storing the sequence of r symbols in a component of the state.

(2) Move the read/write heads back to their original position, meanwhile updating the block of r cells in accordance with the step of TM_1 being simulated.

(3) Move the read/write heads to the leftmost cell of an adjacent block of r cells as determined by the move of TM_1 being simulated.

All of this requires no more than $3r$ steps. Thus TM_2 accepts L and is $c \cdot S$-space-bounded and $c \cdot T$-time-bounded for $c = 3r$. ■

11.3 On the number of tapes of a Turing machine

Theorem 11.4

If a language $L \subseteq \Sigma^*$ is accepted by an S-space-bounded (non-)deterministic Turing machine having p tapes, then there is a single-tape, S-space-bounded, (non-)deterministic Turing machine which accepts L.

Proof

Let $TM_1 = (p, V_1, \Sigma, B_1, Q_1, q_1, F_1, M_1)$ be a (non-)deterministic Turing machine accepting L. We construct a (non-)deterministic Turing machine $TM_2 = (1, V_2, \Sigma, B_2, Q_2, q_2, F_2, M_2)$ to accept L.

Let $V_2 = \Sigma \cup (V_1 \times \{\text{mark}, \text{space}\})^p$. That is, we look at the single tape of TM_2 as being divided into p tracks. Each track represents a tape of TM_1. The track stores the contents of the tape and also the position of the read/write head, by marking the cell on which the read/write head is located.

TM_2 simulates a step of TM_1 by making a sweep along the tape from the leftmost marked cell to the rightmost one, meanwhile storing the contents of the marked cells as a component of its state. Then TM_2 moves back to the leftmost marked cell, meanwhile updating the tracks. Updating includes changing the symbol in the marked cell of a track and also updating the markings themselves in accordance with movement of the read/write heads of TM_1, as determined by the step of TM_1 being simulated.

TM_2 accepts its input if and only if TM_1 accepts this input. If TM_1 is S-space-bounded, then TM_2 obviously is S-space-bounded as well. ∎

Corollary 11.3

If $L \subseteq \Sigma^*$ is accepted by a T-time-bounded (non-)deterministic Turing machine with p tapes, then L can also be accepted by a single-tape (non-)deterministic Turing machine, which is time bounded by $\lambda n[(T(n))^2]$.

Proof

Consider the machines TM_1 and TM_2 from the proof above. In any accepting computation of TM_2 of an input of length n, the leftmost and the rightmost marked cells are no farther than $T(n)$ cells apart. Thus simulating a single step of TM_1 requires no more than $cT(n)$ steps of TM_2. Since there are $T(n)$ steps of TM_1 to be simulated, TM_2 requires no more than $O(T^2(n))$ steps. ∎

It can be shown that the above result is optimal, reducing the number of tapes indeed requires squaring of the time consumed (see Vitany, 1984).

11.4 Relationships between space and time

The details of the relation between time and space complexity are as yet unknown. In most instances, however, space is a more powerful resource than time.

Theorem 11.5

$$DTIME(f) \subseteq NDTIME(f) \subseteq DSPACE(f) \subseteq NDSPACE(f)$$

Proof

The inclusions $DTIME(f) \subseteq NDTIME(f)$ and $DSPACE(f) \subseteq NDSPACE(f)$ are immediate, since a deterministic Turing machine is a non-deterministic Turing machine as well. It remains to show that $NDTIME(f) \subseteq DSPACE(f)$.

Let $L \in NDTIME(f)$ and let TM_1 be an f-time-bounded non-deterministic Turing machine accepting L. TM_1 can access no more than $f(n)+1$ cells of any tape in a computation of length $f(n)$. Thus TM_1 is $\lambda n[f(n)+1]$ space-bounded.

At any time during a computation, the number of alternative continuations of this computation is bounded by a constant m, which depends on TM_1, but not on the computation itself. Thus a k-step computation of TM_1 can be specified by a string of length k over some finite alphabet Γ, specifying for each moment of time which alternative to take.

We construct a deterministic Turing machine TM_2 accepting L. TM_2 tries to find an accepting computation of length k of TM_1 on input $w \in \Sigma^*$ by systematically enumerating all strings of length k over the alphabet Γ and given such a string $\gamma \in \Gamma^*$ of length k, performing the computation of TM_1 induced by γ. As soon as an accepting computation is found, TM_2 stops and accepts the input w. The above search procedure is repeated for successive values of k until an accepting computation is found.

On input $w \in L$ of length n this procedure needs

(1) space to represent the strings γ, which is bounded by $f(n)$,

(2) space to perform the computation specified by γ, which is bounded by $f(n)+1$.

By the linear space compression Theorem 11.1, we can redesign TM_2 to operate within space f. (The possibility of enumerating the computations of a non-deterministic algorithm in this way has already been used in the proof of Lemma 9.1 on page 216.) ■

Theorem 11.6

$$NDSPACE(f) \subseteq \bigcup_{c>0} DTIME(\lambda n[c^{f(n)}])$$

Proof

Let TM_1 be an f-space-bounded non-deterministic Turing machine accepting $L \subseteq \Sigma^*$. Let $c_0 \vdash \cdots \vdash c_t$ be an accepting computation on input $w \in L$ of length n. We may assume that no instantaneous description (ID) occurs more than once in the above computation. Also, if the ID $c=((\alpha_1,\ldots,\alpha_p),q,(\beta_1,\ldots,\beta_p))$ occurs in this computation, then the space used by c (see Section 10.3 for definitions), is at most

$f(n)$. Thus the length of $\alpha_i\beta_i$ is bounded by $f(n)$ for all i, $2 \leqslant i \leqslant p$. The string $\alpha_1\beta_1 = w$ in all the *ID*s occurring in the computation, because tape 1 is assumed to be a read-only tape. The number of different such space-bounded *ID*s is $n \cdot |Q| \cdot |V|^{2pf(n)}$, which is bounded by $d^{f(n)}$ for some d.

We construct a deterministic Turing machine TM_2 to accept L. This machine TM_2 checks whether or not w belongs to L in the following way:

(1) It constructs a list of all IDs of TM_1 which use space at most k. The initial value of k is 1. An *ID* can be represented in $O(k)$ tape cells, and there are no more than d^k such *ID*s.

(2) It marks the initial *ID active*, and all other *ID*s *untried*. In the search process described below, three marks are used, namely *used*, *untried* and *active*.

   (a) As yet, it is not known whether or not an *untried ID* is reachable from the initial *ID*.
   (b) A *used ID* is reachable from the initial *ID*, and its successor *ID*s have already been determined.
   (c) An *active ID* is reachable from the initial *ID*, but its successor *ID*s have not yet been determined.

(3) TM_2 considers each *active ID* in turn. For each *active ID* c

   (a) TM_2 checks to see whether or not c is an accepting *ID*. If so, TM_2 terminates and accepts w,
   (b) otherwise, TM_2 changes the mark of c from *active* to *used*, and marks all *untried* successor *ID*s of c as *active*, and then repeats the process starting at step (a).
   If there are no more *active* IDs, TM_2 sets $k = k + 1$, and restarts the whole procedure from step (1) onwards, for this new value of k.

The above procedure takes $O(k \cdot d^k)$ for steps (1) and (2). Because there are no more than d^k *active ID*s, and no more than d^k successor *ID*s, and because at each step (*b*) of the search procedure described under (3) at least one *ID* is marked *used*, step (3) takes $O(d^{3k})$ steps.

Thus the whole process, consisting of steps (1), (2) and (3), requires $O(d_1^k)$ for some $d_1 > 0$. Because $k \leqslant f(n)$ for any $w \in L$ of length n, we conclude that TM_2 is $\lambda n[c^{f(n)}]$ time-bounded. ■

In the proofs of the theorems above, we have used a search procedure for ever-increasing values of a parameter k. It is not possible to compute $f(n)$, and then only use the value $k = f(n)$, because we know nothing about how much time or space is needed to compute $f(n)$ in the first place. Therefore we must avoid computing $f(n)$. Using the construction above, we do not even have to assume that the complexity bounding function f is computable.

By the preceding two theorems we have the following corollary.

Corollary 11.4

$$DTIME(f) \subseteq NDTIME(f) \subseteq DSPACE(f)$$
$$\subseteq NDSPACE(f) \subseteq \bigcup_{c>0} DTIME(\lambda n[c^{f(n)}]) \quad ■$$

We will see in Section 11.6 that there is a language which belongs to $DTIME(\lambda n[2^{f(n)}])$, but does not belong to $DTIME(f)$. Therefore, at least one of the above inclusions must be strict. It is as yet unknown *which* of the inclusions of the above corollary are strict.

As regards the relation between deterministic and non-deterministic space complexity, the best result is Savitch's theorem discussed below. It requires space-bounding functions which are sufficiently well-behaved in the sense defined below.

Definition 11.4

A function $S: N \to N$ is called

(1) *space constructible*, if there is a deterministic Turing machine TM which is S-space-bounded, and for each n there is some input of length n on which TM actually uses exactly $S(n)$ tape cells on one of its tapes.

(2) *fully space constructible*, if the Turing machine uses exactly $S(n)$ cells on each input of length n. ■

Theorem 11.7

If $S: N \to N$ is a fully space constructible function, and $(\forall n)[S(n) \geq \log_2(n)]$, then

$$NDSPACE(S) \subseteq DSPACE(\lambda n[(S(n))^2]).$$

Proof

Let S be a function as specified above, and let $L \subseteq \Sigma^*$ be accepted by an S-space-bounded non-deterministic Turing machine TM_1. There is a $c > 0$ such that, if TM_1 accepts an input $w \in L$ of length n, it can do so via an accepting computation of length at most $c^{S(n)}$, because there are no more than $c^{S(n)}$ different instantaneous descriptions (IDs) using space at most $S(n)$.

We describe an algorithm to see whether or not $w \in L$, and then discuss how to implement this algorithm as a deterministic Turing machine.

Let the predicate R be defined as follows:

$$R \triangleq \lambda c_1 c_2 i[\text{there is a computation } c_1 \vdash \cdots \vdash c_2 \text{ of length at most } 2^i]$$

The following recursive procedure *TEST* computes the predicate *R*.

```
procedure TEST(ID_1, ID_2, i)
  if i = 0 then
    if (ID_1 = ID_2 ∨ ID_1 ⊢ ID_2) then
      return (true)
    else return (false)
    fi
  else for_each ID using space at most S(n) do
           if (TEST(ID_1, ID, i − 1) ∧ TEST(ID, ID_2, i − 1))
             then return (true)
           fi
       od;
       return (false)
  fi
end_of_procedure
```

The algorithm computing the predicate $\lambda w[w \in L]$ is as follows:

```
input: w
method:
  n := length(w);
  m := log_2(c^{S(n)});
  ID_0 := the initial ID of TM_1 with input w;
  for_each accepting ID_f of TM_1 using space at most S(n) do
    if R(ID_0, ID_f, m) then
      terminate and accept w
    fi
  od;
  reject w
```

The predicate R is evaluated using the recursive procedure *TEST*. Because the value of the third parameter of *TEST* decreases by 1 with each recursive call of *TEST*, there are never more than m pending calls.

A Turing machine can implement the above algorithm by storing activation records of each call of *TEST* on a work tape. An activation record is used to store all the information relevant for a particular call of the procedure. In this case, this information consists of the values of the three input parameters and the value of the local variable *ID*.

To store an *ID* on a work tape, we need $O(S(n))$ cells to store the contents of the work tapes i, $2 \leqslant i \leqslant p$, and $O(n)$ cells to store the contents of the input tape 1. The latter, however does not change. We only need to store the position of the read/write head, for which we need $O(\log_2(n))$ cells. By assumption, $\log_2(n) = O(S(n))$. Therefore an *ID*, and hence an activation record, can be stored using $O(S(n))$ cells.

Because there are never more than $m = O(S(n))$ pending calls of *TEST*, the algorithm as a whole requires $O((S(n))^2)$ space. By the linear space compression Theorem 11.1, we can redesign TM_2 such that it is $\lambda n[(S(n))^2]$ space-bounded. ■

A similar relation between deterministic and non-deterministic time is not known. The best result known is stated in Corollary 11.4,

$$NDTIME(f) \subseteq \bigcup_{c>0} DTIME(\lambda n[c^{f(n)}]).$$

The power of non-determinism relative to time as a measure of complexity is the most prominent open problem of complexity theory. The remaining chapters are devoted to a discussion of this problem.

It will be argued that it is reasonable to consider problems which can be solved in time polynomial in the size of the input, as problems whose solution is feasible. This class of problems is denoted by **P**. A great many problems of practical importance are not known to belong to **P**, although it is easy to show that they belong to **NP**, the class of problems which can be solved by a *non-deterministic* algorithm in time bounded by a polynomial in the size of the input.

The question of whether or not **P** is equal to **NP** is a particular case of the general question about the power of non-determinism, and is of great practical importance.

11.5 Gödel numbering of Turing machines

In the following we need the technique of diagonalization in order to prove properties of the complexity hierarchy. Hence we define a Gödel numbering of Turing machines, a corresponding universal Turing machine, and corresponding termination predicates.

A Turing machine is an 8-tuple $TM = (p, V, \Sigma, B, Q, q, F, M)$, where the machine specification M is a subset of $Q \times V^p \times V^p \times V^p \times Q$. For the purpose of Gödelization, we will fix

- the number p of tapes,
- and the input alphabet Σ.

To represent the action items of M, we have to represent the states and the symbols of the tape alphabet as strings over a finite alphabet. It suffices to assume that the states and tape symbols are indexed by natural numbers, and represented by their indices written in decimal notation. We will also assume that the indexing is such that the initial state q_0 is always indexed by 0, and similarly, that the blank symbol is indexed by 0.

Thus we can represent a Turing machine by writing down the action items comprising the machine specification M in some arbitrary order, followed by the set of finite states also written down in some arbitrary order. The result is a string over some finite alphabet W, $V \subseteq W$.

EXAMPLE 11.1

Let $p = 2$, $\Sigma = \{a, b\}$. Consider the Turing machine

$$TM = (p, V, \Sigma, B, Q, q, F, M),$$

where

- $V = \{a, b, B\}$,
- $Q = \{p, q, r\}$,
- $F = \{p, r\}$ and
- $M = \{(q, (a, B), (B, B), (R, S), p), (p, (b, B), (B, B), (S, S), r)\}$.

Let us assume the indexing in Table 11.1 Then the above Turing machine is represented by

$$(0, (1, 0), (0, 0), (R, S), 1), (1, (2, 0), (0, 0), (S, S), 2)\{1, 2\}$$

This is a string over an alphabet W consisting of the digits 0 to 9, the direction indicators R, L and S, left and right parentheses, left and right braces and comma.

The representation is not unique. We can use a different

Table 11.1 Assigning indices to the states and tape symbols of a Turing machine.

State	Index	Tape symbol	Index
q	0	B	0
p	1	a	1
r	2	b	2

indexing of Q and a different order of writing M and F. But the machine is uniquely determined by the representation.

Finally, we represent the symbols of W by groups of digits, thus obtaining a number written in decimal.

EXAMPLE 11.2

We continue Example 11.1. Choose the representation of the symbols of W shown in Table 11.2. Then the machine of the example above is represented by

$$20004020014000214020 \cdots 10214002213001400231$$

According to the above, not every string of decimal digits represents a Turing machine. A string which does not satisfy the formation rules, such as 202031, which stands for (({}, or 12345, which denotes no string over W, do not represent Turing machines. Therefore we add the proviso that any string of decimal digits which is ill-formed according to the description above represents the Turing machine

$$(p, \Sigma \cup \{B\}, \Sigma, B, \{q, r, s\}, q, \{s\}, \{(q, B^p, B^p, S^p, r)\}),$$

an arbitrary but fixed Turing machine.

Thus we have established a Gödel numbering of p-tape non-deterministic Turing machines with input alphabet Σ. This will be our standard Gödel

Table 11.2 Representing symbols as sequences of digits.

Symbol	Representation	Symbol	Representation
0	00	S	10
1	01	R	11
2	02	L	12
3	03	(	20
4	04	)	21
5	05	{	30
6	06	}	31
7	07	,	40
8	08		
9	09		

numbering scheme. TM_x will denote the Turing machine having index x with respect to this standard Gödel numbering.

This Gödel numbering satisfies the padding Lemma 7.1. That is, given an index m_0 we can effectively obtain ever larger indices m such that TM_{m_0} and TM_m are equivalent in that $L(TM_{m_0}) = L(TM_m)$.

These indices can be obtained by adding action items of the form (q, B^p, B^p, S^p, q). For each action item added, we choose a state q which is new, i.e. does not already occur in the machine specification. Because the states occurring in these added action items cannot be reached from the initial state, the accepting computations are in no way influenced. They take exactly the same time and space.

We also need a Gödel numbering of deterministic Turing machines. This can be obtained in the same way. But this time we add the extra proviso that a string of decimal digits denoting a Turing machine which is not deterministic according to our standard numbering stands for the Turing machine

$$(p, \Sigma \cup \{B\}, \Sigma, B, \{q, r, s\}, q, \{s\}, \{(q, B^p, B^p, S^p, r)\}).$$

We turn now to the universal Turing machine.

Let us assume that $\Sigma \cap \{0, 1, \ldots, 9\} = \varnothing$. The input to the universal Turing machine U is a string xw, with x a natural number written in decimal, and $w \in \Sigma^*$. The universal Turing machine U must accept w if and only if TM_x accepts w. On the work tapes, U uses codes for the symbols of the tape alphabet, as described in the proof of Theorem 11.3. The code used is implicitly given in the index x of the machine to be simulated. The universal machine U has $p+2$ tapes and proceeds as follows:

(1) Compute from x the machine specification M_x and store this on tape $p+1$. Tape $p+2$ is used to store the current state of TM_x, and is therefore initialized to represent zero.

(2) Simulate TM_x using the remaining input w as input for TM_x. To execute a step of TM_x, U searches on tape $p+1$ an action item (q_1, u, v, D, q_2) such that q_1 is the current state as recorded on tape $p+2$, and $u = (u_1, \ldots, u_p)$ corresponds to the currently read symbols.

(3) Having found such an action item, U updates the contents of tapes 1 to p, shifts the read/write heads as required, and updates tape $p+2$ to represent the new state q_2.

Let m be the length of x. Computing the machine specification from x takes $O(m)$ steps. Searching for an adequate action item takes $O(m^2)$ steps. Updating the work tapes also requires $O(m)$ steps. Thus the universal machine U is at most a factor m^2 slower than TM_x itself.

It requires but a simple adaptation of this universal machine U to make it into a machine computing a resource-bounded termination predicate, bounded with either time or space as a measure of complexity. (This is analogous to the way we adapted our universal SAL program to compute the Kleene termination predicate in Section 6.2.)

Definition 11.5

(1) Let TTT denote the Termination predicate for Turing machines with respect to Time as a measure of complexity:

$$TTT \triangleq \lambda mxt[\text{there is an accepting computation of } TM_m \text{ on input } x \text{ of at most } t \text{ steps}].$$

(2) Let TTS denote the Termination predicate for Turing machines with respect to Space as a measure of complexity:

$$TTS \triangleq \lambda mxt[\text{there is an accepting computation of } TM_m \text{ on input } x, \text{ which uses space at most } t].$$

∎

Let $C_m^t \triangleq \lambda x[\mu t[TTT(m,x,t)]]$. Note that C_m^t is *not the same* as the time complexity associated with the Turing machine TM_m as defined in Definition 11.1 (page 270). According to this definition we use worst-case complexity. That is, T_{TM_m} satisfies

$$T_{TM_m}(n) = \begin{cases} \uparrow & \text{if } (\forall x)[\text{if } length(x) = n \text{ then } C_m^t(x)\uparrow] \\ \max\{C_m^t(x) \mid C_m^t(x)\downarrow \text{ and } length(x) = n\} & \text{otherwise} \end{cases}$$

This property looks problematic; it is not obvious how to decide that the maximum has been found. Indeed, worst-case time complexity, as defined in Definition 11.1, is not always a partial recursive function, unless the input alphabet Σ is a singleton.

Assume that $|\Sigma| \geqslant 2$, $a, b \in \Sigma$, and $a \neq b$. Let $L \subseteq \Sigma^*$ be recursively enumerable but not recursively decidable. Define

$$L_1 \triangleq \{aw \mid w \in L\} \cup \{b\omega \mid \omega \in \Sigma^*\} \cup \{\varepsilon\}$$

(ε denotes the empty word.) Clearly L_1 is recursively enumerable. Furthermore, L_1 is not recursively decidable, because $L \leqslant_m L_1$ via $\lambda w[aw]$.

Let TM be a Turing machine which accepts L_1. Because L_1 is recursively enumerable, such a Turing machine exists. The worst-case time complexity T_{TM} of this Turing machine is a total function. Assume now that T_{TM} is also a recursive function. Then the following algorithm accepts L.

input: w
method:

$n := length(aw)$;
$m := T_{TM}(n)$;

if (*there is an accepting computation of* TM *on input* aw
of at most m *steps*) **then** *accept*(w)
else *reject*(w)
fi

It follows, contrary to our assumption on L, that L is recursively decidable. Therefore T_{TM} is not a partial recursive function.

A similar argument shows that the worst-case space complexity need not be a partial recursive function.

In conclusion, there is no recursive enumeration of the set of all worst-case complexity functions, neither for time nor for space. This makes the hierarchy results obtained via diagonalization non-constructive. This undesirable property can be avoided by taking some measure of complexity which allows a recursive indexing. Examples of such measures of complexity are

(1) the measures obtained from the termination predicates

(a) $C^t_m \triangleq \lambda x[\mu t[TTT(m, x, t)]]$.

(b) $C^s_m \triangleq \lambda x[\mu t[TTS(m, x, t)]]$.

These measures are called the *natural* measures of time and space complexity.

(2) the best-case complexity, defined by

$$T^b_m(n) = \begin{cases} \uparrow & \text{if } (\forall x)[\text{if } length(x) = n \text{ then } C^t_m(x)\uparrow] \\ \min\{C^t_m(x) \mid C^t_m(x)\downarrow \text{ and } length(x) = n\} & \text{otherwise} \end{cases}$$

11.6 Density of the time and space hierarchies

Let us first establish that there exist arbitrarily complex predicates and that in consequence the time and space hierarchies are infinite.

Theorem 11.8

For any total recursive function $t: N \to N$, there exists a set which cannot be accepted by any t-time-bounded Turing machine.

Proof

Let TM_i be a recursive enumeration of Turing machines with a single-letter input alphabet $\Sigma = \{a\}$. Then the worst-case, best-case, and natural time complexity measures coincide.

Define L as follows:

$$L \triangleq \{a^n \mid TTT(n, a^n, t(n)) = \text{FALSE}\}$$

Clearly L is a recursively decidable subset of Σ^*. Assume that L is accepted by a t-time-bounded Turing machine, and let m be the index for this machine.

Then

$a^m \in L$ if and only if

$TTT(m, a^m, t(m)) = \text{TRUE}$ (by definition of TTT) if and only if

$a_m \notin L$ (by definition of L).

The result for space is obtained similarly, using the space termination predicate TTS. ■

The above theorem implies the existence of an infinite hierarchy of complexity classes of ever-growing complexity. The theorem gives no information about the density of this hierarchy. The hierarchy has some rather strange properties, to be discussed in Section 11.8.

For now, we consider well-behaved bounding functions, in the sense defined below. For such functions the hierarchy can be shown to be dense.

Definition 11.6

(We repeat part of Definition 11.4).

(1) A function $S: N \to N$ is called

 (a) *space constructible* if there is a deterministic Turing machine TM which is S-space-bounded, and for each n there is some input of length n on which TM actually uses exactly $S(n)$ tape cells on one of its tapes;

 (b) *fully space constructible* if the Turing machine uses exactly $S(n)$ cells on each input of length n.

(2) A function $T: N \to N$ is called

 (a) *time constructible* if there is a deterministic Turing machine TM which is T-time-bounded, and for each n there is some input of length n on which TM actually makes exactly $T(n)$ computation steps;

 (b) *fully time constructible* if the Turing machine takes exactly $T(n)$ computation steps on each input of length n. ■

To obtain the hierarchy result we diagonalize over the S-space-bounded Turing machines. To do this it must be possible to compute the value $S(|x|)$ using no more than $S(|x|)$ space, where $|x|$ denotes the length of the string x. It must also be possible to allocate an amount $S(|x|)$ of space to the machine being simulated. If S is fully space constructible, there is a Turing machine which can do this.

Theorem 11.9

For every S-space-bounded Turing machine P there exists an S-space-bounded Turing machine Q which accepts the same set and does not loop on any finite amount of tape.

Proof

There is some constant d such that the number of different instantaneous descriptions of P using no more than k cells on any tape is at most equal to d^k.

The Turing machine Q simulates P, simultaneously checking that more than k tape cells are used, or less than d^k steps are performed. To do this, Q uses an extra tape to count the number of steps of P, which it has simulated.

More precisely, Q operates as follows.

(1) Mark one cell of each tape as *usable*, and set the counter to zero. Thus, Q assumes that P will not use more than k, initially set to 1, cells on any tape.

(2) Simulate the steps of P and count the number of steps simulated. Whenever a cell is needed which is not yet marked *usable*, Q marks an extra cell where this is required, and also marks an extra cell as *usable* on its counter tape, thus effectively increasing the length of the counter. Q also resets the counter to 0.

(3) If P halts, then so does Q, accepting if and only if P accepts its input.

(4) If the counter itself produces an overflow, that is, requires a cell not yet marked *usable*, then this computation of P does not terminate, P loops on a finite amount of tape. In this case Q terminates rejecting its input.

Clearly, Q does not loop on any finite amount of tape, uses no more space than P, and accepts the same set as P does. ■

Theorem 11.10

Let S_1 be any function from N to N, and S_2 a fully space constructible function such that $(\forall k)(\exists m)(\forall n)[\text{if } n \geqslant m \text{ then } S_2(n) \geqslant kS_1(n)]$. Thus S_2 increases more than linearly in S_1. Then there is a language in

$DSPACE(S_2)$ which is not in $DSPACE(S_1)$. In other words, $DSPACE(S_1)$ is strictly included in $DSPACE(S_2)$.

Proof

By Theorem 11.4 (page 277), it suffices to diagonalize over the class of single-tape Turing machines with input alphabet $\{0, 1, \ldots, 9\}$.

We construct an S_2-space-bounded Turing machine M which disagrees with any S_1-space-bounded Turing machine on at least one input.

On input w of length n, M begins by marking $S_2(n)$ cells on one of its work tapes. Since S_2 is fully space constructible, there exists a Turing machine that uses exactly $S_2(n)$ cells on any input of length n. Thus marking the $S_2(n)$ cells can be done by simulating this Turing machine.

M now starts simulating the Turing machine TM_w, on input w. If in the course of this simulation, M attempts to scan an unmarked cell, then M halts and accepts w. Otherwise, M accepts w only if it can complete the simulation using no more than $S_2(n)$ tape cells, and finds that TM_w rejects w.

The language $L(M)$ accepted by M belongs to $DSPACE(S_2)$ but not to $DSPACE(S_1)$.

(1) $L(M) \in DSPACE(S_2)$ is immediate from the construction of M.

(2) Assume that $L(M) \in DSPACE(S_1)$. Then there is an S_1-space-bounded Turing machine M_1 which accepts $L(M)$.

By the preceding theorem, we may assume that M_1 does not loop on any finite amount of tape. Since our Gödel numbering of Turing machines satisfies the padding lemma, there exist arbitrarily large x such that TM_x is S_1-space-bounded, uses the same tape alphabet as M_1, and accepts the same set, i.e. $L(TM_x) = L(M)$.

Owing to the representation of the symbols of the tape alphabet as strings of decimal digits, the simulation of any TM_x by M_1 requires no more than $\lambda n[kS_1(n)]$ cells, for some $k \geqslant 1$. By assumption, there is an m such that $(\forall n)$[if $n \geqslant m$ then $S_2(n) \geqslant kS_1(n)$].

Let w be the least integer such that $w \geqslant m$. Let TM_w be S_1-space-bounded, use the same tape alphabet as M_1, and accept the same set, that is, $L(TM_w) = L(M)$.

If $w \in L(TM_w)$, then M has sufficient space to complete the simulation of TM_w on input w. It finds that TM_w accepts w, and therefore M rejects w.

If $w \notin L(TM_w)$, then M accepts w. Because TM_w does not loop on any finite amount of tape, the simulation of TM_w by M either requires more than $S_2(|w|)$ space, in which case M

accepts w, or can be completed. Since TM_w rejects w, M accepts w.

Thus we have a contradiction. Consequently, $L(M) \notin DSPACE(S_1)$. ■

Regarding the density of the deterministic time hierarchy, we have only a weaker result. The problem is that in order to diagonalize over all multi-tape Turing machines we have to simulate a machine with an arbitrary number of tapes on a fixed machine, which of course has a fixed number of tapes. Keeping track of the head positions on all these tapes slows the machine down – see Corollary 11.3 (page 278).

Theorem 11.11

If T_2 is a fully time constructible function and $(\forall k)(\exists m)(\forall n)[\text{if } n \geqslant m \text{ then } T_2(n) \geqslant k(T_1(n))^2]$, then there exists a language which can be accepted by a deterministic T_2-time-bounded Turing machine, but not by a deterministic T_1-time-bounded Turing machine.

Proof

We diagonalize over single-tape Turing machines. By Corollary 11.3 any language accepted by a T_1-time-bounded multi-tape Turing machine can also be accepted by a single-tape $\lambda n[(T_1(n))^2]$ time-bounded Turing machine.

We construct a multi-tape T_2-time-bounded Turing machine M. On input w, M simulates the computation of TM_w on input w.

In order to ensure that M is T_2-time-bounded, M executes, simultaneously with simulating TM_w, the steps of a Turing machine that uses exactly $T_2(n)$ steps on every input of length n. Such a machine exists because T_2 is fully time constructible. It has some fixed number of tapes, and we can assume that M has a sufficient number of tapes to accommodate them all.

If the simulation of TM_w has not been completed after $T_2(n)$ steps of M, M halts and accepts w. Otherwise, M accepts w if and only if TM_w rejects w.

Obviously, $L(M)$ is in $DTIME(T_2)$. Assume that $L(M)$ is in $DTIME(T_1)$. Then there is a T_1 time-bounded multi-tape Turing machine and hence a $\lambda n[(T_1(n))^2]$ time-bounded single-tape Turing machine M_1 accepting $L(M)$. Because the Gödel numbering of Turing machines satisfies the padding lemma, there exist arbitrarily large x such that TM_x is equivalent to M_1 and executes exactly the same number of steps as M_1.

Hence there will be an index w such that M can carry the simulation of TM_w to completion within $T_2(|w|)$ steps. In this case $w \in L(M)$ if and only if $w \notin L(M_1)$, contradicting our assumption on M_1. Hence $L(M) \in DTIME(T_2) - DTIME(T_1)$. ■

The above result could be sharpened if we were able to simulate a multi-tape Turing machine more efficiently on a k-tape Turing machine. If $k=1$, and we must simulate the multi-tape Turing machine on a single-tape Turing machine, this is not possible. If we may use more than one tape, we can do better.

The proof is rather involved, but it can be shown that any language which can be accepted by a p-tape t-time-bounded deterministic Turing machine, can also be accepted by a $\lambda n[t(n)\log(t(n))]$ time-bounded Turing machine with two tapes.

Diagonalizing over two-tape Turing machines thus gives the following theorem.

Theorem 11.12

If T_2 is a fully time constructible function such that

$$(\forall k)(\exists m)(\forall n)[\text{if } n \geqslant m \text{ then } T_2(n) \geqslant kT_1(n)\log(T_1(n))]$$

then there exists a language which can be accepted by a T_2-time-bounded deterministic Turing machine, but cannot be accepted by a T_1-time-bounded deterministic Turing machine using any number of tapes. ■

We turn now to the hierarchies for non-deterministic Turing machines. In a proof by diagonalization as in the case of Theorem 11.10, we specify a machine which accepts a word w if and only if TM_w rejects w. If this is to work, it must be possible to determine whether or not TM_w accepts w.

If $L \in NDTIME(f)$, there is an f-time-bounded non-deterministic Turing machine M which accepts L. For every $w \in L$ there is an accepting computation of length at most $f(|w|)$; but there is no method known which can establish in at most $f(|w|)$ non-deterministic steps that M *rejects* w. For non-deterministic space-bounded complexity classes, such a method does exist.

Consider some S-space-bounded non-deterministic Turing machine M. Let U denote the set of all IDs of M which use no more than $S(n)$ space for some n. The set U can be enumerated without repetitions if $S(n)$ is given. Let $ID_0, ID \in U$ and consider the problem to determine whether or not $ID_0 \vdash^* ID$.

Define a sequence $W_0, W_1, \ldots, W_i, \ldots$ of subsets of U as follows.

$$W_0 \triangleq \{ID_0\}$$

$$W_{i+1} \triangleq \{ID' | (\exists ID_i \in W_i)[ID_i \vdash ID']\} \cup W_i$$

The sets W_i cannot be represented in $S(n)$ space, but their cardinalities *can* be represented in $S(n)$ space. Because U can be enumerated without repetitions, it suffices to know $card(W_i)$ in order to non-deterministically test membership in W_i.

Consider the following non-deterministic algorithms.

```
Algorithm check_yes
  input: ID, i
  output: z{z = 1 if ID ∈ W_i}
  method:
    {assume that ID ∈ W_i and check this assumption}
    guess a computation ID_0 = c_0 ⊢ c_1 ⊢ ··· ⊢ c_t, where t ≤ i and each c_j ∈ U;
    if c_t ≠ ID then enter a non-terminating computation
    else z := 1
    fi
```

```
Algorithm check_no
  input: ID, i, card(W_i)
  output: z{z = 0 if ID ∉ W_i}
  method:
    {assume ID ∉ W_i and check this assumption}
    count := 0;
    for each c ∈ U do
      if c ≠ ID then
        either
          check_yes(c, i);
          count := count + 1
        or
          {assume c ∉ W_i and proceed}
        od
      fi
    od;
    if count ≠ card(W_i) then enter a non-terminating computation
    else z := 0
    fi
```

The correctness of the above algorithm *check_no* depends on the fact that the elements of U can be enumerated without repetitions, so that each reachable instantaneous description is counted exactly once.

The cardinality of W_{i+1} can be computed non-deterministically from $card(W_i)$ as follows.

```
Algorithm increment
  input: i, card(W_i)
  output: r{r = card(W_{i+1})}
  method:
    r := 0;
```

```
for each c∈U do
  either
    check_yes(c, i + 1);
    r := r + 1
  or
    count := 0;
    for each d∈U do
      either
        check_yes(d, i);
        if d⊢c then enter a non-terminating computation
        else count := count + 1
        fi
      or
        {proceed}
      od
    od;
    if count ≠ card(W_i) then enter a non-terminating computation
    fi
  od
od
```

Let REL_i be the relation computed by the above algorithm *increment*. REL_i satisfies

if $((i, c), r) \in REL_i$ and $c = card(W_i)$ **then** $r = card(W_{i+1})$.

Note that in the preceding three algorithms we never need to store more than two instantaneous descriptions and no more than two numbers each at most $card(U)$. Thus these algorithms can be implemented in $O(S(n))$ space.

Theorem 11.13

Let S be fully space constructible and such that $S(n) \geqslant \log(n)$ for all n. For every $L \in NDSPACE(S)$ there is an S-space-bounded non-deterministic Turing machine $(2, V, \Sigma, B, Q, q_0, F_1, M)$ which accepts L and such that there is a subset $F_0 \subseteq Q$ with $F_1 \cap F_0 \neq \varnothing$ and for every $w \in \Sigma^*$ we have

(1) $w \in L$ if and only if there is an $S(|w|)$ space-bounded terminating computation $((\varepsilon, \varepsilon), q_0, (w, \varepsilon)) \vdash^* ((\alpha_1, \alpha_2), q, (\beta_1, \beta_2))$ with $q \in F_1$, and

(2) $w \notin L$ if and only if there is an $S(|w|)$ space-bounded terminating computation $((\varepsilon, \varepsilon), q_0, (\omega, \varepsilon)) \vdash^* ((\alpha_1, \alpha_2), q, (\beta_1, \beta_2))$ with $q \in F_0$.

Proof

Let $L \in NDSPACE(S)$ and TM_1 an S-space-bounded non-deterministic Turing machine which accepts L. We construct a second Turing acceptor which satisfies the conditions of the theorem and uses the three preceding algorithms. TM_2 has two special states q_y and q_n, and we define $F_1 \triangleq \{q_y\}$ and $F_0 \triangleq \{q_n\}$.

Let $ID_0 \triangleq ((\varepsilon, \varepsilon), q_0, (w, \varepsilon))$ and let the sets U and W_i be defined as above, that is, U is the set of all instantaneous descriptions using no more than $S(|w|)$ workspace, and the sets W_i contain precisely those instantaneous descriptions reachable from ID_0 in at most i steps. For some $j \leqslant card(U)$ we have $W_j = W_{j+k}$ for all $k \geqslant 0$; let W denote this set W_j. Then $ID_0 \vdash^* ID$ if and only if $ID \in W$.

Because S is fully space constructible, the elements of U can be generated without repetitions in some fixed order, say alphabetically. The position of the read head on the input tape can be represented by a number of size $\log(n) \leqslant S(n)$, where $n = |w|$. Thus, the elements of U can each be represented using $O(S(n))$ cells.

TM_2 operates as follows.

(1) It computes $card(W)$ iteratively as follows.

```
i:= 0; card:= 1;
repeat
   card:= increment(i, card);
   i:= i + 1
until card remains constant
```

Now $card = card(W)$.

(2) It checks whether or not an accepting instantaneous description is reachable using a variant of the algorithm *check_no*.

```
count:= 0;
for each ((α1, α2), q, (β1, β2))∈U do
   either
      check_yes(((α1, α2), q, (β1, β2)), i);
      if q∈F then goto state q_y else count:= count + 1 fi
   or
      {proceed}
   od
od;
if count = card then goto state q_n
else enter a non-terminating computation
fi
```

Because no more than two elements of U and two numbers less than or equal to $card(U)$ are needed simultaneously, the Turing machine TM_2 needs only $O(S(n))$ cells; by the linear tape compression

Theorem 11.1 (page 272) we can modify TM_2 such that it is S-space-bounded. ■

Corollary 11.5

$NDSPACE(S)$ is closed with respect to complement if S is a fully space constructible function such that $S(n) \geqslant \log(n)$ for all n. ■

Theorem 11.13 can be used to show that the non-deterministic space hierarchy is dense.

Theorem 11.14

Let S_1 and S_2 be fully space constructible functions such that $(\forall n)[S_1(n) \geqslant \log(n)]$ and $(\forall k)(\exists m)(\forall n)[\text{if } n \geqslant m \text{ then } S_2(n) \geqslant kS_1(n)]$. Then there is a language in $NDSPACE(S_2)$ which is not in $NDSPACE(S_1)$. In other words, $NDSPACE(S_1)$ is strictly included in $NDSPACE(S_2)$.

Proof

By Theorem 11.4, it suffices to diagonalize over the class of single-tape Turing machines with input alphabet $\{0, 1, \ldots, 9\}$. We may assume that all these Turing machines satisfy Theorem 11.13 because the construction described there is uniform and can be executed to transform any given Turing acceptor.

We construct an S_2-space-bounded Turing machine M which disagrees with any S_1-space-bounded Turing machine on at least one input.

On input w of length n, M begins by marking $S_2(n)$ cells on one of its work tapes and then starts simulating an arbitrarily chosen computation of Turing machine TM_w, on input w.

If in the course of the simulation, M attempts to scan an unmarked cell, then M enters a non-terminating computation which does not consume any new space.

Let q_y^w and q_n^w denote the special accepting and rejecting states of TM_w. M accepts w if and only if in the course of the simulation it encounters the state q_n^w.

The language $L(M)$ accepted by M belongs to $NDSPACE(S_2)$ but not to $NDSPACE(S_1)$.

(1) $L(M) \in NDSPACE(S_2)$ is immediate from the construction of M.

(2) Assume that $L(M) \in NDSPACE(S_1)$. Then there is an S_1-space-bounded Turing machine M_1 which accepts $L(M)$.

Since our Gödel numbering of Turing machines satisfies the padding lemma, there exist arbitrarily large x such that TM_x satisfies the requirements of Theorem 11.13, is S_1-space-bounded, uses the same tape alphabet as M_1, and accepts the same set, i.e. $L(TM_x) = L(M)$.

Owing to the representation of the symbols of the tape alphabet as strings of decimal digits, the simulation of any TM_x by M requires no more than $\lambda n[kS_1(n)]$ cells, for some $k \geqslant 1$. By assumption, there is an m such that $(\forall n)$[if $n \geqslant m$ then $S_2(n) \geqslant kS_1(n)$].

Let w be the least integer such that $w \geqslant m$, and TM_w satisfies the requirements of Theorem 11.13, is S_1-space-bounded uses the same tape alphabet as M_1, and accepts the same set, that is, $L(TM_w) = L(M)$.

If $w \in L(TM_w)$, then no computation of TM_w can enter state q_n^w, therefore M does not have an accepting computation; consequently M rejects w, $w \notin L(M)$.

If $w \notin L(TM_w)$, then there is an $S_1(|w|)$ space-bounded computation of TM_w which at some moment brings TM_w into the special rejecting state q_n^w. The simulation of this computation requires no more than $S_2(|w|)$ space and can therefore be completed by M. Consequently M has an accepting computation and therefore $w \in L(M)$.

Thus we have a contradiction. Consequently, $L(M) \notin NDSPACE(S_1)$.

■

11.7 Translational lemmas

Translational lemmas provide a tool to prove that $DTIME(f)$ is strictly included in $DTIME(g)$ for functions f and g which do not differ enough to make the hierarchy theorems of the preceding section applicable.

The general idea is as follows. Assume that f is slightly larger than g and that $DTIME(f) \subseteq DTIME(g)$. The translational lemma for time will then give functions f_1 and g_1 which differ more than f and g such that $DTIME(f_1) \subseteq DTIME(g_1)$. This process can be repeated until functions f_i and g_i are obtained, for which $DTIME(f_i) \subseteq DTIME(g_i)$ contradicts Theorem 11.11 or 11.12 (pages 292 and 293). Consequently, the assumption was false, and we conclude that $DTIME(g)$ is strictly included in $DTIME(f)$.

Lemma 11.1

If $NDSPACE(S_1) \subseteq NDSPACE(S_2)$ and S_1, S_2 and f are fully space constructible and furthermore $(\forall n)[S_2(n) \geqslant n$ and $f(n) \geqslant n]$, then $NDSPACE(S_1 \cdot f) \subseteq NDSPACE(S_2 \cdot f)$.

Proof

Let $L \in NDSPACE(S_1 \cdot f)$ and M_1 an $S_1 \cdot f$-space-bounded Turing machine which accepts L. Define L^p, the padded version of L, as follows:

$$L^p \triangleq \{wa^i | M_1 \text{ accepts } w \text{ in space } S_1(|w| + i)\}$$

where $|w|$ denotes the length of w and a is a new symbol, i.e. not in the tape alphabet of M_1.

If $w \in L$ then M_1 accepts w in space $S_1(f(|w|))$. Thus, $wa^i \in L^p$ if $|wa^i| = f(|w|)$. Consequently, $\{wa^i | w \in L \wedge |wa^i| = f(|w|)\} \subseteq L^p$.

Let $L^p \in NDSPACE(S_1)$. An S_1-space-bounded Turing machine accepting L^p might operate as follows:

- On input wa^i, it reserves a block of $S_1(|wa^i|)$ cells. This can be done because S_1 is fully space constructible.
- Now the simulation of M_1 on input w commences.
- wa^i is accepted if and only if M_1 accepts w within the allotted space of $S_1(|wa^i|)$ cells.

By assumption, $NDSPACE(S_1) \subseteq NDSPACE(S_2)$, therefore $L^p \in NDSPACE(S_2)$ as well. In consequence there is an S_2-space-bounded Turing machine M^p which accepts L^p.

Finally, we construct an $S_2 \cdot f$-space-bounded Turing machine M_2 which accepts L. On input w it operates as follows.

- First, M_2 marks off a block of $f(|w|)$ cells. This is possible because $S_2(f(|w|)) \geqslant f(|w|)$. In these cells the input w is copied, and then the remaining cells are filled with as. There are cells remaining because $f(|w|) \geqslant |w|$.
- M_2 now simulates M^p on input wa^i. M_2 accepts w if and only if M^p accepts wa^i without using more than $S_2(|wa^i|) = S_2(f(|w|))$ cells.

By assumption, M^p is S_2-space-bounded, therefore M_2 is $S_2 \cdot f$-space-bounded. Also, M_2 accepts w if and only if $wa^i \in L^p$ if and only if $w \in L$.

■

Note that the same argument applies both to deterministic and non-deterministic machines, whence under the same conditions on S_1, S_2 and f, $DSPACE(S_1) \subseteq DSPACE(S_2)$ implies $DSPACE(S_1 \cdot f) \subseteq DSPACE(S_2 \cdot f)$.

Lemma 11.2

If $NDTIME(T_1) \subseteq NDTIME(T_2)$ and T_1, T_2 and f are fully time constructible non-decreasing functions such that

$$(\forall n)[T_1(n) \geqslant n \wedge T_2(n) \geqslant n \wedge f(n) \geqslant n]$$

then $NDTIME(T_1 \cdot f) \subseteq NDTIME(T_2 \cdot f)$.

Proof

Let $L \in NDTIME(T_1 \cdot f)$ and let M_1 be a $T_1 \cdot f$ time-bounded non-deterministic multi-tape Turing machine accepting L. Define L^p as follows:

$$L^p \triangleq \{wa^i | M_1 \text{ accepts } w \text{ in time } T_1(|w| + i)\}.$$

As in the preceding lemma, we have

$$\{wa^i | w \in L \wedge f(|w|) = |w| + i\} \subseteq L^p.$$

Clearly, L^p can be accepted in $O(T_1)$ time. Thus $L^p \in NDTIME(T_1)$, which by assumption is a subset of $NDTIME(T_2)$. Let M^p be a non-deterministic $T_2(n)$-time-bounded Turing machine which accepts L^p.

We construct a non-deterministic $T_2 \cdot f$-time-bounded Turing machine M_2 which is to accept L and operates as follows:

- On input w M_2 appends i *a*s to w, i such that $f(|w|) = |wa^i|$. This can be done in $f(|w|)$ moves.
- Now M_2 executes a sequence of moves of M^p and checks to see whether according to this chosen sequence of moves M^p accepts wa^i. If this is the case then M_2 accepts w. Since M^p is T_2-time-bounded, this requires no more than $T_2(|wa^i|) = T_2(f(|w|))$ steps.

Consequently, M_2 accepts L in $O(T_2)$ time, therefore $L \in NDTIME(T_2 \cdot f)$. ■

Note that the above argument also applies to deterministic machines. Therefore, under the same conditions on T_1, T_2 and f, we have that $DTIME(T_1) \subseteq DTIME(T_2)$ implies $DTIME(T_1 \cdot f) \subseteq DTIME(T_2 \cdot f)$.

The preceding two lemmas can be used to show the existence of dense hierarchies. Before these results are stated, the technique will be illustrated in an example.

In the following example and theorems, we are concerned with classes such as $NDSPACE(\lambda n[n^k])$. To avoid a lot of λs, we denote these as $NDSPACE(n^k)$.

EXAMPLE 11.3

We want to prove that $NDSPACE(n^2) \neq NDSPACE(n^3)$.

This is an immediate consequence of the hierarchy Theorem 11.14 for non-deterministic space. The result can also, albeit more clumsily, be proved using the hierarchy Theorem 11.10

for deterministic space, the translational Lemma 11.1 for non-deterministic space and Savitch's Theorem 11.7. This latter proof runs as follows.

Trivially $NDSPACE(n^2) \subseteq NDSPACE(n^3)$, so it remains to show that $NDSPACE(n^3) \nsubseteq NDSPACE(n^2)$.

Assume that it is true that $NDSPACE(n^3) \subseteq NDSPACE(n^2)$. By Lemma 11.1, this inclusion implies the inclusions:

(1) $NDSPACE(n^6) \subseteq NDSPACE(n^4)$, by letting $f \triangleq \lambda n[n^2]$

(2) $NDSPACE(n^9) \subseteq NDSPACE(n^6)$, by letting $f \triangleq \lambda n[n^3]$

Thus, by Savitch's Theorem 11.7 we have:

$$\begin{aligned} DSPACE(n^9) \subseteq NDSPACE(n^9) &\subseteq NDSPACE(n^6) \\ &\subseteq NDSPACE(n^4) \subseteq DSPACE(n^8) \end{aligned}$$

Because $(\forall k)(\exists m)(\forall n)[\text{if } n \geqslant m \text{ then } n^9 \geqslant kn^8]$, the density theorem for deterministic space, Theorem 11.10, implies that $DSPACE(n^9) \nsubseteq DSPACE(n^8)$.

Consequently, our assumption that $NDSPACE(n^3) \subseteq NDSPACE(n^2)$ must be false, and $NDSPACE(n^2)$ must be strictly included in $NDSPACE(n^3)$.

In Example 11.3, we used the hierarchy Theorem 11.10 for deterministic space, the translational Lemma 11.2 for non-deterministic space and Savitch's Theorem 11.7 to show strict inclusion. As there is no time analog of Savitch's theorem, we need some separation theorem for non-deterministic time-bounded complexity classes in order to apply the translational lemma for non-deterministic time. This is provided by the following theorem of Cook.

Theorem 11.15

For all integer constants $k \geqslant 1$, $NDTIME(n^k)$ is strictly included in $NDTIME(n^{6k})$. ■

The proof of this theorem is too involved to be presented here, it can be found in Cook (1973).

Using the above theorem and the translational Lemma 11.2, we can prove a hierarchy theorem for non-deterministic time.

Theorem 11.16

For any positive integers s and t such that $s/t > 1$, $NDTIME(n^{s/t})$ is strictly included in $NDTIME(n^{(s+1)/t})$.

Proof

Trivially, $NDTIME(n^{s/t}) \subseteq NDTIME(n^{(s+1)/t})$.

It remains to show that the reverse inclusion does not hold.

Assume that $NDTIME(n^{(s+1)/t}) \subseteq NDTIME(n^{s/t})$. Then by the translational Lemma 11.2 for non-deterministic time we have

$$(0) \quad NDTIME(n^{(s+1)(s+0)}) \subseteq NDTIME(n^{s(s+0)}) \quad \text{by letting } f(n) = n^{(s+0)t}$$

$$(1) \quad NDTIME(n^{(s+1)(s+1)}) \subseteq NDTIME(n^{s(s+1)}) \quad \text{by letting } f(n) = n^{(s+1)t}$$

.

.

.

$$(i) \quad NDTIME(n^{(s+1)(s+i)} \subseteq NDTIME(n^{s(s+i)}) \quad \text{by letting } f(n) = n^{(s+i)t}$$

.

.

.

$$(6s-1) \quad NDTIME(n^{(s+1)(6s-1)}) \subseteq NDTIME(^{s(6s-1)}) \quad \text{by letting } f(n) = n^{(6s-1)t}$$

Furthermore, for $i \geqslant 1$, we have

$$NDTIME(n^{s(s+i)}) \subseteq NDTIME(n^{(s+1)(s+i-1)}),$$

because $s(s+i) \leqslant (s+1)(s+i-1)$ for all $i \geqslant 1$.
By the above inclusions we have:

$$\begin{aligned}
NDTIME(n^{6s^2}) &= NDTIME(n^{s(6s)}) \\
&\subseteq NDTIME(n^{(s+1)(6s-1)}) \\
&\subseteq NDTIME(n^{s(6s-1)}) \\
&\subseteq NDTIME(n^{(s+1)(6s-2)}) \\
&\quad . \\
&\quad . \\
&\quad . \\
&\subseteq NDTIME(n^{(s+1)(s+0)}) \\
&\subseteq NDTIME(n^{s(s+0)}) \\
&= NDTIME(n^{s^2})
\end{aligned}$$

Thus:

$$NDTIME(n^{6s^2}) \subseteq NDTIME(n^{s^2})$$

This contradicts Cook's Theorem that $NDTIME(n^m)$ is strictly included in $NDTIME(n^{6m})$. Hence our assumption is wrong and in consequence $NDTIME(n^{s/t})$ is strictly included in $NDTIME(n^{(s+1)/t})$. ■

11.8 Strange properties

In the preceding sections we have seen various properties of problems and their complexities which are more or less in agreement with intuition. In this section two rather counter-intuitive properties will be discussed.

The first is that there are arbitrarily large gaps in the complexity hierarchy; that is, there exist functions f and g such that f and g differ by an arbitrary amount, but nevertheless, $DTIME(f) = DTIME(g)$. Thus, although more resources have been made available, no more problems can be solved.

Definition 11.7

A statement having a parameter n, which is to take values from N, is said to be

(1) true *almost everywhere* (a.e) if it is true for all but a finite number of values of n;

(2) true *infinitely often* (i.o.) if it is true for infinitely many values of n. Note that both a statement and its negation may be true i.o., though at most one of them may be true a.e. ■

Theorem 11.17

Given any total recursive function f such that $(\forall n)[f(n) \geqslant n]$, there exist arbitrarily large functions g such that $DTIME(g) = DTIME(\lambda n[f(g(n))])$. In other words there are no languages which can be accepted by an $f \cdot g$-time-bounded Turing machine, but cannot be accepted by a g-time-bounded Turing machine.

Proof

Let h be an arbitrary recursive function. We construct a function g which is larger than h. Define g as follows:

$$g(0) = h(0)$$

$$g(x) = \mu n[n \geqslant h(x) \wedge \neg(\exists i)[0 \leqslant i \leqslant x \wedge n \leqslant T_{TM_i}(x) \leqslant f(n)]]$$

Let $L \in DTIME(\lambda n[f(g(n))])$ and TM_z an $f \cdot g$-time-bounded deterministic Turing machine accepting L. Then $T_{TM_z}(n)\downarrow$ implies $T_{TM_z}(n) \leqslant f(g(n))$.

If L is finite, then $L \in DTIME(\lambda n[n]) \subseteq DTIME(g)$.

Assume that L is infinite. Then there are infinitely many $n \geqslant z$ such that $T_{TM_z}(n)\downarrow$, and thus $T_{TM_z}(n) \leqslant f(g(n))$. But thanks to the definition of g, we have, for all these values,

$$\neg [g(n) \leqslant T_{TM_z}(n) \leqslant f(g(n))].$$

Therefore $T_{TM_z}(n) \leqslant f(g(n))$ implies $T_{TM_z}(n) \leqslant g(n)$.

In consequence, $T_{TM_z}(n) \leqslant g(n)$ almost everywhere. By the linear speed-up theorem (Theorem 11.2 on page 274) we have $L = L(TM_z) \in DTIME(g)$.

Thus $DTIME(f \cdot g) \subseteq DTIME(g)$. $DTIME(g) \subseteq DTIME(f \cdot g)$ is immediate, since $(\forall n)[g(n) \leqslant f(g(n))]$ by our assumption about f. ■

Note that the function g defined above is recursive, if we use a measure of complexity which allows a recursive indexing, such as the natural time complexity.

The second property concerns languages whose complexity is undefined. In Section 11.1, we noted that it is not obvious that there must exist a least-complex algorithm. The time-bounding functions might be incomparable, for example.

Below we will construct a language over a single-letter alphabet, which does not have a least-complex algorithm. Moreover, every algorithm can be speeded up by an arbitrary factor. The set is constructed via diagonalization. There is no 'natural' set known such that any algorithm accepting the set can be speeded up arbitrarily.

Theorem 11.18

For every total recursive function $r: N \rightarrow N$ there exists a language L such that if TM_i accepts L then there is another Turing machine TM_j which also accepts L, and furthermore $r(T_{TM_j}(n)) \leqslant T_{TM_i}(n)$ almost everywhere.

Proof

Let $\bar{r} \triangleq \lambda x[\max\{x^2, r(x)\}]$, and define the function $h: N \rightarrow N$ as follows:

$$h(0) = 2$$

$$h(n+1) = \bar{r}(h(n)) \qquad \text{for } n \geqslant 0$$

Clearly h is a total recursive function. An index for h can be obtained from an index for r.

Using diagonalization, we define a set L such that

(1) $(\forall i)[\text{if } L(TM_i) = L \text{ then } T_{TM_i}(n) \geqslant h(n \dot{-} i) \text{ a.e.}]$

(2) $(\forall k)(\exists j)[L(TM_j) = L \wedge T_{TM_j}(n) \leqslant h(n \dot{-} (k \dot{-} 1)) \text{ a.e.}]$

From these two statements the theorem follows directly. Suppose $L = L(TM_i)$. Let k in statement (2) have the value $i+2$. Then there is a machine TM_j which accepts L and $T_{TM_j}(n) \leqslant h(n \dot{-} (i+1)) = h(n \dot{-} i \dot{-} 1)$. Then we have

$$r(T_{TM_j}(n)) \leqslant \bar{r}(T_{TM_j}(n)) \leqslant \bar{r}(h(n \dot{-} i \dot{-} 1)) = h(n \dot{-} i)$$
$$\leqslant T_{TM_i}(n) \text{ almost everywhere.}$$

L is obtained by diagonalization over a class of Turing machines which includes all machines for which $T_{TM_i}(n) \geqslant h(n \dot{-} i)$ a.e. is false. In consequence, if L is accepted by TM_i, then TM_i is outside this class, and therefore $T_{TM_i}(n) \geqslant h(n \dot{-} i)$ a.e.

We begin by defining a recursive enumeration of this class of machines. Define σ as follows:

$$\sigma(0) = \mu j(T_{TM_j}(0) < h(0)]$$
$$\sigma(n) = \mu j[T_{TM_j}(n) < h(n \dot{-} j) \wedge (\forall i)[\text{if } 0 \leqslant i < n \text{ then } j \notin \sigma(i)]] \text{ for } n \geqslant 1$$

(1) σ is a total function.
Consider the Turing machine whose specification consists of the following action items:

$$(q_0, (B, B^{p-1}), (B, B^{p-1}), S^p, q_1)$$
$$(q_0, (a, B^{p-1}), (B, B^{p-1}), S^p, q_1)$$

$F = \{q_1\}$ and the machine has a single-letter input alphabet $\Sigma = \{a\}$. The above machine accepts Σ^*, and every computation consists of a single step. By adding action items $(q_j, B^p, B^p, S^p, q_j)$ for $j = 2, 3, \ldots$ we get ever larger indices for Turing machines which behave exactly like the one above. Thus there is always a value of j which satisfies all the requirements.

(2) If $T_{TM_i}(n) \geqslant h(n \dot{-} i)$ a.e. is false, then there are infinitely many values such that $T_{TM_i}(n) < h(n \dot{-} i)$. Let $m_1, m_2, \ldots$ denote these values. If i is not in the range of σ, then $\sigma(m_j) < i$ for all $j \geqslant 1$. Because σ is injective, we have a contradiction. Thus any machine for which $T_{TM_i}(n) \geqslant h(n \dot{-} i)$ a.e. is false is enumerated by σ.

Using σ, L is defined as follows:

$$a^x \in L \text{ if and only if } a^x \notin L(TM_{\sigma(x)}).$$

By Church's thesis, L is recursively decidable. Since L is constructed by diagonalization, we have $L \neq L(TM_{\sigma(i)})$ for all $i \geqslant 0$.

Therefore, if $L = L(TM_i)$ then i does not belong to the range of σ, and consequently $T_{TM_i}(n) \geqslant h(n \dot{-} i)$ a.e., and L satisfies property (1).

It remains to construct a machine such that property (2) is satisfied. Let k be given.

We construct a machine M which accepts L. M must simulate the computation of $TM_{\sigma(x)}$ on input a^x. To do so, it must compute $\sigma(x)$, and this requires checking whether $T_{TM_j}(m) < h(m \dot{-} j)$ for $m < n$. Direct simulation of TM_j can be used only if $h(m \dot{-} j) \leqslant h(n \dot{-} k + 1)$. Direct simulation is not possible if $m \dot{-} j > n \dot{-} k$, in other words if $j < k + m \dot{-} n$, or $j < k$.

For these small values of j, we incorporate the required information in the machine specification, which can be achieved by using extra components of the state vector (see Section 11.2).

Let $n_1 \triangleq \max\{n | \sigma(n) \leqslant k\}$, and incorporate in the machine specification of M a table giving the values of $\sigma(i)$ for $i = 0, 1, \ldots, n_1$, and also a table specifying whether or not $a^x \in L$ for $x = 0, 1, \ldots, n_1$. Note that it is not possible to effectively determine the value of n_1; nevertheless n_1 exists. Because there is but a finite amount of information to be stored, this can be done in the finite machine specification. It is, however, not possible, and also not necessary in our case, to constructively determine an index for the machine M, because the value of n_1 cannot be computed.

For $x > n_1$, the Turing machine M operates by direct simulation of $TM_{\sigma(x)}$ on input a^x. Because σ is injective, we have $(\forall x > n_1)[\sigma(x) > k]$. Thus to find $\sigma(x)$ for $x > n_1$ it suffices to evaluate

$$(\mu j > k)[T_{TM_j}(x) < h(x \dot{-} j) \wedge (\forall i)[\text{if } 0 \leqslant i < x \text{ then } j \neq \sigma(i)]].$$

Then we only have to evaluate $T_{TM_j}(m) < h(m \dot{-} j)$ for values m and j such that $m < x$ and $j > k$. Since $m < x < x + j \dot{-} k$ and thus $m \dot{-} j < x \dot{-} j$, this can be achieved within our time bound $h(x \dot{-} k)$. In all, x^2 of such evaluations are necessary.

Having determined $\sigma(x)$, M checks to see whether or not $TM_{\sigma(x)}$ accepts a^x. This requires no more than $h(x \dot{-} \sigma(x)) < h(x \dot{-} k)$ steps of $TM_{\sigma(x)}$. Our machine M requires s^2 times as many steps, where s is the size of the machine specification of $TM_{\sigma(x)}$. This is the slow-down introduced by the universal Turing machine (see Section 11.5). Also, $s = O(\log(x))$.

In all, M requires no more than $(x^2 + (\log(x))^2)h(x \dot{-} k)$, which is bounded above by $h(x \dot{-} k + 1)$ a.e.

To see this, observe that $h(n) \geqslant 2^{2^{n \dot- 1}}$, and $n^2 + (\log(n))^2 \leqslant 2^{2^{(n \dot- k) \dot- 1}}$ a.e. Therefore,

$$(n^2 + (\log(n))^2)h(n \dot- k) \leqslant 2^{2^{(n \dot- k) \dot- 1}} h(n \dot- k) \leqslant (h(n \dot- k))^2$$

$$\leqslant \bar{r}(h(n \dot- k)) = h(n \dot- (k \dot- 1)).$$

Thus M has sufficient time, its index satisfies property (2) above, whence we are done. ■

The theorems above are phrased using time as a measure of complexity. For space as a measure of complexity, the proofs run similarly. These strange properties are actually valid for any measure of complexity which allows a recursive indexing and a recursive termination predicate.

The theorems have been found in *abstract complexity theory*, an effort to create a machine independent theory of complexity.

In this abstract setting, we begin with a recursive indexing $\phi_0, \phi_1, \ldots$ of the partial recursive functions. Then a *measure of complexity* is an indexed set of partial recursive functions $\Phi_0, \Phi_1, \ldots$ satisfying the following axioms:

Axiom 1 $dom(\phi_i) = dom(\Phi_i)$, $i \geqslant 0$

Axiom 2 the function $M \triangleq \lambda ixm[\text{if } \Phi_i(x) \leqslant y \text{ then } 1 \text{ else } 0]$ is a total recursive function.

Thus our worst-case complexity measures *are not* complexity measures in the sense defined above. Many others are, in particular best-case complexity and natural time and space complexity.

In this general setting, a set A is called *speedable* if for all i such that $W_i = A$ and for all total recursive functions h, there is an index j such that $W_j = A$ and for infinitely many x we have $\Phi_i(x) > h(x, \Phi_j(x))$. Otherwise A is not speedable. As stated above, it can be shown in this general setting that speedable sets exist.

An important goal of the theory is to find characterizations of properties such as speedability in recursion-theoretic terms; i.e. without reference to any complexity measure. One of the results obtained in this direction is the following characterization of speedability.

Property: A recursively enumerable set A is not speedable if and only if there exists a total recursive function δ such that

(1) $(\forall i)[W_i \cap \bar{A} = W_{\delta(i)} \cap \bar{A}]$ and

(2) $(\forall i)[\text{if } W_i \subseteq A \text{ then } W_{\delta(i)} \text{ is finite}]$.

($\bar{A}$ denotes the complement of A, i.e. $N - A$.)

11.9 Exercises

11.1 Find a $\lambda n[n^2]$-time-bounded, $\lambda n[n]$-space-bounded deterministic Turing machine which accepts the language $L \triangleq \{a^m b^m | m \geqslant 1\}$. Also construct a $\lambda n[n \log(n)]$-time-bounded deterministic Turing machine which accepts L.

11.2 This exercise requires the following definition:

Definition 11.8

Let $f: N \to N$ be any function. Then $\inf_{n \to \infty} f(n)$ denotes the limit as n goes to infinity of the greatest lower bound of $f(n), f(n+1), \ldots$. ■

Let M be an S-space-bounded two-tape deterministic Turing machine such that there is no constant upper bound on the amount of tape used. The head on the input tape may only move to the right or remain stationary. Prove that

$$\inf_{n \to \infty} \frac{S(n)}{\log n} > 0.$$

11.3 Prove that

(a) every time-constructible function is also space-constructible.

(b) if S is space-constructible then $\lambda n[c^{S(n)}]$ is time-constructible for a certain positive integer c.

11.4 Prove that every language L accepted by a T-time-bounded k-tape deterministic Turing machine can also be accepted by a $\lambda n[T(n) \log(T(n))]$ time-bounded two-tape Turing machine.

11.5 Let f and g be time-constructible functions with the property

$$\inf_{n \to \infty} \frac{f(n) \log(f(n))}{g(n)} = 0.$$

Prove that there exists a language which can be accepted by a g-time-bounded Turing machine, but cannot be accepted by an f-time-bounded Turing machine.

11.10 References

Studies of space and time hierarchies for Turing machines can be found in Hartmanis and Stearns (1964, 1965) and in Hartmanis *et al.* (1965). For the non-deterministic Turing machine hierarchies see Cook (1973) and Ibarra (1972).

The theorem relating the space complexity of deterministic and non-deterministic machines is due to Savitch (1970).

The result that non-deterministic space is closed with respect to complement has been found in 1987 by Neil Immerman (1987) and by Robert Szelepcsenyi (1987). A special case concerns $\lambda n[n]$-space-bounded non-deterministic Turing acceptors. These are also called linear bounded automata; they accept all and only context sensitive languages. The LBA problem is the question whether context sensitive languages are closed with respect to complement. The LBA problem has been open since 1964. Immerman's 1987 theorem answered the question affirmatively. Hartmanis and Hunt (1974) gives an overview of the LBA problem.

A general reference on complexity issues giving a fairly complete survey is Wagner and Wechsung (1986).

The axiomatic approach to complexity theory was developed by Blum (1967). In this paper the speed-up theorem and the compression theorem are also mentioned. The gap theorem is discussed in Borodin (1972). The property characterizing speedable sets is from Soare (1982).

Chapter 12
Tractable Problems

We will now concentrate on complexity classes consisting of languages of 'low' complexity. The intention is that these classes consist of problems whose solutions are feasible in practice, or are more or less on the boundary between being feasible and being infeasible.

First we will argue that it is reasonable to consider 'solvable by a polynomial-time-bounded Turing machine' as the formal meaning of 'feasibility in practice'.

12.1 The case for polynomial time

A problem is considered tractable if it can be solved by some 'sufficiently efficient' algorithm. The meaning of the term 'sufficiently efficient' evidently depends on the circumstances. If you are interested in solving just one single instance of a problem, and having solved that instance, you will abandon the problem forever, there is little to be said about what should rightly be called acceptable. In this case the decision about what is acceptable depends on everything but the problem itself.

When classifying problems as tractable or intractable, we envisage another situation. An indefinite number of instances of the problem must be solved, and the size of these instances is in principle unbounded. These assumptions are necessary to make 'tractability' a meaningful property. If the number of instances were bounded, because their size were bounded or

Table 12.1 Polynomial and non-polynomial time complexity functions.

Time complexity function	Size of instance			
	10	20	40	80
$\lambda n[n]$	10^{-5} seconds	2×10^{-5} seconds	4×10^{-5} seconds	8×10^{-5} seconds
$\lambda n[n^2]$	10^{-4} seconds	4×10^{-4} seconds	16×10^{-4} seconds	64×10^{-4} seconds
$\lambda n[n^{10}]$	2.8 hours	118.5 days	3.3×10^2 years	3.4×10^3 centuries
$\lambda n[2^n]$	10^{-3} seconds	1 second	12.7 days	3.8×10^8 centuries
$\lambda n[3^n]$	5.9×10^{-2} seconds	0.97 hours	3.9×10^3 centuries	4.7×10^{22} centuries

for some other reason, all those instances could be solved once and for all and subsequently a table of the solutions constructed. The program to solve the problem could then consist of a table-lookup routine, and the table of solutions constructed earlier would be included in the program as fixed data. Clearly this approach is unrealistic; therefore we insist on an unrestricted number of instances of unrestricted size.

A problem is generally considered to be tractable if it can be solved by an algorithm with a time-complexity function which is bounded by a polynomial in the size of the problem instance, mainly because of the rate of growth of the cost involved. The distinction between polynomial-time algorithms and algorithms requiring more than polynomial time is displayed especially well when considering instances of large problems.

In Table 12.1 a number of time-complexity functions are compared. It is assumed that $f(n)$ time steps are executed for a problem instance of size n and that each step takes 1 μs (10^{-6} seconds).

A salient characteristic is the growth rate:

- a $\lambda n[2^n]$-time-bounded algorithm takes ten times as long as a $\lambda n[n^2]$-time-bounded algorithm on instances of size 10
- on instances of size 20 the $\lambda n[2^n]$-time-bounded-algorithm takes 2500 times as long as the $\lambda n[n^2]$-time-bounded algorithm.

Note that even if the time requirements of an algorithm are bounded by a polynomial, this does not imply that such an algorithm is always efficient

Table 12.2 Growth of the size of the largest problem instance solvable in a given amount of time.

Time complexity function	Maximal size of problem instance	
	Solvable in K seconds	Solvable in $2K$ seconds
$\lambda n[n]$	n_1	$2 \times n_1$
$\lambda n[n^2]$	n_2	$1.41 \times n_2$
$\lambda n[n^{10}]$	n_3	$1.07 \times n_3$
$\lambda n[2^n]$	n_4	$n_4 + 1$
$\lambda n[3^n]$	n_5	$n_5 + 0.6$

in practice. A $\lambda n[n^{10}]$-time-bounded algorithm is far worse than a $\lambda n[2^n]$-time-bounded algorithm for values of n up to say 40, and already requires an intolerably long time on instances of size 20. The same effect occurs if constants of proportionality are taken into account. In the long run, however, polynomials will always be on the winning side, of course.

In Table 12.2 the same functions are compared in a different way. Suppose that the algorithms execute at every time step one out of some given set of operations, that all operations have equal cost, and that you have a fixed amount of money available, sufficient to pay for say K operations. It is possible to compute the maximum size of a problem instance to solve for which sufficient funds are available. Table 12.2 shows how this maximal size increases if you double the amount of money you are willing to spend. If the algorithm is polynomial-time-bounded, the new maximum is a constant (depending on the degree of the polynomial) times the old maximum. If on the other hand, you have a $\lambda n[2^n]$-time-bounded algorithm, the maximum is increased by only 1. The same effect occurs for non-polynomial functions such as $\lambda n[n^{\log(n)}]$, although it is less pronounced.

One should not equate polynomial-time-bounded algorithms with good, efficient algorithms. As can be seen in Table 12.1, running a $\lambda n[n^{10}]$-time-bounded algorithm can take a lot of time. Furthermore, when considering time bounds, we calculate with the worst-case values. The algorithm may very well prove to run far faster in the cases to which it is applied in practice.

To classify problems on the basis of their difficulty, worst-case analysis is most appropriate, because the intrinsic difficulty of a problem is sure to emerge in the worst case, but not necessarily in the average case. On the other hand, the results of a worst-case analysis are not always useful in predicting the behaviour of an algorithm as observed in actual practice.

An example of this phenomenon is the *simplex* algorithm for solving linear programming problems. The linear programming problem consists of the following:

Given an $m \times n$ integer matrix A, an n-vector $b \in Z^n$ and a linear cost function $c(x) \triangleq c_1x_1 + \cdots + c_nx_n$ with integer coefficients, find a value $x \in Q^n$ which minimizes the cost function $c(x)$ and satisfies the conditions $A(x) = b$ and $x \geqslant 0$.

The simplex algorithm is an algorithm which starts with an initial value x_0 which satisfies the constraints, and then modifies this value until the minimum of $c(x)$ has been obtained. In the worst case, the number of iterations is exponential in n. Yet this algorithm has an impressive record of running very fast in practice.

There exist algorithms for the linear programming problem which run in polynomial time in the worst case. The first of these is the so-called 'ellipsoid algorithm' developed by Khachian (1979). Although this algorithm is polynomial-time-bounded and should therefore be considered a 'good' algorithm in the formal sense of the word, it is not a useful algorithm in practice. Tests have shown that the algorithm in its current form is not of practical importance for real-world problems, and is not useful for large or small linear programming problems.

More recently, Karmarkar (1984) published a polynomial-time-bounded algorithm, claiming that its behaviour in practice is better than that of the simplex method, at least for large problems.

12.2 Some complexity classes and their simplest relations

In this section we will define the complexity classes which play a key role in our discussion of tractable and intractable problems.

Definition 12.1

L consists of all languages which can be accepted by deterministic Turing machines in space bounded by the logarithm of the size of the input, i.e. $\log_2|w|$, where $|w|$ denotes the length of the string w.

NL consists of all languages which can be accepted by nondeterministic Turing machines in space bounded by the logarithm of the size of the input.

P consists of all languages which can be accepted by deterministic Turing machines in time bounded by polynomials in the size of the input. The polynomials may depend on the languages to be accepted. Membership of **P** is our formal equivalent of tractability.

NP consists of all languages which can be accepted by nondeterministic Turing machines in time bounded by polynomials in the size of the input. The polynomials may depend on the languages to be accepted. The membership of **NP** is our formal equivalent of the tractability of the problem to verify whether a proposed solution actually is a solution. (See Section 9.4 for a discussion of the relation between non-determinism and verifying whether a proposed candidate actually is a solution.)

PSPACE consists of all languages which can be accepted by deterministic Turing machines in space bounded by polynomials in the size of the input. The polynomials may depend on the languages to be accepted.

NPSPACE consists of all languages which can be accepted by non-deterministic Turing machines in space bounded by polynomials in the size of the input. The polynomials may depend on the languages to be accepted. ■

The complexity classes defined above, can be compared using the inclusion relation. Most of the inclusions are obvious, e.g. $\mathbf{P} \subseteq \mathbf{NP}$, or they are easily proved, see the theorem below.

The crucial question is whether or not particular inclusions, foremost among them $\mathbf{P} \subseteq \mathbf{NP}$, are strict. This has proved to be an extremely difficult problem, which is still largely unresolved. It is widely believed that all these classes actually are different and thus that the inclusions are strict, but that it will require entirely new mathematical techniques to prove it.

It is also not known whether or not $\mathbf{L} = \mathbf{NL}$. Savitch (1970) constructed a set which is in **NL** and furthermore, if it is also in **L**, then $DSPACE(S)$ is equal to $NDSPACE(S)$ for all bounding functions S such that for all natural numbers n, $S(n) \geqslant \log(n)$.

Theorem 12.1

$$\mathbf{L} \subseteq \mathbf{NL} \quad \text{and} \quad \mathbf{P} \subseteq \mathbf{NP} \subseteq PSPACE = NPSPACE$$

Proof

- $\mathbf{L} \subseteq \mathbf{NL}$ and $\mathbf{P} \subseteq \mathbf{NP}$.
  Clearly, a deterministic Turing machine is a non-deterministic Turing machine as well.
- $\mathbf{NP} \subseteq NPSPACE$.
  Clearly, the number of tape cells used is at most equal to the number of steps executed.
- $PSPACE = NPSPACE$.
  This follows directly from Savitch's theorem (Theorem 11.7) and the fact that the square of a polynomial is itself a polynomial. ■

Little is known about the relation between the space-bounded and the time-bounded classes. By Theorem 11.6 we have

$$NDSPACE(f) \subseteq \bigcup_{c>0} DTIME(\lambda n[c^{f(n)}])$$

Substitution of $f \triangleq \lambda n[\log(n)]$ in the above formula, gives the following corollary.

Corollary 12.1

$$\mathbf{L} \subseteq \mathbf{NL} \subseteq \mathbf{P} \subseteq \mathbf{NP} \subseteq PSPACE = NPSPACE \qquad \blacksquare$$

By Theorem 11.7, $\mathbf{NL} \subseteq DSPACE(\lambda n[(\log(n))^2])$; thus by Theorem 11.10 **NL** is strictly included in *PSPACE*.

The preceding theorems are summarized in Figure 12.1.

As has been elaborated in Section 12.1, **P** can be identified with the class of problems whose solution is feasible. There are many problems of practical importance for which no feasible solution is known, that is, they cannot be shown to be in **P**. Sometimes they can be shown to be in **NP**. Examples of such problems can be found in Chapter 13. As discussed in Section 9.4, this means that the corresponding *candidate verification problem* is feasible, i.e. belongs to the class **P**.

The theorem below justifies identifying **NP** with the class of all problems whose corresponding candidate verification problems are feasible, i.e. in **P**. It is also useful as a tool in showing that some given language belongs to **NP**.

Theorem 12.2

A language A belongs to **NP** if and only if there exist a polynomial p and a polynomially decidable predicate R such that:

$$x \in A \text{ iff } (\exists y)[|y| \leqslant p(|x|) \wedge R(x, y)].$$

(To say that R is a polynomially decidable predicate means that there is a polynomial q, and a q-time-bounded deterministic Turing machine which computes R – that is, accepts an input w if and only if $R(w) = \text{TRUE}$. Without loss of generality, we may assume that this Turing machine halts on all inputs.)

Proof

- (*If*) Assume that such a polynomial p and polynomially decidable predicate R exist. We construct a non-deterministic

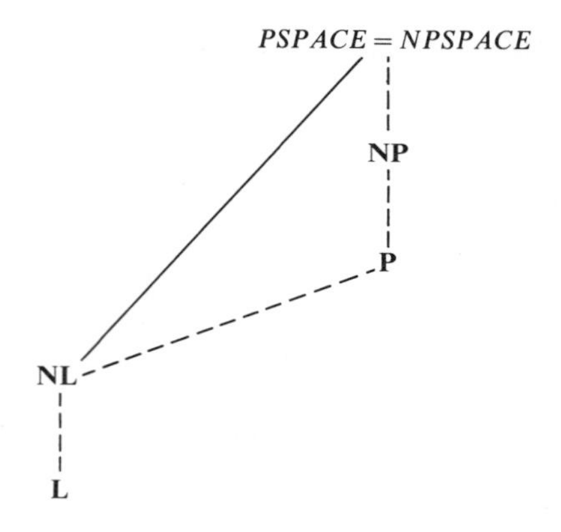

Figure 12.1 The classes **L**, **NL**, **P**, **NP**, *PSPACE* and *NPSPACE* with respect to $\subseteq$ (solid lines represent strict inclusions; dashed lines inclusions not known to be strict).

Turing machine M which is to accept A. On input w, it operates as follows:

(1) M non-deterministically selects a string y such that $|y| \leqslant p(|w|)$.

(2) M checks whether $R(w, y)$ holds, by simulating the Turing machine which computes the predicate R. If so, M accepts w, and otherwise M enters a non-terminating computation.

- (*Only if*) Assume that A is accepted by the p-time-bounded non-deterministic Turing machine M_A, where p is some polynomial.

  At any time during a computation of M_A, the number of alternative continuations of this computation is bounded by a constant m, which depends on M_A, but not on the computation itself. Thus a k-step computation of M_A can be specified by a string of length k over some finite alphabet Γ, specifying for each moment of time which alternative to take.

  Thus we define the predicate R as follows:

  $R(w, \gamma) = \text{TRUE}$ iff γ specifies an accepting computation of the Turing machine M_A on input w.

  This predicate R is clearly polynomially decidable. Also

  $$w \in A \text{ iff } (\exists y)[|y| \leqslant p(|w|) \wedge R(w, y)]$$

  (The possibility of enumerating the computations of a non-determinstic Turing machine in this way has already been used in Theorem 11.5). ■

12.3 Turing transducers and oracle acceptors

In the preceding section we have seen that there are many questions of the type 'is class **X** strictly included in class **Y**?' to which the answer is still unknown. More or less the same is true for questions of the type 'what is the complexity of problem **A**?'

Often, an algorithm solving **A** is known. This algorithm can be implemented as a Turing machine, and the complexity of this Turing machine can then be determined. But in the typical case, we are not able to show that this algorithm and Turing machine are optimal; thus, although we know the complexity of a particular solution, we do not know the complexity of the problem **A** itself. For instance, from the Turing machine implementation of the algorithm it follows that the problem **A** is in **NP**, furthermore, we cannot find a polynomial-time-bounded deterministic Turing machine solving **A**, and are also unable to show that no such machine exists.

Let A and B be languages, and **X** a complexity class. Little or no conclusion can be drawn from the statement 'A and B both belong to **X**'. This only means that the complexities of A and B have a common upper bound, as determined by the definition of the class **X**.

In the absence of knowledge about the actual complexities of A and B, we need some means to compare the complexity of A to the complexity of B. To this purpose we can use *reduction.*

In Chapters 7 and 8 we have used Karp and Cook reducibilities in their most general form, i.e. in Karp reducibility any function $f \in REC_1^1$ was permitted, and in Cook reducibility any effective computational scheme was permitted.

Let $A \subseteq \Sigma^*$ and $B \subseteq \Delta^*$ be recursively decidable languages, and let $A \leqslant B$ via the function $h: \Sigma^* \rightarrow \Delta^*$. To see whether $w \in A$, we can first compute $h(w)$ and then determine whether $h(w) \in B$. If the overhead of computing $h(w)$ and the difference in size between w and $h(w)$ can be neglected, we can say that the complexity of A is less than or equal to the complexity of B. To make sure that the problem of evaluating the reducing function is not more difficult than solving the problems to be compared, we must choose an adequate set L of resource-bounded reducing functions.

EXAMPLE 12.1

Assume that we want to compare the complexities of the languages in *NPSPACE.*

It does not make sense to define a Karp type many-to-one reducibility relation by taking L equal to the class of all elementary functions. (Here L denotes the set of all functions which may be used as reducing functions, see Definition 7.1.)

The reducibility relation thus defined would partition the set of all elementary subsets of $\{0, 1, \ldots, 9\}^*$ (i.e. subsets of $\{0, 1, \ldots, 9\}^*$

such that the characteristic functions of the corresponding subsets of **N** are elementary) into three equivalence classes:

- two singleton classes $[\varnothing]_\equiv$ and $[\{0,1,\ldots,9\}^*]_\equiv$
- one class $[\{1\}]_\equiv$ containing all the remaining elementary subsets of $\{0,1,\ldots,9\}^*$

The latter class $[\{1\}]_\equiv$ certainly contains all sets with feasibly computable characteristic functions, because the computation of a non-elementary function requires more than a hyperexponential amount of time (see Section 4.5). A non-elementary characteristic function can therefore not be considered to be feasibly computable.

In what follows we will use three types of resource-bounded reducibilities:

- polynomial-time-bounded Cook reducibility
- polynomial-time-bounded Karp reducibility
- log-space-bounded Karp reducibility.

To compute the reducing functions, we need Turing machines which produce output, that is, we use Turing machines as *transducers*, and not as *acceptors*. If Definitions 10.4–10.6 are used, the space needed to represent the output is counted as space used by the transducer. However, we do *not* wish to count the space needed to represent the output as *workspace* of the transducer.

Definition 12.2

A p-tape non-deterministic *Turing transducer* is a $p+2$-tape non-deterministic Turing machine, which uses tape 1 as a *read-only* input tape, and tape 2 as a *write-only* output tape.

Formally, this means that all the action items of the Turing machine are of the following form:

$$(q_1,(x,y,\ldots),(x,z,\ldots),(D_1,D_2,\ldots),q_2)$$

where q_1 and q_2 are states, x, y, and z are tape symbols, D_1 is L, R, or S, and D_2 is R or S.

The *workspace* used by this transducer is the space used on the tapes 3 to $p+2$. ■

Let A and B be recursively enumerable languages, and suppose that

$A \leqslant_L B$ via a function h, such that h can be computed by a $\lambda n[\log(n)]$-space-bounded Turing transducer.

Suppose we test whether $w \in A$ by computing $h(w)$ and testing whether $h(w) \in B$. According to the above, the space needed to represent the output $h(w)$ is completely disregarded. Neither the Turing transducer computing h nor the Turing acceptor accepting B is charged for this space.

By Theorem 11.6, a computation of an f-space-bounded nondeterministic Turing transducer on an input of size n is at most $c^{f(n)}$ steps long; thus, the size of the output is $O(c^{f(n)})$. It does not seem reasonable, in a context of space-bounded computations, to completely disregard this space.

We can, however, avoid the need to represent the entire output, by changing the transducer in such a way that it does not compute $h(w)$, but instead computes the ith digit of $h(w)$, given w and i.

Lemma 12.1

Let h be the function computed by an f-space-bounded nondeterministic Turing transducer. There exists a Turing machine which, given w and i, where $i \leqslant |h(w)|$, computes the ith symbol of $h(w)$, and uses no more than $\max\{f(|w|), \log(i)\}$ tape cells.

Proof

We construct a Turing machine M which simulates the Turing transducer T computing h. It does not give, but only counts the digits written on the output tape by T. When the ith digit is produced, the M halts, writing this digit on the output. More precisely, M operates as follows:

(1) Initialize a counter.

(2) Simulate the steps performed by T. If T produces a digit, that is, if an action item of the form

$$(q_1, (x, y, \ldots), (x, z, \ldots), (D_1, D_2, \ldots), q_2)$$

with $D_2 = R$ is used, then M increments the counter, without producing any output on tape 2.

(3) If the counter reaches the value i, then M halts, writing the ith digit (z according to the action item above), on tape 2. Otherwise, M continues the simulation of T. ■

Corollary 12.2

The above Turing machine M is f-space-bounded.

Proof

By Theorem 11.6, $|h(w)| = O(c^{f(|w|)})$. Thus $i = O(c^{f(|w|)})$, and consequently $\log(i) = O(f(|w|)$. Therefore, the Turing machine M is f-space-bounded. ■

By the above lemma and its corollary, it is legitimate to disregard the space required to represent the output of a Turing transducer, and we can characterize resource-bounded reducing functions using resource-bounded Turing transducers, no matter whether the resource is time or space.

Definition 12.3

Let $A \subseteq \Sigma^*$ and $B \subseteq \Delta^*$, Σ and Δ being arbitrary alphabets.

(1) $A \leqslant_P^K B$ (pronounced A is polynomial-time Karp reducible to B), if there exists a function $f: \Sigma^* \to \Delta^*$ such that:

(a) f can be computed by a deterministic Turing transducer time-bounded by a polynomial in the size of the input,

(b) $(\forall w \in \Sigma^*)[w \in A \text{ iff } f(w) \in B]$.

(2) $A \leqslant_L B$ (pronounced A is log-space reducible to B), if there exists a function $f: \Sigma^* \to \Delta^*$ such that:

(a) f can be computed by a $\lambda n[\log(n)]$-space-bounded deterministic Turing transducer,

(b) there is a constant c such that $(\forall w \in \Sigma^*)[|f(w)| \leqslant c|w|]$,

(c) $(\forall w \in \Sigma^*)[w \in A \text{ iff } f(w) \in B]$. ■

The relations $\leqslant_P^K$ and $\leqslant_L$ are Karp reducibility relations in the sense of Definition 7.1. Let K_P be the class of all functions computable by polynomial-time-bounded deterministic Turing transducers, and K_L the class of all functions f-computable by $\lambda n[\log(n)]$-space-bounded deterministic Turing transducers, which also satisfy requirement (2)(b) that is,

$$(\exists c)(\forall w \in \Sigma^*)[|f(w)| \leqslant c|w|].$$

Then both K_P and K_L contain the identity function and are closed with respect to function composition.

The class of functions which can be computed by $\lambda n[\log(n)]$-space-bounded deterministic Turing transducers is not closed with respect to function composition, the restriction to functions of which the size of the function value is at most linear in the size of the function argument is added to ensure that the resulting class K_L *is* closed with respect to function composition.

We turn now to Cook reducibility relations. In recursion theory, Cook reducibility was defined using *oracle* computations (see Sections 8.1 and 8.2). We assumed an assignment statement $\langle$*identifier*$\rangle := \chi(\langle$*identifier*$\rangle)$, where χ is some arbitrary function $\chi: N \to N$, in some mysterious way evaluated by an *oracle*, denoted by χ. The resulting model of computation is SAL($^\chi$**N**).

We need an equivalent concept in terms of Turing machines. To this purpose, we define *oracle Turing acceptors*. The general idea is that an oracle Turing acceptor is like an ordinary Turing acceptor, as defined in Definition 10.7, which has one of its tapes designated as the *oracle query tape*, and also has some of its states designated as *oracle evaluation states*. The machine specification of a non-deterministic oracle Turing acceptor consists of ordinary action items, and of *oracle evaluation action items*. The latter have the form (q, q_1, q_2), where q is an oracle evaluation state, and q_1 and q_2 are ordinary states.

An oracle acceptor behaves like an ordinary Turing acceptor until it reaches an oracle evaluation state. From there, it goes in a single step to either q_1 or q_2, dependent on whether or not the word written on the oracle query tape does or does not belong to a set X, particular to the oracle. In this step nothing changes as regards the contents of the tapes and the position of the read/write heads.

Definition 12.4

A non-deterministic oracle Turing acceptor is an 11-tuple

$$T = (p, V, \Sigma, B, Q, q_0, F, M, Q_o, M_o, i_o),$$

where

(1) $p \geqslant 1$ is the number of tapes of the acceptor;

(2) tape i_o, $1 \leqslant i_o \leqslant p$, is the *oracle query* tape;

(3) V is a finite non-empty alphabet, consisting of all the symbols that can be printed on the tape squares;

(4) Σ is a non-empty subset of V, which specifies the input alphabet;

(5) B is blank symbol of the Turing machine;

(6) Q is a finite non-empty set, whose elements are the *ordinary* states of the machine;

(7) Q_o is a finite, possibly empty set, $Q \cap Q_o = \varnothing$, the elements of Q_o are called the *oracle evaluation* states;

(8) $q_0 \in Q$ is the initial state;

(9) F is a subset of Q, its elements are called *accepting states*;

(10) $M \subseteq Q \times V^p \times V^p \times D^p \times (Q \cup Q_o)$, where $D = \{S, L, R\}$, and $M_o \subseteq Q_o \times Q \times Q$, the elements of M are called *ordinary* action items, those of M_o are called *oracle* action items, the set $M \cup M_o$ is called the machine specification.

The acceptor is called deterministic if for each $q \in Q$ and $u \in V^p$ there is at most one ordinary action item (q, u, v, d, q_n) in M, and for each $q \in Q_o$ there is at most one oracle action item (q, q_1, q_2) in M_o. ■

Note that a non-deterministic oracle Turing acceptor may actually be an ordinary non-deterministic Turing acceptor, because the set Q_o of oracle evaluating states may be empty.

Computations of the acceptor are defined using instantaneous descriptions as in Definition 10.5.

Definition 12.5

Let $T = (p, V, \Sigma, B, Q, q_0, F, M, Q_o, M_o, i_o)$ be a non-deterministic oracle Turing acceptor, and

$$c_1 = ((\alpha_1, \alpha_2, \dots, \alpha_p), p, (\beta_1, \beta_2, \dots, \beta_p))$$

and

$$c_2 = ((\gamma_1, \gamma_2, \dots, \gamma_p), q, (\delta_1, \delta_2, \dots, \delta_p))$$

*ID*s of the Turing acceptor *T*. Assume that the oracle evaluates the characteristic function of the language $X \subseteq V^*$.

Then *ID* c_1 is transformed into *ID* c_2 in a single-step, notation $c_1 \vdash_T c_2$ if

(1) there is an ordinary action item by which this follows according to Definition 10.5 or

(2) there is an oracle evaluation action item (p, q_1, q_2), $\alpha_i\beta_i \in X$ and $q = q_1$, or $\alpha_i\beta_i \notin X$ and $q = q_2$, and in either case $\alpha_i = \gamma_i$, and $\beta_i = \delta_i$ for all i, $1 \leqslant i \leqslant p$.

The language accepted by the non-deterministic oracle Turing acceptor is defined as follows:

$$L(T) \triangleq \{\omega \in \Sigma^* \mid \text{there is a computation } ((\varepsilon, \dots, \varepsilon), q_0, (\omega, \varepsilon, \dots, \varepsilon)) \vdash_T^* ((\alpha_1, \alpha_2, \dots, \alpha_p), q, (\beta_1, \beta_2, \dots, \beta_p)) \text{ such that } q \in F\}$$

where $u \vdash_T^* v$ means that *ID* u can be transformed into *ID* v by a computation of $k \geqslant 0$ steps.

An oracle Turing acceptor whose oracle evaluates the characteristic function of the language A is called an A-oracle Turing acceptor. ■

Using resource-bounded oracle Turing acceptors, we can define resource-bounded Cook reducibilities.

Definition 12.6

Let $A \subseteq \Sigma^*$ and $B \subseteq \Delta^*$, Σ and Δ being arbitrary alphabets.

$A \leqslant_P^C B$, pronounced 'A is polynomial-time Cook reducible to B', if there exists a deterministic B-oracle Turing acceptor in time bounded by a polynomial in the size of the input, which accepts the language A, that is, a deterministic oracle Turing acceptor which accepts A and whose oracle evaluates the characteristic function of B.

■

In Definitions 12.3 and 12.6 we have defined three relations, $\leqslant_P^K$, $\leqslant_L$, and $\leqslant_P^C$. These relations are reflexive and transitive. Thus they induce equivalence relations and partial orderings on the equivalence classes, as explained in Section 7.1.

In Definition 7.2 the concepts *completeness* and *hardness* have been defined. Those definitions also apply to the present cases. For instance, a set $L \subseteq \Sigma^*$ is $\leqslant_P^K$-hard for **NP** if $(\forall X \in \mathbf{NP})[X \leqslant_P^K L]$. This set L is $\leqslant_P^K$-complete for **NP** if it is $\leqslant_P^K$-hard for **NP** and if furthermore $L \in \mathbf{NP}$ (similarly for the other reducibility relations and other complexity classes).

The following theorem summarizes some properties of the above reducibility relations. For the greater part it is a rephrasing of properties proved earlier.

Theorem 12.3

(1) if $A \leqslant_L B$ then $A \leqslant_P^K B$.

(2) if $A \leqslant_P^K B$ then $A \leqslant_P^C B$.

(3) if $A \leqslant_P^K B$ then $\bar{A} \leqslant_P^K \bar{B}$.

(4) $A \leqslant_P^C B$ if and only if $\bar{A} \leqslant_P^C B$.

(5) if $A \leqslant_P^C B$ and $B \in \mathbf{P}$ then $A \in \mathbf{P}$.

(6) if $A \leqslant_P^C B$ and $B \in \mathbf{NP}$ then $A \in \mathbf{NP}$. ■

Part (5) of the above theorem illustrates the importance of the concept of *completeness*. If we can show that some given set A is $\leqslant_P^C$-complete for **NP** and that A also is a member of **P**, then $\mathbf{P} = \mathbf{NP}$. This follows, because if the $\leqslant_P^C$-complete set A is in **P**, then for any other set $X \in \mathbf{NP}$ we have $X \leqslant_P^C A$ and thus $x \in \mathbf{P}$ by part (5) of the theorem.

The polynomial-time-bounded reducibility relations have the drawback that all the classes **L**, **NL**, and **P** are in a single $\equiv_P^C$ equivalence class and also in a single $\equiv_P^K$ equivalence class, if, at least, we disregard the pathological cases, i.e. the empty set and the universal set.

They have the further drawback that the reducing function, may use a polynomial amount of space. This is somewhat excessive, and almost all known reductions use only an amount of space logarithmic in the size of the input. The logarithmic space reducibility relation permits discussing a sub-polynomial hierarchy, i.e., it distinguishes between languages complete for **P**, and languages complete for **NL**. Thus, it is useful to consider in particular the reducibility relation $\leqslant_L$.

12.4 Polynomial-time and log-space reducibility

In this section we concern ourselves with the question whether complexity classes are closed with respect to reducibilities. Also, it will be shown that $\leqslant_P^K$ and $\leqslant_P^C$ are different relations.

Definition 12.7

Let **K** be some class of languages and let $\leqslant$ be some reducibility relation. The class **K** is *closed* with respect to the relation $\leqslant$ if from $A \in \mathbf{K}$ and $B \leqslant A$ it follows that $B \in \mathbf{K}$. ■

Lemma 12.2

The classes **P** and *PSPACE* are closed with respect to $\leqslant_P^C$.

Proof

Consider the class **P**.

Let $A \in \mathbf{P}$ and $B \leqslant_P^C A$. Then there are polynomials p and p_A and an m-tape, p-time-bounded deterministic A-oracle Turing acceptor TM which accepts B, and an m_A-tape, p_A-time-bounded deterministic Turing machine TM_A which accepts A, and halts on all inputs.

We construct an $(m + m_A)$-tape deterministic Turing machine TM_B which accepts B. It operates as follows.

TM_B simulates the deterministic A-oracle Turing acceptor using its tapes 1 to m. If this acceptor enters an oracle evaluation state, TM_B copies the contents of the oracle query tape to its tape $m + 1$, and then simulates TM_A using its tapes $m + 1$ to $m + m_A$. When the computation of TM_A has terminated, TM_B resumes the simulation of the oracle acceptor in a state corresponding to the result of the computation of TM_A. The machine TM_B halts and accepts its input if and only if TM halts and accepts its input.

On an input w of size n, an accepting computation of TM consists of no more than $p(n)$ steps. No more than $p(n)$ oracle queries can occur, and the contents of the query tape are always of size at most $p(n)$. Answering a query requires no more than $p_A(p(n))$ steps. Thus, at most $p(n) + p(n).p_A(p(n))$ steps need to be simulated by TM_B. Therefore TM_B is q-time-bounded for some polynomial q, and consequently $B \in \mathbf{P}$.

Consider the class *PSPACE*. In this case, the above construction is still valid. The oracle acceptor is p-space-bounded because it is p-time-bounded, and TM_A is p_A-space-bounded by assumption. Thus TM_B requires no more than $\max\{p(n), p_A(p(n))\}$ space on an input of size n, and is therefore q-space-bounded for some suitably chosen polynomial q, and consequently $B \in PSPACE$. ■

The construction in the above proof is not valid for complexity classes defined by non-deterministic resource-bounded Turing acceptors.

EXAMPLE 12.2

Let $A \in \mathbf{NP}$ and $B \leqslant_P^C A$ via the following deterministic polynomial-time-bounded oracle Turing acceptor

$$T = (2, \{a, b, B\}, \{a, b\}, B, \{p, q_f\}, p, \{q_f\}, M, \{q\}, M_o, 2)),$$

where

$$\begin{aligned} (1) \quad M = \{&(p, (a, B), (a, a), (R, R), p), \\ &(p, (b, B), (b, B), (R, S), p), \\ &(p, (B, B), (B, B), (S, S), q)\}, \end{aligned}$$

$$(2) \quad M_o = \{(q, p, q_f)\}.$$

If we construct an acceptor TM_B as in the above proof, TM_B must go to the accepting state q_f if the word w on tape 2, the oracle query tape, *is not* in A. It is not known how to do this in polynomial time.

No alternative construction is known, thus it is as yet unknown whether or not the class **NP** is closed with respect to $\leqslant_P^C$.

Lemma 12.3

NP is closed with respect to $\leqslant_P^K$.

Proof

Let $A \in \mathbf{NP}$ and $B \leqslant_P^K A$ via h. Then there are polynomials p and p_A, a p-time-bounded deterministic Turing transducer TM computing h, and a p_A-time-bounded non-deterministic Turing acceptor TM_A which accepts A.

We construct a non-deterministic Turing machine which accepts B. On input w, TM_B starts simulating TM to compute $h(w)$. When this computation has terminated, TM_B simulates TM_A on input $h(w)$. TM_B accepts w if and only if TM_A accepts $h(w)$.

If $w \in L(TM_B)$ then there is an accepting computation of TM_B of length at most $p(|w|) + p_A(p(|w|))$. Therefore TM_B is q-time-bounded for some polynomial q, and accepts B, whence $B \in \mathbf{NP}$. ■

By Lemmas 12.2 and 12.3 and properties (1) and (2) of Theorem 12.3, we have the following corollary.

Corollary 12.3

The classes **P**, **NP**, *PSPACE* and *NPSPACE* are closed with respect to $\leqslant_P^K$ and $\leqslant_L$. ■

Lemma 12.4

The classes **L** and **NL** are closed with respect to $\leqslant_L$.

Proof

Let $A \in \mathbf{NL}$ and $B \leqslant_L A$ via the function h. Then there is a $\lambda n[\log(n)]$-space-bounded deterministic Turing transducer which computes h, and a $\lambda n[\log(n)]$-space-bounded non-deterministic Turing machine TM_A which accepts A.

By Lemma 12.1 there is a $\lambda n[\log(n)]$-space-bounded deterministic Turing machine **TM**, which given w and i, computes the ith digit of $h(w)$.

We construct a $\lambda n[\log(n)]$-space-bounded non-deterministic Turing machine TM_B which accepts B. It operates by simulating TM_A on input $h(w)$. It does not compute the whole of $h(w)$, but instead uses the machine TM which computes the digits of $h(w)$. TM_B uses a separate counter to keep track of the position of the read/write head on tape 1 of TM_A.

On input w, TM_B simulates TM_A on input $h(w)$. By definition, there is a constant c such that $(\forall w)[|h(w)| \leqslant c|w|]$. Thus TM_A uses $O(\log(|w|))$ space on input $h(w)$. Consequently, TM_B is $\lambda n[\log(n)]$-space-bounded, whence $B \in \mathbf{NL}$.

The same construction can be used for the class L. ■

Combining the preceding closure properties, we have the following theorem.

Theorem 12.4

The classes **L**, **NL**, **P**, **NP**, *PSPACE* and *NPSPACE* are closed with respect to $\leqslant_L$. ■

The preceding results find application in deducing statements such as:

If A is $\leqslant_L$-complete for **P**, and $A \in \mathbf{NL}$, then $\mathbf{P} \subseteq \mathbf{NL}$, and consequently $\mathbf{P} = \mathbf{NL}$.

This follows by Theorem 12.4, because for each $B \in \mathbf{P}$ we have $B \leqslant_L A$ by the completeness of A, therefore $B \in \mathbf{NL}$ by Theorem 12.4, and consequently $\mathbf{P} \subseteq \mathbf{NL}$.

As long as questions such as 'does $\mathbf{P} \subseteq \mathbf{NL}$ hold?' remain unsolved, these types of conditional statements are the best we can offer.

The most important statements of this type are:

(1) If A is $\leqslant_P^K$-complete for **NP**, and $A \in \mathbf{P}$, then $\mathbf{P} = \mathbf{NP}$.

(2) If A is $\leqslant_L$-complete for **NL**, and A actually is in **L**, then $\mathbf{L} = \mathbf{NL}$.

Although these statements are meaningful in themselves, they only have substance if complete languages can be shown to exist. This can indeed be done, and we will show in the next section that certain resource-bounded versions of the Halting Problem are complete. Apart from these more or less artificial problems, many practical problems have been shown to be complete. Some of these problems will be considered in the following chapters.

We turn now to the relations $\leqslant_P^C$ and $\leqslant_P^K$. In the theorem below we consider a language A such that *not* $A \leqslant_P^K \bar{A}$. Since $A \leqslant_P^C \bar{A}$ for all A by Theorem 12.3 it follows that $\leqslant_P^K$ and $\leqslant_P^C$ are different in general.

Theorem 12.5

There exists a language A such that *not* $A \leqslant_P^K \bar{A}$.

Proof

Let $h: N \rightarrow N$ be the function recursively defined by:

$$\begin{cases} h(0) = 1 \\ h(x) = 2^{h(x-1)}, \text{ for all } x \geqslant 1. \end{cases}$$

Let Σ be an arbitrary alphabet containing at least two symbols, say 0 and 1, and let $H, H_0: \Sigma^* \rightarrow N$ be the functions:

$$H \triangleq \lambda x[\mu y[h(y) > |x|] \dot{-} 1]$$

$$H_0 \triangleq \lambda x[\text{if } h(H(x)) = |x| \text{ then } 1 \text{ else } 0]$$

Both H and H_0 are computable in polynomial time.

Below, we construct a language A, such that *not* $A \leqslant_P^K \bar{A}$. To do this we need a Gödel numbering of deterministic Turing transducers. We will assume a numbering as described in Section 11.5, the function computed by the ith Turing transducer will be denoted by ϕ_i.

The language A is defined as follows.

Let $x \in \{0, 1\}^*$.

(1) if $x = \varepsilon$ then $x \in A$,

(2) if $H_0(x) \neq 1$ or $x \notin \{0^n | n \geqslant 1\}$ then $x \notin A$,

(3) otherwise, $H_0(x) = 1$ and $x \in \{0^n | n \geqslant 1\}$, and therefore x is a sequence of $h(H(x))$ 0s, i.e. $x = 0^{h(H(x))}$.

Let $\langle i, j \rangle = H(x)$, where $\langle\,\rangle: N^2 \rightarrow N$ is the standard

Cantor numbering (see Section 5.1), and let TTT denote the Termination predicate for Turing machines with respect to Time as a measure of complexity (see Definition 11.5). Then the predicate TTT satisfies

if $TTT(i, x, 2^{|x|}) \neq 1$ then
  $x \notin A$
else
  if $|\phi_i(x)| > h(H(x) \dot{-} 1)$ then
    $x \notin A$
  else
    $x \in A$ if and only if $\phi_i(x) \in A$
  fi
fi

Assume that $A \leqslant_P^K \bar{A}$ via the function $\phi_i \in P$, and furthermore assume that the ith deterministic Turing transducer, which computes ϕ_i, is p-time-bounded, for some polynomial p. Choose the natural number j such that $2^{h(\langle i,j \rangle)} > p(h(\langle i,j \rangle))$.

Consider the element $x_0 \triangleq 0^{h(\langle i,j \rangle)}$. We have $H(x_0) = \langle i,j \rangle$, $x_0 = 0^{h(H(x_0))}$, and $H_0(x_0) = 1$. Thus part (3) of the above definition of A applies.

Let $y_0 \triangleq \phi_i(x_0)$. Our assumption that $A \leqslant_P^K \bar{A}$ via ϕ_i implies that we should have $x_0 \in A$ if and only if $y_0 \notin A$. However,

(1) if $|y_0| > h(H(x_0) \dot{-} 1) = h(\langle i,j \rangle \dot{-} 1)$ then $x_0 \notin A$. Since by our assumption on j we have

$$p(h(\langle i,j \rangle)) < 2^{h(\langle i,j \rangle)} = h(\langle i,j \rangle + 1)$$

we have $|y_0| < h(\langle i,j \rangle + 1)$. Thus

$$h(\langle i,j \rangle \dot{-} 1) < |y_0| < h(\langle i,j \rangle + 1).$$

This implies that $y_0 \notin A$, unless $y_0 = 0^{h(\langle i,j \rangle)}$. But then $y_0 = x_0 \notin A$. Hence, if $|y_0| > h(H(x_0) \dot{-} 1)$ we have $x_0 \in A$ if and only if $y_0 \in A$.

(2) If $|y_0| \leqslant h(H(x_0) \dot{-} 1)$ we have $x_0 \in A$ if and only if $y_0 \in A$ by the definition of A.

Therefore $x_0 \in A$ if and only if $y_0 = \phi_i(x_0) \in A$. This contradicts our assumption that $A \leqslant_P^K \bar{A}$ via ϕ_i. Thus, *not* $A \leqslant_P^K \bar{A}$. ■

Corollary 12.4

The polynomial-time Karp and Cook reducibility relations are different. ■

It can be shown that $2^{|x|}$ steps are required to determine whether a given x belongs to the language A defined in the above proof. Thus, $A \in DTIME(\lambda n[2^n])$. Hence $\leqslant_P^K$ and $\leqslant_P^C$ differ at least on the 'higher' complexity classes. It is not known whether or not these relations differ on **NP**.

12.5 Complete languages

The Halting Problem, represented in the set $K_0 = \{\sigma_1^2(x, y) | \phi_x(y)\downarrow\}$, is $\leqslant_m$- and $\leqslant_T$-complete for the class of all recursively enumerable sets (see Theorem 7.4 and Corollary 8.2). In this section, we will consider resource-bounded analogs of the Halting Problem. For each of the classes **NL**, **P**, **NP**, *PSPACE* and *NPSPACE*, there is a resource-bounded analogue of the Halting Problem, which is $\leqslant_L$-complete for the class in question. Practical problems, which are complete for these classes, will be discussed in the next two chapters.

By 'resource-bounded analogue of the Halting Problem' we mean a language of the following type:

$\{x¢w¢a^m |$ input w is accepted by Turing machine TM_x without using more than $b(m)$ space (more than $b(m)$ computation steps), and m is sufficiently large, dependent on the bound b and the type of resource, time or space$\}$.

We will assume the Gödel numbering of Turing acceptors, as discussed in Section 11.5. The Gödel numbering concerns non-deterministic Turing acceptors with two tapes, the read-only input tape, and a single work tape, and with some fixed input alphabet Σ, which we assume to have at least two letters.

Definition 12.8

$U_{\mathbf{NL}} \triangleq \{x¢w¢a^m | x \geqslant 0,\ w \in \Sigma^*,\ TM_x$ accepts w using at most $\log(|w|)$ space, and $m \geqslant |w|^{|x|}\}$. ■

Lemma 12.5

The language $U_{\mathbf{NL}}$ is in **NL**.

Proof

We construct a Turing machine M with four work tapes, which accepts $U_{\mathbf{NL}}$ by simulating a computation of TM_x on input w. Let $v = x¢w¢a^m$ be the input of M. The four tapes of M are used as follows.

(1) The input of TM_x is read from the input tape of M. Tape 1 is used to store the value of a pointer which specifies the position

of the read head on the input tape of TM_x. This requires $\log(|w|) \leqslant \log(|v|)$ cells.

(2) Tape 2 specifies the action item of TM_x which is the currently active item in the simulation of TM_x. The action item is specified by the value of a pointer, which determines the position within x at which the action item is represented. This also requires no more than $\log(|v|)$ tape cells.

(3) Tape 3 consists of three tracks:

(a) an upper track which is used to store the contents of TM_x's work tape

(b) a middle track which is used to mark the position of the read/write head of TM_x's work tape

(c) a lower track, which is used to measure the number of cells used by TM_x.

The symbols of the tape alphabet of TM_x are coded as strings over the tape alphabet of M. No more than $|x|$ symbols are needed per symbol. On the lower track, we only mark cells as either *used* or *unused*, such that the number of cells marked *used* is equal to the number of cells actually used at that point of time. Initially, a block of $\log(|w|)$ cells is marked off to limit the space which TM_x may use.

In all, tape 3 uses no more than $|x| \log(|w|) \leqslant \log(|v|)$ cells.

(4) Tape 4 specifies the state of TM_x which is the current state in the simulation of TM_x. This also requires no more than $\log(|v|)$ cells.

M starts to check that $m \geqslant |w|^{|x|}$. For this purpose, it computes $\log(|w|)$ and $\log(m)$ and verifies that $\log(m) \geqslant |x| \log(|w|)$. This can be done using no more than $\log(|v|)$ space. If m is not sufficiently large, M halts and rejects its input, otherwise M properly initializes the contents of the tapes 1 to 4, and then begins to simulate a computation of TM_x on input w. A step in the computation of TM_x is simulated as follows:

Let q be the current state, and a and b the current symbols on the input and the work tape. M moves the read head on its input tape to some position in x, and checks to see that this and adjacent cells contain the description of an action item $(q,(a,b),(a,c),(D_1,D_2),p)$. If so, it updates tape 4, updates the tracks of tape 3, and updates the contents of tape 1.

Next M checks to see whether p is a final state. If this is the case M halts accepting w, otherwise the simulation is continued.

If something happens to go wrong in the course of this, e.g. there is no action item when one is expected, or the tape used by TM_x grows beyond the allotted $\log(|w|)$ cells, M halts, rejecting its input w.

Obviously the non-deterministic Turing machine M accepts $U_{\mathbf{NL}}$ and uses no more than $\log(|v|)$ tape cells. In other words, $U_{\mathbf{NL}} \in \mathbf{NL}$.

■

Lemma 12.6

The language $U_{\mathbf{NL}}$ is $\leqslant_L$-hard for **NL**.

Proof

For every language $A \in \mathbf{NL}$ there exists a $\lambda n[\log(n)]$-space-bounded non-deterministic Turing machine, which accepts A. Let z be an index for this machine.

Define the function f as follows.

$$f(w) \triangleq z ¢ w ¢ a^m, \text{ where } m = |w|^{|z|}.$$

Then $(\forall w \in \Sigma^*)[w \in A \text{ iff } f(w) \in U_{\mathbf{NL}}]$.

The function f can be computed by a $\lambda n[\log(n)]$-space-bounded deterministic Turing transducer. The transducer operates as follows:

- Write $z¢$ to the output. This requires no space, because $z¢$ is a constant. Therefore writing this constant to the output can be programmed in the machine specification of the Turing transducer.
- Copy the input w to the output.
- Write $¢$ to the output.
- Let the work tape of the transducer consist of $|z|+1$ tracks. (Remember that z is a constant.) The uppermost track is used to count up to $|w|$, thereby marking off a block of $\log(|w|)$ cells of the work tape. The transducer now counts from 0 to $|w|^{|z|}$, interpreting the tape as a length $|z|$ counter, each of $|z|$ tracks representing a radix $|w|$ digit. The transducer sends a symbol 'a' to the output at each step of the counter.
- When the counter overflows, the transducer halts.

Thus $A \leqslant_L U_{\mathbf{NL}}$ via the function f, and consequently $U_{\mathbf{NL}}$ is $\leqslant_L$-hard for **NL**. ■

Combining the preceding two lemmas we have:

Theorem 12.6

The Halting Problem for non-deterministic $\lambda n[\log(n)]$-space-bounded Turing machines, as represented in the language $U_{\mathbf{NL}}$, is $\leqslant_L$-complete for **NL**. ■

Let us now define resource-bounded versions of the Halting Problem which are complete for other complexity classes, in particular for **P** and **NP**.

Definition 12.9

(1) $U_{\mathbf{P}} \triangleq \{x¢k¢w¢a^m \mid x, k \geqslant 0,\ w \in \Sigma^*$, input w is accepted by the deterministic Turing machine TM_x using no more than $|w|^k$ steps, and $m \geqslant |w|^k\}$.

(2) $U_{\mathbf{NP}} \triangleq \{x¢k¢w¢a^m \mid x, k \geqslant 0,\ w \in \Sigma^*$, input w is accepted by the non-deterministic Turing machine TM_x using no more than $|w|^k$ steps, and $m \geqslant |w|^k\}$.

■

Lemma 12.7

The language $U_{\mathbf{P}}$ is in the class **P**.

Proof

We construct a Turing machine M to accept $U_{\mathbf{P}}$. The machine will have five tapes, which are used as follows:

- Tapes 1 and 2 consist of 2 tracks each. The upper track of tape 1 contains a copy of x, the upper track of tape 2 contains a string of $|w|^k$ as. The lower tracks contain markers. Tape 1 is used to represent the machine specification, and tape 2 is used to count to $|w|^k$ to limit the number of steps of TM_x to be simulated.
- Tape 3 is used to represent the state of TM_x used as the current state in the simulation.
- Tapes 4 and 5 consist of two tracks each. Tape 4 represents the input tape of TM_x, its upper track contains w, its lower track is used to mark the position of TM_x's read head. Tape 5 represents the work tape of TM_x, the upper track represents the contents of TM_x's work tape, the lower track determines the position of the read/write head.

M begins to check that $m \geqslant |w|^k$. Computing $|w|^k$ takes

$$O((\log(|w|))^k) = O(|w|^k) = O(|v|)$$

steps, where $v = x¢k¢w¢a^m$ is M's input. If $m < |w|^k$, M halts and rejects its input. Otherwise, M properly initializes the tapes, and begins to simulate a computation of TM_x on input w.

Let p be the current state of TM_x, a the current symbol from its input tape, and b that from its work tape. To simulate a single step of the computation of TM_x on input w, M takes the following actions.

(1) Find an action item $(p,(a,b),(a,c),(D_1,D_2),q)$ of TM_x which can be used. If there is no usable action item, M halts, and rejects its input $x¢k¢w¢a^m$.

(2) Otherwise, M updates the tapes 3, 4 and 5 as specified by the action item. Next M checks to see whether the new state q is a final state. If so, M halts, and accepts its input $x¢k¢w¢a^m$.

(3) Otherwise, M shifts the marker on tape 2 one place to the right. If the upper track of the marked cell does not contain the letter a, then M halts, and rejects its input $x¢k¢w¢a^m$. Otherwise, M continues the simulation.

Actions (1) and (2) require $O(|x|^2)$ steps each. Since action (3) requires a constant number of steps, the simulation of a single step of TM_x takes at most $O(|x|^2)$ steps.

M accepts its input if TM_x accepts its own input w in $|w|^k$ steps at most. Thus the total number of steps required for M to accept its input v is $O(|v|^3)$, and hence $U_{\mathbf{P}}$ is in the class **P**. ■

Lemma 12.8

The language $U_{\mathbf{P}}$ is $\leqslant_L$-hard for **P**.

Proof

For every $A \in P$ there is some deterministic Turing machine TM_{z_A} which accepts A in polynomial time, i.e. there is some k_A such that any $x \in A$ is accepted according to a computation of no more than $|x|^{k_A}$ steps. Define the function f as follows.

$$f(w) \triangleq z_A ¢ k_A ¢ w ¢ a^m, \text{ where } m = |w|^{k_A}.$$

The function f can be computed by a $\lambda n[\log(n)]$-space-bounded deterministic Turing transducer. It operates like the transducer described in Lemma 12.6. Now we use the fact that both z_A and k_A are constants, which allows us to output the required number of padding symbols while using only an amount of space logarithmic in the size of w. Thus $A \leqslant_L U_{\mathbf{P}}$ via f, and consequently $U_{\mathbf{P}}$ is $\leqslant_L$-hard for **P**. ■

Combining the preceding two lemmas we have:

Theorem 12.7

The Halting Problem for polynomial-time-bounded deterministic Turing machines, as represented in the language $U_{\mathbf{P}}$, is $\leqslant_L$-complete for **P**. ■

Theorem 12.8

The Halting Problem for polynomial-time-bounded non-deterministic Turing machines, as represented in the language $U_{\mathbf{NP}}$, is $\leqslant_L$-complete for **NP**.

Proof

The proof is very similar to the proof of the preceding theorem. To show that $U_{\mathbf{NP}}$ itself belongs to the class **NP**, a non-deterministic Turing machine is constructed which will accept $U_{\mathbf{NP}}$. This machine differs from the machine in Lemma 12.7, in that the action item of TM_x to be used is non-deterministically selected. The proof that $U_{\mathbf{NP}}$ is $\leqslant_L$-hard for **NP** is practically identical to the proof of Lemma 12.8. ■

Definition 12.10

$U_{PSPACE} \triangleq \{x¢k¢w¢a^m | x, k \geqslant 0,\ w \in \Sigma^*$, input w is accepted by the deterministic Turing machine TM_x using no more than $|w|^k$ tape cells, and $m \geqslant |w|^k\}$. ■

The Halting Problem for polynomial-space-bounded deterministic Turing machines, as represented in the set U_{PSPACE} defined above, is $\leqslant_L$-complete for the class *PSPACE*. This follows by an argument similar to those above. It is only required to mark off a block of $|w|^k$ cells and to reject the input if more space is needed during simulation than has been allotted to the machine TM_x being simulated.

The preceding results are summarized in the following theorem.

Theorem 12.9

For each of the classes **L**, **NL**, **P**, **NP**, and *PSPACE*, there exists a language which is $\leqslant_L$-complete for the complexity class in question. (By items (1) and (2) of Theorem 12.3, these are also $\leqslant_P^K$- and $\leqslant_P^C$-complete for the class in question.) ■

The complete languages referred to in the preceding theorem are constructed using resource-bounded variants of the Halting Problem. Many natural and practical problems have been found which are $\leqslant_L$-, $\leqslant_P^K$-, or $\leqslant_P^C$-complete for **NL**, **P** and **NP**. In the chapters to follow we will consider some of these problems and show that they indeed are complete.

12.6 Exercises

12.1 Prove that $\mathbf{P} = \mathbf{NP}$ if $\mathbf{P} = PSPACE$.

12.2 Prove that sorting n integers is a problem that belongs to $\mathbf{P}$.

12.3 *QBF-SAT* is the following problem.

$$QBF\text{-}SAT \triangleq \{(Q_1x_1)(Q_2x_2)\cdots(Q_kx_k)F \mid F \text{ is a propositional formula, } Q_i \text{ is either } \forall \text{ or } \exists \text{ for all } i, 1 \leqslant i \leqslant k, \text{ and there is an assignment } v{:}(\text{var}(F) - \{x_1,\ldots,x_k\}) \rightarrow \{0,1\} \text{ such that } v((Q_1x_1)(Q_2x_2)\cdots(Q_kx_k)F) = 1\}$$

Prove that *QBF-SAT* is in *PSPACE*.

12.4 In a search problem Π each instance $\mathbf{I}$ has an associated solution set $S_\Pi(\mathbf{I})$ and for a given $\mathbf{I}$ we are to find an element of $S_\Pi(\mathbf{I})$.

The enumeration problem based on a search problem Π reads: 'Given $\mathbf{I}$, what is the cardinality of $S_\Pi(\mathbf{I})$?' or 'How many solutions does Π have?'

A *non-deterministic counting* Turing machine is a standard non-deterministic Turing machine with an auxiliary output device which 'magically' prints in binary notation on a special output tape the number of accepting computations induced by the input. The machine is f-time-bounded if the length of the *longest* computation induced by an input of size n is bounded by $f(n)$.

Let $\#P$ denote the set of all enumeration problems Π whose solution can be computed by a non-deterministic counting polynomial-time-bounded Turing machine.

The enumeration problem corresponding to the Hamiltonian Circuit problem (*UHC*, see Definition 13.5), asks for the number of Hamiltonian circuits in the given graph; similarly the enumeration problem corresponding to the satisfiability problem *SAT* asks for the number of assignments which make the given propositional expression true.

Prove that the enumeration problems corresponding to the Hamilton circuit problem and the satisfiability problem both belong to $\#P$.

12.5 Let C be a complexity class and L a problem. If L is complete for C and C contains an intractable problem, then L is intractable. Prove this.

12.6 If $L \in \mathbf{NP}$ and L is complete for a class C with respect to $\leqslant_P^K$, then C is a subset of $\mathbf{NP}$. Prove this.

12.7 Let X be any class of problems. The classes $\mathbf{P}^X$ and $\mathbf{NP}^X$ are defined as follows:

(a) $\mathbf{P}^X = \{L | (\exists M \in X)[L \leqslant_{\mathbf{P}}^{C} M]\}$
(b) $\mathbf{NP}^X = \{L | (\exists M \in X)[L \leqslant_{\mathbf{NP}}^{C} M]\}$

where $L \leqslant_{\mathbf{NP}}^{C} M$ if and only if there exists a polynomial-time-bounded non-deterministic Turing acceptor which accepts L and whose oracle evaluates the characteristic function of M. Prove that $\mathbf{P}^{\mathbf{NP}} \subseteq \mathbf{NP}^{\mathbf{NP}}$.

12.8 Define sets Σ_k^p, Π_k^p and Δ_k^p as follows:

(a) $\Sigma_0^p = \Pi_0^p = \Delta_0^p = \mathbf{P}$
(b) $\Delta_{k+1}^p = \mathbf{P}^{\Sigma_k^p}$
$\Sigma_{k+1}^p = \mathbf{NP}^{\Sigma_k^p}$
$\Pi_{k+1}^p = co\text{-}\Sigma_{k+1}^p$.
for all $k \geqslant 0$.

Prove the following inequalities:

(i) $\Delta_k^p \subseteq (\Pi_k^p \cap \Sigma_k^p)$.
(ii) $(\Pi_k^p \cup \Sigma_k^p) \subseteq \Delta_{k+1}^p$.

(All the above inclusions are strict. The resulting hierarchy is called the *polynomial hierarchy*, it is an analogue of Kleene's *arithmetical hierarchy*. More details can be found in Rogers (1967), Stockmeyer (1977) and Hermes (1961).)

12.7 References

As already mentioned, the basic reference regarding Turing reducibility and complete sets is Post (1943). The pioneer papers about complete languages, sets or problems, are Cook (1973) and Karp (1972). In the latter paper the importance of the concept '**NP**-completeness' is clearly demonstrated.

Classification of problems by time-bounded Turing machines is due to Hartmanis and Stearns (1965). Classification by tape complexity is from Hartmanis *et al.* (1965).

Finally we mention that complexity classes can also be characterized by logical means. For further references to this topic see Wagner and Wechsung (1986).

Chapter 13
The Classes P, NP, and *co*-NP

In this chapter we concentrate on the classes **P**, **NP** and *co*-**NP**. In Sections 13.1 and 13.2 a number of practical problems will be shown to be $\leqslant_P^C$-complete for **NP**.

By a 'practical problem' we mean a problem which does not originate directly from the theory of algorithms, as do the resource-bounded versions of the Halting Problem discussed in Section 12.5, but one which is an abstraction of a problem occurring in practice. Such problems are concerned with formulae, graphs, functions, and partially ordered sets, and are abstractions of problems concerned with finding sets of values which satisfy given constraints, finding optimal routes for transport, finding optimal task schedules and so forth.

Problems will be specified as in the example below.

EXAMPLE 13.1

$CLIQUE \triangleq \{((V,E),k) \mid$ where (V,E) is an undirected graph, k an integer, and (V,E) contains at least one k-clique as a subgraph$\}$.

The name of the set, *CLIQUE*, is also used as a name for the problem. Thus we say the '*CLIQUE* problem', when we should say the 'membership problem of the language *CLIQUE*'.

Concepts such as k-clique, which are treated as known in this chapter, are mostly defined in Chapter 1; otherwise the concepts are explained in the text.

The precise string representation of objects such as graphs is not specified for every problem. It is simply assumed that the representation is concise.

To be precise, it is assumed that the alphabet consists of the printable ASCII characters; thus, we have letters, digits, punctuation marks, brackets, braces, parentheses and so forth. It is assumed further that:

- Integers are represented in decimal notation, without leading zeros.
- Variables are indexed, and x_i is represented by x_{r_i}, where r_i is the decimal representation of i without leading zeros.
- Directed and undirected graphs with n vertices and m edges are represented by a list of n vertices and m edges, using the format
$$\langle \{x_1, x_2, \ldots, x_n\}, [(x_{i_1}, x_{i_2}), \ldots, (x_{i_m}, x_{i_{m+1}})] \rangle$$
for directed graphs, and the format
$$\langle \{x_1, x_2, \ldots, x_n\}, [\{x_{i_1}, x_{i_2}\}, \ldots, \{x_{i_m}, x_{i_{m+1}}\}] \rangle$$
for undirected graphs, where the variables are represented as described above.
- Numerical expressions, logical formulae and such are represented in the obvious way, using operation symbols, variables and parentheses.

It is customary to say 'problem X is **NP**-complete' instead of 'problem X is $\leqslant_P^C$-complete for **NP**'. Because we use both time-bounded and space-bounded reducibility relations, it is necessary to specify which of these reducibility relations is meant. When we use 'problem X is **NP**-complete' this is meant to be an abbreviation of 'problem X is $\leqslant_P^C$-complete for **NP**'.

13.1 The satisfiability problem

A great many practical problems originating from very different scientific disciplines have been shown to be $\leqslant_P^C$-complete for the class **NP**. From an intuitive point of view these complete problems are as difficult as any other problem in **NP**. In addition, an efficient solution for a single **NP**-complete problem A, i.e. a polynomial-time-bounded algorithm solving A, gives an efficient solution for all problems in **NP**, and thereby shows that $\mathbf{P} = \mathbf{NP}$.

The above may be rephrased as follows. If we classify problems solvable by polynomial-time-bounded algorithms as computable in practice, then we may think of the problems for **NP** as being on, or slightly over, the current border line of what can actually be computed at the moment in an 'efficient way' by a sequential computer.

The point in proving that a particular problem X is $\leqslant_P^C$-complete for the class **NP** is twofold:

(1) It provides one more way of rephrasing the general question 'is **P** equal to **NP**?' as a particular instance of it, namely, 'does X belong to **P**?'

(2) Although it does not sharply determine the complexity of problem X, it is a very strong hint that problem X is intractable, because it is at least as complex as any other problem in **NP**.

From a practical point of view, it probably does not pay to search for an efficient algorithm solving a problem X which is $\leqslant_P^C$-complete for **NP**. It is better to spend time on finding some properties of the problem instances you will encounter which simplify the problem sufficiently, but do not hold in general. If you cannot find such properties, then it is best to settle for some non-optimal solution which you can determine efficiently.

The problem to be shown $\leqslant_P^C$-complete for **NP** in this section concerns propositional formulae. In the next section problems from other disciplines will be considered.

Definition 13.1

Let $\{x_i | i = 1, 2, \ldots\}$ be the set of propositional variables. A propositional formula is an expression built from propositional variables and the propositional connectives NOT($\neg$), AND($\wedge$) and OR($\vee$), according to the following syntax rules:

$$\begin{aligned}\langle \textit{prop. formula} \rangle ::= {} & \langle \textit{prop. variable} \rangle | \\ & \neg(\langle \textit{prop. formula} \rangle) | \\ & (\langle \textit{prop. formula} \rangle \wedge \langle \textit{prop. formula} \rangle) | \\ & (\langle \textit{prop. formula} \rangle \vee \langle \textit{prop. formula} \rangle)\end{aligned}$$ ■

The propositional connectives have their usual meanings, i.e. in accordance with the truth functions in Table 13.1. Thus, for every assignment of truth values to the propositional variables occurring in a propositional formula, the truth value of the formula is well defined.

Definition 13.2

(1) Let F be a propositional formula. The set of propositional variables occurring in F is denoted by $\text{var}(F)$. An assignment of truth values to the variables occurring in F is a function $v\colon \text{var}(F) \to \{0, 1\}$. The truth value of F under this assignment, as determined by Table 13.1, is denoted by $v(F)$.

Table 13.1 Truth-functional definition of propositional connectives; 1 denotes the truth value TRUE, 0 denotes FALSE.

x	y	$\neg x$	$x \wedge y$	$x \vee y$
0	0	1	0	0
0	1	1	0	1
1	0	0	0	1
1	1	0	1	1

(2) A propositional formula F is *satisfiable* if there is at least one assignment $v\colon \text{var}(F) \to \{0, 1\}$, such that $v(F) = 1$.

(3) A propositional formula F is *falsifiable* if there is at least one assignment $v\colon \text{var}(F) \to \{0, 1\}$, such that $v(F) = 0$. ■

Because the binary propositional connectives are associative, there is no need to write all the parentheses required by Definition 13.1. The number of parentheses can be further reduced by assigning priorities to the connectives. We will use the customary priorities:

$\neg$ has the highest priority,
$\wedge$ has the highest but one, and
$\vee$ has the lowest.

Definition 13.3

(1) A *literal* is either a propositional variable or the negation of a propositional variable. The literal $\neg x_i$ is also denoted by $\bar{x}_i$.

(2) A propositional formula is in *conjunctive normal form*, abbreviated CNF, if it is of the format $C_1 \wedge C_2 \wedge \cdots \wedge C_n$ for some $n \geqslant 1$. The C_is are called clauses and are of the format $L_1 \vee L_2 \vee \cdots \vee L_{m_i}$, where each L_j is a literal.

(3) A propositional formula is in k-conjunctive normal form (k-CNF), if it is in conjunctive normal form and all clauses contain at most k literals. ■

EXAMPLE 13.2

(1) The formula $F \triangleq x_1 \wedge (x_2 \vee x_3)$ is in conjunctive normal form. There are two clauses, namely x_1 and $(x_2 \vee x_3)$. $\text{var}(F) = \{x_1, x_2, x_3\}$.

There is an assignment $v\colon \text{var}(F) \to \{0, 1\}$, for example, $v(x_1) = 1$, $v(x_2) = 1$, and $v(x_3) = 0$, which makes the propositional formula F true, i.e. such that $v(F) = 1$. Thus F is satisfiable.

There also is an assignment $v:\text{var}(F) \rightarrow \{0, 1\}$, such that $v(F) = 0$, for instance $v(x_1) = 0$, $v(x_2) = 1$, and $v(x_3) = 0$. Thus F is falsifiable too.

(2) The formula $x_1 \wedge (x_2 \vee x_3) \vee x_1$ is not in conjunctive normal form.

In logic, one is interested in tautologies, i.e. formulae which are true for every assignment of truth values to the propositional variables occurring in the formula, and also in methods to prove that a given formula is a tautology. Such methods consist of applying rules of inference, i.e. rules which allow the deduction of new formulae from a set of already deduced or given formulae. Three such rules of inference are as follows:

Modus ponens Formula B may be deduced from the formulae $A \rightarrow B$ and A.

Resolution Given formulae $A \vee C_1 \vee C_2 \vee \cdots \vee C_k$ and $\neg A \vee D_1 \vee D_2 \vee \cdots \vee D_m$ we may deduce the formula $C_1 \vee C_2 \vee \cdots \vee C_k \vee D_1 \vee D_2 \vee \cdots \vee D_m$. This formula is called the *resolvent* of the two given formulae.

Unit resolution This is a special case of the preceding rule of inference. From a unit clause, i.e. a literal, say A, and a clause $\neg A \vee B_1 \vee B_2 \vee \cdots \vee B_m$ the clause $B_1 \vee B_2 \vee \cdots \vee B_m$ may be deduced.

As a special case, the *empty clause*, denoted by $\square$, may be deduced from the formulae A and $\neg A$, where A is a literal.

The rule *modus ponens* is a well-known rule of inference, and it is used in many logical theories. *Resolution* is an equally well-known rule, which finds application in many mechanical theorem-proving procedures. The rule is used in *refutation*-type arguments. This means that to prove a formula F, the formula $\neg F$ is converted to conjunctive normal form, and then the resolution rule is applied repeatedly until the empty clause $\square$ is obtained as a resolvent.

Any assignment $v:\text{var}(F) \rightarrow \{0, 1\}$ which satisfies $\neg F$ also satisfies every resolvent constructed. If $\square$ is a resolvent, then there is no assignment such that $v(\neg F) = 1$, and consequently the formula F is valid.

The inference rule 'resolution' is sound, meaning that if the empty clause can be deduced from formula F, then F is not satisfiable, i.e. it is a contradiction. The rule is also complete for the propositional calculus, meaning that if F is a propositional formula which is not satisfiable, then the empty clause can be deduced from F by *resolution*.

A set of clauses is called provably false by *(unit-)resolution* if the inference rule *(unit-)resolution* allows us to deduce the empty clause from the given clause.

In this section, we will consider the problem of determining whether a given propositional formula is satisfiable. The other concepts introduced above will be used in Chapter 14, where practical problems $\leqslant_L$-complete for **P** or for **NL** are discussed.

Definition 13.4

$SAT \triangleq \{F \mid F$ is a satisfiable propositional formula$\}$.

$CNF\text{-}SAT \triangleq \{F \mid F$ is a satisfiable propositional formula which is in conjunctive normal form$\}$.

$3\text{-}CNF\text{-}SAT \triangleq \{F \mid F$ is a satisfiable propositional formula which is in conjunctive normal form, and each clause contains at most three literals$\}$. ∎

Lemma 13.1

SAT, $CNF\text{-}SAT$ and $3\text{-}CNF\text{-}SAT$ are in **NP**.

Proof

This follows easily by Theorem 12.2. Let F be a propositional formula. An assignment $\phi: \text{var}(F) \rightarrow \{0,1\}$ can be represented by a list $((v_1, a_1), \ldots, (v_r, a_r))$, where v_1 to v_r are the variables occurring in F, and a_1 to a_r are the values from $\{0,1\}$ assigned to these variables. Thus $|\phi| = O(|F|)$. Given ϕ and F it can be verified in polynomial, even linear, time that $v(F) = 1$. Hence, by Theorem 12.2, SAT, $CNF\text{-}SAT$ and $3\text{-}CNF\text{-}SAT$ are in **NP**. ∎

Below, it will be shown that $CNF\text{-}SAT$ is $\leqslant_L$-complete for **NP**. The proof is rather involved.

Lemma 13.2

$CNF\text{-}SAT$ is $\leqslant_L$-hard for **NP**.

Proof

Let $A \subseteq \Sigma^*$ be in **NP**. Then there is a polynomial p and a single-tape non-deterministic Turing machine TM_A which accepts A.

Below, we describe how to transform an element $w \in \Sigma^*$ into a propositional formula F_w in conjunctive normal form, such that $w \in A$ if and only if F_w is satisfiable. Because the transformation can be carried out by a $\lambda n[\log(n)]$-space-bounded Turing transducer, it follows that $A \leqslant_L CNF\text{-}SAT$.

Because the construction works for an arbitrary language A, it follows that $A \leqslant_L CNF\text{-}SAT$ for all $A \in$ **NP**. Consequently, $CNF\text{-}SAT$ is $\leqslant_L$-hard for **NP**.

We turn now to the construction of the formula F_w. An accepting computation of TM_A is a sequence of length $m \leqslant p(|w|)$ of instantaneous descriptions $ID_1 \vdash \cdots \vdash ID_m$.

By convention, the IDs give a succinct description of the contents of the tape; they contain no superfluous Bs (B is the blank symbol). To construct the formula F_w it is easier to be less succinct, and specify the contents of all the tape cells which TM_A might conceivably visit during an accepting computation. This is a group of $2p(|w|)+1$ adjacent cells; the cell on which the read/write head is located at the start of the computation, and its $p(|w|)$ neighbouring cells to the left and to the right. Thus the contents of the tape are specified by a string over the tape alphabet Γ of length $2p(|w|)+1$. The position of the read/write head is similarly specified by a string over the alphabet $\{0, 1\}$ of length $2p(|w|)+1$, the intention being that the cell on which the read/write head is located is marked by a 1, whereas the other entries are 0.

A computation can be represented in this way by a $p(|w|)$ vector $\mathbf{S}$ specifying the current state of the machine, a $p(|w|) \times (2p(|w|)+1)$ matrix H specifying the position of the read/write head, and a $p(|w|) \times (2p(|w|)+1)$ matrix $\mathbf{C}$ specifying the contents of the tape (see Table 13.2).

The Turing machine TM_A accepts w if and only if it is possible to fill the vector $\mathbf{S}$ with elements of the state set Q, the matrix H with elements of $\{0, 1\}$, and the matrix $\mathbf{C}$ with elements of the tape alphabet Γ, such that the resulting vector and matrices together represent an accepting computation of TM_A on input w.

To represent the state vector $\mathbf{S}$ using propositional variables, the $p(|w|)$ vector $\mathbf{S}$ is extended to a $p(|w|) \times |Q|$ matrix S defined as follows:

$$S_{i,p} \triangleq \text{if } \mathbf{S}_i = p \text{ then } 1 \text{ else } 0$$

For example, let $Q = \{p, q, r, s\}$; then

$$\begin{array}{ccccc} & \mathbf{S} & & S \\ & & & \begin{array}{ccccc} p & q & r & s & t \end{array} \\ \text{if} & \begin{pmatrix} p \\ r \\ q \\ q \\ t \\ t \end{pmatrix} & \text{then} & \begin{pmatrix} 1 & 0 & 0 & 0 & 0 \\ 0 & 0 & 1 & 0 & 0 \\ 0 & 1 & 0 & 0 & 0 \\ 0 & 1 & 0 & 0 & 0 \\ 0 & 0 & 0 & 0 & 1 \\ 0 & 0 & 0 & 0 & 1 \end{pmatrix} \end{array}$$

Similarly, instead of the $p(|w|) \times (2p(|w|)+1)$ matrix $\mathbf{C}$, a $p(|w|) \times$

Table 13.2 Matrices representing a computation.

S: state	H: head position	**C**: tape contents
q_0	$0\cdots010\cdots0$	$B\cdots Bab\cdots x$
.	$.\cdots...\cdots.$	$.\cdots...\cdots.$
q	$0\cdots000\cdots1$	$B\cdots aba\cdots B$

$(2p(|w|)+1)\times|\Gamma|$ matrix C is used, which is defined as follows:

$$C_{i,j,\gamma} \triangleq \text{if } \mathbf{C}_{i,j}=\gamma \text{ then } 1 \text{ else } 0$$

The Turing machine TM_A accepts $w\in\Sigma^*$ if and only if the matrices S, H and C can be filled with elements from the set $\{0,1\}$ such that the resulting matrices represent an accepting computation of TM_A. It remains to specify this as the requirement that a propositional formula F_w is satisfiable.

Let $I\triangleq\{i|1\leqslant i\leqslant p(|w|)\}$ and $J\triangleq\{j|1\leqslant j\leqslant 2p(|w|)+1\}$.

(1) The formula F_w is constructed using the propositional variables:

(a) $S_{i,p}$, for $i\in I$ and $p\in Q$,
(b) $H_{i,j}$, for $i\in I$ and $j\in J$,
(c) $C_{i,j,\gamma}$, for $i\in I$, $j\in J$ and $\gamma\in\Gamma$.

Thus a valuation $v:\text{var}(F_w)\to\{0,1\}$ specifies a particular value of the matrices S, H and C.

(2) The formula F_w to be constructed is a conjunction $F_1\wedge F_2\wedge F_3\wedge F_4\wedge F_5$, and each of the formulae F_i expresses a particular property of the matrices.

(a) F_1 expresses that the rows of S and H contain precisely one 1, and similarly that for each i and j precisely one of the variables $C_{i,j,\gamma}$ is 1.

Below, a formula $U(V)$ is defined which is satisfiable if precisely one of the variables in the set V of propositional variables is 1.

$$U(V)\triangleq\left(\bigvee_{v\in V} v\right)\bigwedge_{v,w\in V,\,v\neq w}(\neg v\vee\neg w)$$

For example, $U(\{x,y,z\})=(x\vee y\vee z)\wedge(\neg x\vee\neg y)\wedge(\neg x\vee\neg z)\wedge(\neg y\vee\neg z)$. Thus,

$$F_1\triangleq\bigwedge_{i\in I}U(\{S_{i,p}|p\in Q\})\wedge$$

$$\bigwedge_{i \in I} U(\{H_{i,j} | j \in J\}) \wedge \bigwedge_{i \in I, j \in J} U(\{C_{i,j,\gamma} | \gamma \in \Gamma\}).$$

(Note that F_1 is in conjunctive normal form.)

(b) The formula F_2 expresses that the first rows correspond to the initial *ID*. Let $w = a_1 \cdots a_m$ and $p(m) = n$. Then

$$F_2 \triangleq S_{1,q_0} \wedge H_{1,n+1} \wedge \bigwedge_{1 \leqslant j \leqslant n} C_{1,j,B} \wedge$$
$$C_{1,n+1,a_1} \wedge C_{1,n+2,a_2} \wedge \cdots \wedge C_{1,n+m,a_m} \wedge$$
$$\bigwedge_{n+m \leqslant j \leqslant 2n+1} C_{1,j,B}$$

(c) The subformula F_3 expresses that contents of a tape cell may change only if the read/write head is located at that cell. For cell j at the ith computation step, this requirement is expressed by the formula 'if $H_{i,j} = 0$ then $C_{i,j,\gamma} = C_{i+1,j,\gamma}$'. Writing this formula in conjunctive normal form we arrive at

$$E_{i,j} \triangleq (H_{i,j} \vee C_{i,j,\gamma} \vee \neg C_{i+1,j,\gamma}) \wedge$$
$$(H_{i,j} \vee \neg C_{i,j,\gamma} \vee C_{i+1,j,\gamma})$$

This must hold for all i and j; thus we arrive at the formula

$$F_3 \triangleq \bigwedge_{i \in I', j \in J} E_{i,j},$$

where $I' = I - \{p(|w|)\}$.

(d) F_4 expresses that any change in state, head position and tape contents corresponds to an action item in the specification of the Turing machine TM_A. Let (p, a, b, D, q) be an action item of A. The requirement is expressed by the formula

if $(S_{i,p} \wedge C_{i,j,a} \wedge H_{i,j})$
then $(C_{i+1,j,b} \wedge H_{i+1,j+d} \wedge S_{i+1,q})$

where d is -1, 0 or $+1$ if D is L, S or R respectively.

More generally, let $(p, a, b_1, D_1, q_1), \ldots, (p, a, b_r, D_r, q_r)$ be all the action items in the specification of Turing machine TM_A having current state p and current tape symbol a. Then the statement

if $(S_{i,p} \wedge C_{i,j,a} \wedge H_{i,j})$

then $\bigvee_{k=1}^{r} (C_{i+1,j,b_k} \wedge H_{i+1,j+d_k} \wedge S_{i+1,q_k})$

expresses that the steps must be in accordance with the specification of TM_A. Let $R_{i,j}$ denote the conjunctive normal form of this statement, i.e.

$$R_{i,j} \triangleq (\neg S_{i,p} \vee \neg C_{i,j,a} \vee \neg H_{i,j} \vee C_{i+1,j,b_1} \vee C_{i+1,j,b_2} \vee \cdots \vee C_{i+1,j,b_r}) \wedge (\neg S_{i,p} \vee \neg C_{i,j,a} \vee \neg H_{i,j} \vee C_{i+1,j,b_1} \vee C_{i+1,j,b_2} \vee \cdots \vee H_{i+1,j+d_r}) \vdots (\neg S_{i,p} \vee \neg C_{i,j,a} \vee \neg H_{i,j} \vee S_{i+1,q_1} \vee S_{i+1,q_2} \vee \cdots \vee S_{i+1,q_r})$$

Finally we define the formula F_4 as follows:

$$F_4 \triangleq \bigwedge_{i \in I, j \in J} R_{i,j}$$

The above would suffice if every accepting or non-accepting computation had a length of exactly $p(|w|)$ steps. We assume that the enumeration

$$(p, a, b_1, D_1, q_1) \cdots (p, a, b_r, D_r, q_r)$$

includes the action item (p, a, a, S, p) whenever p is an accepting state or the specification of TM_A does not have any action items with current state p and current tape symbol a. This ensures that every computation can be extended to last $p(|w|)$ steps, moreover, once an accepting state is reached the computation can be extended without changing this state.

(e) Finally, the formula F_5 will guarantee that the computation reaches a final state.

$$F_5 \triangleq S_{p(|w|),q_1} \vee \cdots \vee S_{p(|w|),q_t}$$

where $q_1 \cdots q_t$ is an enumeration of the final states of TM_A.

Clearly, the formula F_w can be constructed in polynomial time. Its construction requires a finite number of counters which can count up to $\max\{I, J\}$ and therefore require $O(\log(|w|))$ tape cells each. Thus the construction of the formula F_w can be accomplished by an $\lambda n[\log(n)]$-space-bounded Turing transducer. ■

By the preceding two lemmas we have:

Theorem 13.1

CNF-SAT is $\leqslant_L$-complete for **NP**. ■

The formula F_w constructed in the proof of Lemma 13.2 contains clauses with $2p(|w|)+1$ literals. It is possible to reduce the number of literals per clause to three, thus proving that 3-*CNF-SAT* is **NP**-complete.

Theorem 13.2

3-*CNF-SAT* is **NP**-complete.

Proof

Consider a clause $C_1 = S \vee X \vee Y$, where S is a subclause consisting of one or more literals, and X and Y are literals. We construct a new formula $C_2 \triangleq (S \vee Z) \wedge (\neg Z \vee X \vee Y)$, where Z is a new variable not already occurring in C_1.

The formulae C_1 and C_2 are *not* logically equivalent, but C_1 is satisfiable if and only if C_2 is satisfiable.

To see this, assume that C_1 is satisfiable. Thus there is an assignment $v\colon \mathrm{var}(C_1) \to \{0,1\}$ such that $v(C_1)=1$. Therefore $v(S)=1$ or $v(X \vee Y)=1$. It is always possible to extend v to $w\colon(\mathrm{var}(C_1) \cup \{Z\}) \to \{0,1\}$ such that $w(C_2)=1$.

On the other hand, if $\phi\colon \mathrm{var}(C_2) \to \{0,1\}$ is a satisfying assignment, i.e. $\phi(C_2)=1$, then $\phi(S)=1$ or $\phi(X \vee Y)=1$. In other words $\phi(C_1)=1$.

Note that $S \vee Z$ contains one literal less than C_1 does. By repeatedly applying the above technique, we can replace any clause having more than three literals by a formula in conjunctive normal form, all clauses of which contain at most three literals. The length of the resulting formula is linear in the size of the given formula.

This transformation of a given formula can clearly be achieved in polynomial time; careful analysis shows that the transformation can also be implemented on a $\lambda n[\log(n)]$-space-bounded Turing transducer. ■

13.2 Other NP-complete problems

In this section the following problems will be shown to be **NP**-complete.

Definition 13.5

(1) *Clique*. The problem is to determine whether a given undirected graph contains a k-clique, that is, a subgraph having k vertices and such that there is an edge between any two of these vertices:

$CLIQUE \triangleq \{((V,E),k) \mid$ where (V,E) is an undirected graph, k an integer, and (V, E) contains at least one k-clique as a subgraph, that is, there is a subset $W \subseteq V$ with k vertices and such that for all $u, v \in W$ there is an edge between u and $v\}$.

(2) *Vertex cover.* Given an undirected graph (V, E) and a natural number k, the problem is to determine whether there exists a subset $W \subseteq V$ of k nodes, such that every edge in E is incident upon a node in W:

$VERTEX\text{-}COVER \triangleq \{((V, E), k) \mid$ where k is a natural number and (V, E) an undirected graph, and there is a subset $W \subseteq V$ having k nodes, such that for every edge $(u, v) \in E$, $u \in W$ or $v \in W\}$.

(3) *Subgraph isomorphism.* The problem is to determine whether a given undirected graph (V, E) can be embedded in another given undirected graph (W, T):

$U\text{-}SUB\text{-}ISO \triangleq \{((V, E), (W, T)) \mid (V, E)$ and (W, T) are undirected graphs and there is an injective function $f: V \to W$ such that there is an edge between u and v if and only if there is an edge between $f(u)$ and $f(v)$, for all $u, v \in V$, in other words, (V, E) is isomorphic to a subgraph of $(W, T)\}$.

(4) The problem is the same as above, but now for directed graphs:

$D\text{-}SUB\text{-}ISO \triangleq \{((V, E), (W, T)) \mid (V, E)$ and (W, T) are directed graphs and (V, E) is isomorphic to a subgraph of $(W, T)\}$.

(5) The problem is to determine whether a given undirected graph contains a *Hamiltonian circuit*, that is, a simple cycle containing all the vertices of the graph:

$UHC \triangleq \{(V, E) \mid$ where (V, E) is an undirected graph which contains a Hamiltonian circuit, that is, a cycle containing every vertex of the graph exactly once$\}$.

(6) The problem is the same as above, but now for directed graphs:

$DHC \triangleq \{(V, E) \mid$ where (V, E) is a directed graph which contains a Hamiltonian circuit$\}$.

(7) *The travelling salesman problem.* Given is a directed graph, the

vertices standing for cities, and the edges standing for roads which connect the cities. The edges are labelled by natural numbers, which denote the distance. The problem is to determine whether or not there exists a tour which visits every city exactly once, and has a total travel distance less than a given upper bound. (This problem was used in Section 9.3 to illustrate the relation between an optimization problem, its predicate version and its language version.)

$TSP \triangleq \{((V, E), w, k) |$ where $k \in N$, (V, E) is a directed graph, $w: E \rightarrow N$ is a function labelling the edges of the graph, and (V, E) has a subgraph (V, T) which is a Hamiltonian circuit and $\sum_{e \in T} w(e) \leqslant k\}$.

(8) *Zero–one linear inequalities.* Given an $m \times n$ matrix $\mathbf{M}$ and an m-dimensional vector $\mathbf{b}$ over the integers, the problem is to determine whether there exists an n-dimensional vector $\mathbf{x}$ over $\{0, 1\}$ such that $\mathbf{Mx} \geqslant \mathbf{b}$:

$\{0, 1\}$-$LINEQ \triangleq \{(\mathbf{M}, \mathbf{b}) | \mathbf{M}$ is an $m \times n$ matrix over Z, $\mathbf{b} \in Z^m$, and there exists an $\mathbf{x} \in \{0, 1\}^n$ such that $\mathbf{Mx} \geqslant \mathbf{b}\}$.

(9) *The equal-execution time-scheduling problem.* Given are a set J of jobs, a partial order on J, a time limit t, and a number p of processors. The problem is to determine whether there is a schedule which assigns the jobs to the processors in such a way that the given ordering of the jobs is satisfied, no more than p processors are used at any moment of time, and all jobs have finished within the given time limit:

$SCHED \triangleq \{((J, \prec), t, p) |$ where $t, p \in N$ and $(J, \prec)$ is a partially ordered set and there is a function $s: J \rightarrow \{1, 2, \ldots, t\}$ such that

(a) $(\forall j_1, j_2 \in J)[\text{if } j_1 \prec j_2 \text{ then } s(j_1) < s(j_2)]$, where $<$ is the customary relation 'less than' on N;

(b) $(\forall i \in \{1, 2, \ldots, t\})[$the number of elements of $\{j | s(j) = i$ is at most $p]\}$. ■

Lemma 13.3

All the problems given in Definition 13.5, *CLIQUE*, *VERTEX-COVER*, *U-SUB-ISO*, *D-SUB-ISO*, *UHC*, *DHC*, *TSP*, $\{0, 1\}$-*LINEQ* and *SCHED* are in **NP**.

Proof

This is obvious from Theorem 12.2. Consider for example the problem *CLIQUE*. The relation $R((V, E), k, W) \triangleq$ [if ($W \subseteq V$ and W has k vertices which together constitute a k-clique) then TRUE else FALSE] is clearly polynomially decidable. Also, the size of W is linear in the size of $((V, E), k)$. Because $((V, E), k) \in CLIQUE$ if and only if $(\exists W)[R((V, E), k, W)]$, Theorem 12.2 implies that *CLIQUE* is in **NP**.

The other problems are dealt with in a similar way. In all cases it is evident that checking whether a given candidate is indeed a solution to the problem can be done in polynomial time; it is also evident that the size of any reasonable candidate is polynomial in the size of the problem instance. ■

It remains to show that the above problems are $\leqslant_L$-hard for **NP**.

Lemma 13.4

CLIQUE is $\leqslant_L$-hard for **NP**.

Proof

We show that $CNF\text{-}SAT \leqslant_L CLIQUE$.

Let $F = F_1 \wedge F_2 \wedge \cdots \wedge F_q$ be a propositional formula in conjunctive normal form. We construct a graph which has a node for each and every occurrence of a literal in F, and has an edge between two nodes if the nodes correspond to occurrences of non-contradictory literals from different conjuncts F_i and F_j.

More precisely, the set of nodes is constructed as follows.

For each i, number the occurrences of literals in the conjunct F_i from 1 to m_i, $F_i = X_{i,1} \vee X_{i,2} \vee \cdots \vee X_{i,m_i}$, and let $K_i \triangleq \{(i, 1), (i, 2), \ldots, (i, m_i)\}$.

For example,

$$\begin{array}{ccccccccccccc} F = (x & \vee & \neg y) & \wedge & (y & \vee & \neg z) & \wedge & (z & \vee & \neg x & \vee & u) \\ | & & | & & | & & | & & | & & | & & | \\ X_{1,1} & & X_{1,2} & & X_{2,1} & & X_{2,2} & & X_{3,1} & & X_{3,2} & & X_{3,3} \end{array}$$

$$K_1 = \{(1, 1), (1, 2)\}$$
$$K_2 = \{(2, 1), (2, 2)\}$$
$$K_3 = \{(3, 1), (3, 2), (3, 3)\}$$

The graph $G_F = (V, E)$ associated with the given propositional formula F is defined as follows.

$$V = \bigcup_{i=1}^{q} K_i$$

$E = \{\{(i,j),(k,m)\} \mid i < k$, and there is an assignment ϕ:var$(F) \to \{0,1\}$ such that $\phi(X_{i,j}) = \phi(X_{k,m}) = 1$, i.e. $X_{i,j}$ and $X_{k,m}$ are not contradictory$\}$.

In the above $\{(i,j),(k,m)\}$ denotes an *undirected edge*, the requirement '$i < k$' instead of '$i \neq k$' makes the representation of these undirected edges unique.

This graph G_F has a clique of size q if and only if the given formula F is satisfiable. This can be seen as follows.

(1) Let ϕ:var$(F) \to \{0,1\}$ be a satisfying assignment, i.e. $\phi(F) = 1$. Then $\phi(F_i) = 1$ for all i, $1 \leqslant i \leqslant q$. Consequently, for each i there is a j such that $\phi(X_{i,j}) = 1$.

Then there is a set $S \subseteq V$, such that $S \subseteq \{(i,j) \mid \phi(X_{i,j}) = 1\}$, and for each i there is precisely one j such that $(i,j) \in S$.

Thus there is an edge between any two nodes from S, and consequently, S spans a q-clique in G_F.

(2) Assume now that $T \subseteq V$ spans a q-clique in the graph G_F. Then $(i,j),(k,m) \in T$ implies $i \neq k$, because there are no edges $((i,j),(i,m))$ in G_F. Thus, for each i, $1 \leqslant i \leqslant q$, there is precisely one j such that $(i,j) \in T$.

Since there is an edge between any of these, it follows that there is an assignment ϕ:var$(F) \to \{0,1\}$ such that $\phi(X_{i,j}) = 1$ for all $(i,j) \in T$. Therefore $\phi(F) = 1$, and F is satisfiable.

It remains to show that the graph G_F can be constructed by a $\lambda n[\log(n)]$-space-bounded Turing transducer.

Let n denote the size of F. A node of G_F is represented using two numbers, and an edge of G_F is represented using four numbers. All of these numbers are less than n, and thus require no more space than $\log(n)$.

Enumerating the set V of nodes of G_F can be achieved by making a single sweep along F using two counters. The initial value of both counters is 1. The second counter is incremented when we pass an $\vee$ sign. When we pass an $\wedge$ sign, the first counter is incremented and the second counter is reset to 1.

The set of edges can be enumerated similarly, using four counters. For a particular vertex (i,j) the set

$$\{((i,j),(k,m)) \mid i < k, \text{ and } X_{i,j} \text{ and } X_{k,m} \text{ are non-contradictory}\}$$

of all edges incident upon (i,j), can be enumerated by making a single sweep along the given formula F.

In summary, G_F can be constructed by a $\lambda n[n^2]$-time-bounded, $\lambda n[\log(n)]$-space-bounded Turing transducer. ■

Lemma 13.5

U-SUB-ISO is $\leqslant_L$-hard for **NP**.

Proof

Determining whether a graph contains a clique is a special case of determining whether an arbitrary given graph can be embedded in another given graph.

A given instance $((V, E), k)$ of the clique problem is transformed into the instance $((W, T), (V, E))$ of the subgraph isomorphism problem as follows.

Let $card(V)$ denote the number of vertices in V.

(1) If $k > card(V)$ then $W = V \cup \{x_{card(V)+1}\}$ and $T = E$.

(2) Otherwise $W = \{x_1, x_2, \ldots, x_k\}$, and $T = \{\{x_i, x_j\} \mid 1 \leqslant i < k$, and $i < j \leqslant k\}$.

This instance $((W, T), (V, E))$ can be constructed using space linear in $\log(card(V))$, independent of k. Thus the reducing function can be computed by a $\lambda n[\log(n)]$-space-bounded Turing transducer. ■

Lemma 13.6

VERTEX-COVER is $\leqslant_L$-hard for **NP**.

Proof

We show that $CLIQUE \leqslant_L VERTEX\text{-}COVER$.

Let $G = (V, E)$ be any undirected graph. The *complement* G^c of G is the undirected graph (V, E^c), where $E^c \triangleq \{\{u, v\} \mid \{u, v\} \notin E\}$, that is, there is an edge between nodes u and v in G^c if and only if there is no edge between u and v in G. Assume that W is a vertex cover for G. By definition, there are no edges between two nodes which both belong to $V - W$. Therefore there is an edge between any two nodes belonging to $V - W$ in the complement G^c of G. Thus we have the following property.

> *Property*: For any undirected graph $G = (V, E)$ and every $W \subseteq V$ W is a vertex cover in G if and only if $V - W$ is a clique in the complement G^c of G.

Consequently $CLIQUE \leqslant_L VERTEX\text{-}COVER$ via the function h which maps the instance $((V, E), k)$ to $((V, E^c), m)$, where m is the number of nodes in V minus k, which obviously can be computed by a $\lambda n[\log(n)]$-space-bounded Turing transducer. ■

The remaining problems from the list in Definition 13.5 are also shown to be $\leqslant_L$-hard for **NP** using the technique of space-bounded reduction. What is reduced to what is shown in Figure 13.1.

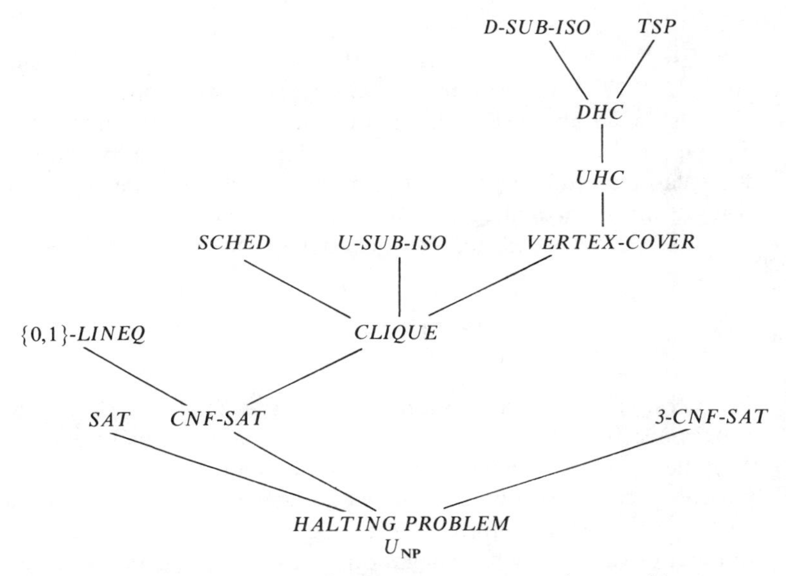

Figure 13.1 Scheme of reductions between various problems.

Of the reductions shown in Figure 13.1, we have only specified

$$CNF\text{-}SAT \leqslant_L CLIQUE,$$

$$CLIQUE \leqslant_L VERTEX\text{-}COVER$$

and

$$CLIQUE \leqslant_L U\text{-}SUB\text{-}ISO.$$

It is clearly true that $U_{\mathbf{NP}} \leqslant_L CNF\text{-}SAT$, where $U_{\mathbf{NP}}$ represents the resource-bounded Halting Problem as specified in Definition 12.9, but we proved in Lemma 13.2 that *CNF-SAT* is $\leqslant_L$-hard for **NP** by a generic transformation showing that $X \leqslant_L CNF\text{-}SAT$ for an arbitrary $X \in \mathbf{NP}$. The remaining proofs will be given as suggested by the reduction scheme in Figure 13.1.

The reader should convince him- or herself that the transformation described can indeed be implemented by a $\lambda n[\log(n)]$-space-bounded Turing transducer. This mostly involves no more than observing that it suffices to keep track of a finite number of indices for variables and such. Generating sets of nodes, edges, etc. can then be done using a finite number of counters, each of which stores the value of a particular index.

Lemma 13.7

$CNF\text{-}SAT \leqslant_L \{0,1\}\text{-}LINEQ$, thus $\{0,1\}\text{-}LINEQ$ is $\leqslant_L$-hard for **NP**.

Proof

Let $F = F_1 \wedge F_2 \wedge \cdots \wedge F_p$ with $\text{var}(F) = \{z_1, z_2, \ldots, z_q\}$ be an instance of *CNF-SAT*. We assume that the conjuncts are nontrivial, that is, there is no variable z such that both z and $\neg z$ occur in some conjunct F_i. Otherwise, this conjunct can be skipped. That a propositional formula in conjunctive normal form is satisfiable can be expressed as a set of linear inequalities, as in the example below.

Consider the formula $(u \vee v \vee w) \wedge (u \vee \neg v) \wedge (v \vee \neg w)$.

$$\begin{array}{ll} u \vee v \vee w & u + v + w \geqslant 1 \\ u \vee \neg v & u + (1 - v) \geqslant 1 \\ v \vee \neg w & v + (1 - w) \geqslant 1 \end{array}$$

Each conjunct gives rise to a formula $\Sigma_j z_j + \Sigma_m (1 - z_m) \geqslant 1$, where j ranges over the indices of the variables which occur as literals in the conjunct, and m ranges over the indices of the variables whose negations occur as literals in the conjunct. Bringing the constants 1 to the right results in a formula $\Sigma_j z_j + \Sigma_m(-z_m) \geqslant 1 - k$, where k is the number of variables whose negations occur as literals in the conjunct. It remains to express these inequalities in matrix notation.

In general, the given instance $F = F_1 \wedge F_2 \wedge \cdots \wedge F_p$ with $\text{var}(F) = \{z_1, z_2, \ldots, z_q\}$ is transformed into an instance $(\mathbf{M}, \mathbf{b})$ of $\{0,1\}$-*LINEQ* as follows:

(1) $\mathbf{M}$ is a $p \times q$ matrix over the integers, where entry m_{ij} satisfies:

$$m_{ij} = \begin{cases} 1 & \text{if variable } z_j \text{ occurs as a literal in } F_i; \\ -1 & \text{if } \neg z_j \text{ occurs as a literal in } F_i; \\ 0 & \text{if neither variable } z_j \text{, nor } \neg z_j \text{ occurs as a literal in } F_i. \end{cases}$$

(2) $\mathbf{b}$ is a p-vector over the integers satisfying $b_i = 1 - k_i$, where k_i is the number of variables z such that the literal $\neg z$ occurs in F_i.

We will now show that $F \in SAT$ if and only if $(\mathbf{M}, \mathbf{b}) \in \{0,1\} - LINEQ$.

(1) Let $\phi : \text{var}(F) \to \{0,1\}$ be a satisfying assignment, and let $x \in \{0,1\}^q$ be defined by $x_i = \phi(z_i)$, $1 \leqslant i \leqslant q$.

Define $u_{i,1}, u_{i,0}, n_{i,1}$ and $n_{i,0}$ as follows:

$u_{i,1}$ is the number of variables z_j such that z_j occurs as a literal in F_i and $\phi(z_j) = 1$,

$u_{i,0}$ is the number of variables z_j such that z_j occurs as a literal in F_i and $\phi(z_j) = 0$,

$n_{i,1}$ is the number of variables z_j such that $\neg z_j$ occurs as a literal in F_i and $\phi(z_j) = 1$,

$n_{i,0}$ is the number of variables z_j such that $\neg z_j$ occurs as a literal in F_i and $\phi(z_j) = 0$.

Note that for all i, $1 \leqslant i \leqslant p$, we have $n_{i,1}+n_{i,0}=k_i$, and since ϕ is a satisfying assignment we also have $u_{i,1}+n_{i,0} \geqslant 1$. Then

$$\begin{aligned}(\mathbf{Mx})_i = u_{i,1}-n_{i,1} = u_{i,1}-(k_i-n_{i,0}) &= u_{i,1}+n_{i,0}-k_i \\ \geqslant 1-k_i = b_i\end{aligned}$$

Consequently, $\mathbf{Mx} \geqslant \mathbf{b}$.

(2) Assume now that $\mathbf{Mx} \geqslant \mathbf{b}$ for some 0, 1-vector $\mathbf{x}$. Let ϕ:var$(F) \to \{0,1\}$ be the assignment $\phi(z_j)=x_j$, and let $u_{i,1}, u_{i,0}, n_{i,1}$ and $n_{i,0}$ be defined as above. Then

$$\begin{aligned}(\mathbf{Mx})_i &\geqslant b_i \\ u_{i,1}-n_{i,1} &\geqslant 1-k_i \\ u_{i,1}-(k_i-n_{i,0}) &\geqslant 1 \\ u_{i,1}+n_{i,0} &\geqslant 1\end{aligned}$$

Thus ϕ is a satisfying assignment, whence F is satisfiable.

It remains to verify that the matrix $\mathbf{M}$ and the vector $\mathbf{b}$ can be constructed by a $\lambda n[\log(n)]$-space-bounded Turing transducer.

Because $k_i < |F|$, the value of b_i can be computed while using no more than $\log(|F|)$ tape cells. The elements of $\mathbf{M}$ can be computed most easily column by column. For each column, this takes $\log(|F|)$ space to store the index of the variable corresponding to the current column, and a single sweep along F.

In all, the instance $(\mathbf{M}, \mathbf{b})$ can be computed by a $\lambda n[\log(n)]$-space-bounded, $\lambda n[n^2]$-time-bounded Turing transducer, and in consequence *CNF-SAT* $\leqslant_L \{0,1\}$-*LINEQ*. ■

By Lemmas 13.3 and 13.7, the problem $\{0,1\}$-*LINEQ* is $\leqslant_L$-complete for **NP**.

The linear inequalities problem comes in different forms. First we may have the requirement $\mathbf{Mx} \leqslant \mathbf{b}$ instead of $\mathbf{Mx} \geqslant \mathbf{b}$. We only have to change the signs of the entries in the matrix $\mathbf{M}$ and the vector $\mathbf{b}$ in the above proof to see that $(\exists \mathbf{x})[\mathbf{Mx} \leqslant \mathbf{b}]$ also is $\leqslant_L$-complete for **NP**.

Another version is $\{0,1\}$-*LP*, zero–one linear programming, where we require $\mathbf{Mx} = \mathbf{b}$.

$\{0,1\}$-*LP* $\triangleq \{(\mathbf{M}, \mathbf{b}) \mid \mathbf{M}$ is an $n \times m$ matrix over the integers and $\mathbf{b}$ is an n-vector over the integers, and there exists an m-vector $\mathbf{x}$ over $\{0,1\}$ such that $\mathbf{Mx} = \mathbf{b}\}$.

Consider the following $p \times q(p+1)$ matrix $\mathbf{M}'$ obtained from the matrix $\mathbf{M}$

in the proof of Lemma 13.7 by extending the rows with a group of q entries all equal to -1, and $p-1$ groups of q entries all equal to 0.

$$\mathbf{M}' = \left(\mathbf{M} \left| \begin{array}{cccc} -1 \cdots -1 & 0 \cdots \ 0 & \cdots & 0 \cdots \ 0 \\ 0 \cdots \ 0 & -1 \cdots -1 & \cdots & 0 \cdots \ 0 \\ 0 \cdots \ 0 & 0 \cdots \ 0 & \cdots & 0 \cdots \ 0 \\ \vdots & \vdots & & \vdots \\ 0 \cdots \ 0 & 0 \cdots \ 0 & \cdots & -1 \cdots -1 \end{array} \right. \right)$$

Formally,

$$m_{ij} = \begin{cases} 1 & \text{if } 1 \leqslant j \leqslant q \text{ and variable } z_j \text{ occurs as a literal in } F_i; \\ -1 & \text{if } 1 \leqslant j \leqslant q \text{ and } \neg z_j \text{ occurs as a literal in } F_i; \\ -1 & \text{if } iq \leqslant j \leqslant (i+1)q; \\ 0 & \text{otherwise.} \end{cases}$$

Consider the product $\mathbf{M}'\mathbf{y}$ where $\mathbf{y} = (\mathbf{x}, \mathbf{u})^{\mathrm{T}}$

$$\mathbf{M}'\mathbf{y} = (\mathbf{M}|\mathbf{C})\left(\frac{\mathbf{x}}{\mathbf{u}}\right)$$

Then $\mathbf{M}'\mathbf{y} = \mathbf{Mx} + \mathbf{Cu}$, and therefore $(\mathbf{Mx})_i - q \leqslant (\mathbf{M}'\mathbf{y})_i \leqslant (\mathbf{Mx})_i$. Consequently, if $(\mathbf{M}'\mathbf{y})_i = b_i$, then $(\mathbf{Mx})_i \geqslant b_i$. On the other hand, if $(\mathbf{Mx})_i \geqslant b$, we can make $(\mathbf{M}'\mathbf{y})_i = b_i$, by properly setting the entries of $\mathbf{u}$. Since $\{0, 1\}$-*LP* clearly is **NP**, it follows that $\{0, 1\}$-*LP* is $\leqslant_L$-complete for **NP**.

If we now relax the restriction on $\mathbf{x}$ in that we only require $\mathbf{x}$ to be a vector over the natural numbers instead of over $\{0, 1\}$, we obtain the problem of integer linear programming, *ILP*:

> *ILP* $\triangleq \{(\mathbf{M}, \mathbf{b}) | \mathbf{M}$ is an $n \times m$ matrix over the integers and $\mathbf{b}$ is an n-vector over the integers, and there exists an m-vector $\mathbf{x}$ over the non-negative integers such that $\mathbf{Mx} = \mathbf{b}\}$.

It is obvious from the above that *ILP* is $\leqslant_L$-hard for **NP**. To apply Theorem 12.2, we need a bound on the size of the solution, if at least a solution exists. In Papadimitriou and Steiglitz (1982) a proof can be found of the following property.

> *Property*: There is a polynomial p such that for any $n \times m$ matrix $\mathbf{M}$ over the integers, and any n-vector $\mathbf{b}$ over the integers, there exists a solution to the equation $\mathbf{Mx} = \mathbf{b}$ if and only if there is an $\mathbf{x}$ such that $\mathbf{Mx} = \mathbf{b}$, and $|x_i| \leqslant p(|\mathbf{M}, \mathbf{b})|)$, for all i, $1 \leqslant i \leqslant m$.

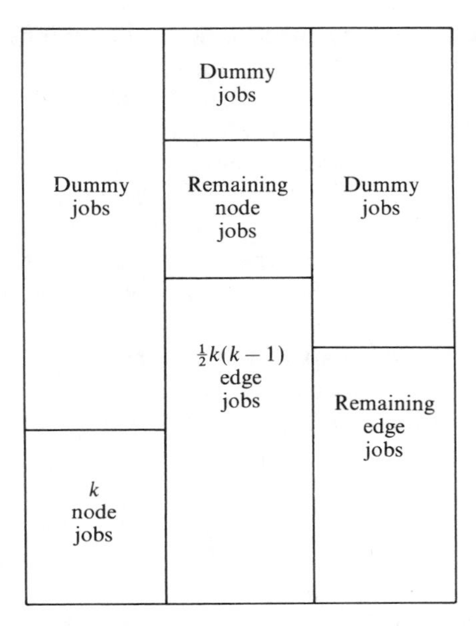

Figure 13.2 The reduction $CLIQUE \leqslant_L SCHED$.

Thus we may apply Theorem 12.2, and conclude that ILP is $\leqslant_L$-complete for **NP**.

If we further relax the restriction so as to allow $\mathbf{x}$ to be a vector over the rationals, the problem $\mathbf{Mx} \geqslant \mathbf{b}$ and also the linear programming problem have been shown to be in **P**. Currently two polynomial-time-bounded algorithms are known, Khachian's algorithm (Khachian, 1979), and Karmarkar's algorithm (Karmarkar, 1984).

Lemma 13.8

$SCHED$ is $\leqslant_L$-hard for **NP**.

Proof

In accordance with the scheme given in Figure 13.1, we show that $CLIQUE \leqslant_L SCHED$, it then follows that $SCHED$ is $\leqslant_L$-hard for **NP**.

Let $((V, E), k)$ be an instance of $CLIQUE$, which is to be transformed into an instance $((J, \prec), t, p)$ of $SCHED$. We use node jobs, edge jobs and dummy jobs, and a time limit $t = 3$. The parameters of the instance are defined so as to make Figure 13.2 a valid schedule for $((V, E), k) \in CLIQUE$.

If (V, E) contains a k-clique, we begin at time $t = 1$ by performing the k node jobs corresponding to the nodes of the clique. At time $t = 2$ we perform the $\frac{1}{2}k(k-1)$ edge jobs corresponding to the edges of the clique, and also the remaining $card(V) - k$ node jobs. At $t = 3$ the remaining $card(E) - \frac{1}{2}k(k-1)$ edge jobs are performed. We must also make sure that only the intended node or edge jobs are executed. This is the purpose of the dummy jobs, of which we have three kinds, to be performed at $t=1$, $t=2$ and $t=3$ respectively.

Thus the instance $((V,E),k)$ is transformed into the instance $((J,\prec),t,p)$ of *SCHED* as follows:

(1) If $card(V)<k$ or $card(E)<\frac{1}{2}k(k-1)$, then $((V,E),k)\notin CLIQUE$ Therefore, we transform into a fixed instance $(J,\prec),t,p)$ such that $((J,\prec),t,p)\notin SCHED$, for example $J=\{x,y\}$, $\prec=\{(x,y)\}$, i.e. x must precede y, $p=1$, and $t=1$.

For this part of the transformation, we only need space to check the condition $card(V)<k \vee card(E)<\frac{1}{2}k(k-1)$. Clearly, no more than $\log(|((V,E),k)|)$ space is required.

(2) If the instance of *CLIQUE* is nontrivial, we proceed as follows:

Let

$$m \triangleq 1+\max\{k, card(V)-k+\tfrac{1}{2}k(k-1), card(E)-\tfrac{1}{2}k(k-1)\}$$
$$D_1 \triangleq \{d_{1,j} \mid 1\leqslant j\leqslant m-k\}$$
$$D_2 \triangleq \{d_{2,j} \mid 1\leqslant j\leqslant m-(card(V)-k+\tfrac{1}{2}k(k-1))\}$$
$$D_3 \triangleq \{d_{3,j} \mid 1\leqslant j\leqslant m-(card(E)-\tfrac{1}{2}k(k-1))\}$$
$$J_V \triangleq \{j_v \mid v\in V\}$$
$$J_E \triangleq \{j_e \mid e\in E\}$$

The instance $((V,E),k)$ of *CLIQUE* is transformed into the instance

$$((J_V\cup J_E\cup D_1\cup D_2\cup D_3,\prec),3,m),$$

where the precedence relation $\prec$ is defined by:

(a) $j_v\prec j_e$ if and only if e is incident upon v,
(b) $d_1\prec d_2$ if and only if $d_1\in D_1$ and $d_2\in D_2$,
(c) $d_2\prec d_3$ if and only if $d_2\in D_2$ and $d_3\in D_3$,
(d) for any jobs $x,y\in J$, we have $x\prec y$ only if this is specified by (a), (b) or (c) above.

A Turing transducer implementing the above transformation needs $\log(|((V,E),k)|)$ tape cells to compute m. Enumerating the set J of jobs also requires $\log(|((V,E),k)|)$ tape cells. Finally, $2\log(|((V,E),k)|)$ cells suffice to enumerate the precedence relation. Consequently, the transformation described above can be implemented by a $\lambda n[\log(n)]$-space-bounded, $\lambda n[n^2]$-time-bounded Turing transducer, whence $CLIQUE\leqslant_L SCHED$, and *SCHED* is $\leqslant_L$-hard for **NP**. ■

Lemma 13.9

$VERTEX\text{-}COVER\leqslant_L UHC$, therefore *UHC* is $\leqslant_L$-hard for **NP**.

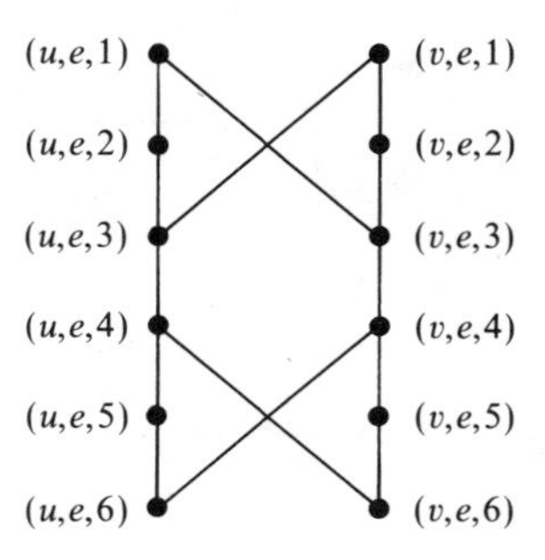

Figure 13.3 Subgraph G_e for an edge $e = \{u, v\}$.

Proof

Let $((V, E), k)$ be an instance of *VERTEX-COVER* which is to be transformed into an instance (W, T) of *UHC*. If $card(V) < k$ then set $W = \{x, y, z\}$ and $T = \{\{x, y\}, \{y, z\}\}$.

Consider now the nontrivial case, when $card(V) \geqslant k$.

(1) For each edge $e \in E$ the graph (W, T) has a subgraph $G_e = (W_e, T_e)$ of the type depicted in Figure 13.3.

Thus for an edge $e = \{u, v\}$ the subgraph $G_e = (W_e, T_e)$ is defined as follows:

$$W_e \triangleq \{(u, e, i), (v, e, i) \mid 1 \leqslant i \leqslant 6\}$$

$$T_e \triangleq \{\{(u, e, i), (u, e, i+1) \mid 1 \leqslant i < 6\} \cup$$

$$\{(v, e, i), (v, e, i+1) \mid 1 \leqslant i < 6\} \cup$$

$$\{\{(u, e, 1), (v, e, 3)\}, \{(u, e, 3), (v, e, 1)\}\} \cup$$

$$\{\{(u, e, 4), (v, e, 6)\}, \{(u, e, 6), (v, e, 4)\}\}\}$$

(2) Subgraphs corresponding to edges incident upon the same node are connected together. For each node $v \in V$, let the edges incident upon v be numbered from 1 to n_v. The numbering is arbitrary; we assume the numbering as implied by the enumeration of E. Thus, for each $v \in V$ the graph (W, T) contains the set

$$T_v \triangleq \{\{(v, e_i, 6), (v, e_{i+1}, 1)\} \mid 1 \leqslant i < n_v\}$$

(see Figure 13.4).

(3) Finally, (W, T) contains the set $\{a_1, a_2, \ldots, a_k\}$ of nodes. These are used to identify a particular vertex cover, and are connected

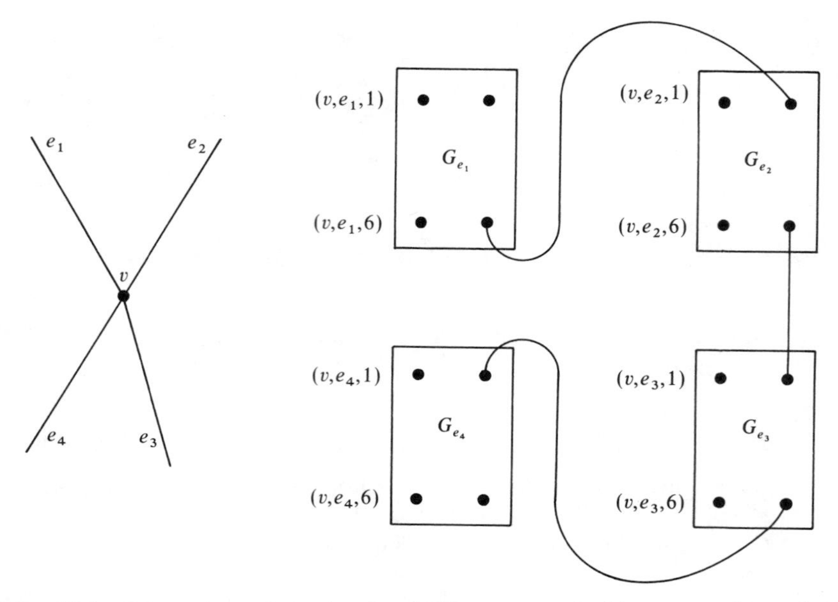

Figure 13.4 Connecting subgraphs G_{e_i} of edges e_1 to e_4 incident upon the node v.

to all the strings of subgraphs corresponding to nodes and the edges incident upon them. That is, T contains the following set T_0 of edges.

$$T_0 \triangleq \{\{a_i,(v,e_1,1)\},\{a_i,(v,e_{n_v},6)\} \mid 1 \leqslant i \leqslant k, v \in V\}$$

In summary, the instance $((V,E),k)$ of $VERTEX\text{-}COVER$ is transformed into the instance (W,T) of UHC defined as follows:

(1) $W = \{a_i \mid 1 \leqslant i \leqslant k\} \cup \left(\bigcup_{e \in E} W_e \right)$

(2) $T = \left(\bigcup_{e \in E} T_e \right) \cup \left(\bigcup_{v \in V} T_v \right) \cup T_0$

Assume that (W,T) has a Hamiltonian circuit. The nodes in the set $\{a_1,a_2,\ldots,a_k\}$ divide the circuit into paths, which begin and end with a node from the set $\{a_1,a_2,\ldots,a_k\}$, but do not pass through such a node. Such a subpath of the circuit necessarily passes through a string of subgraphs, corresponding to the edges from E that are incident upon a particular node v. Since the Hamiltonian circuit must pass all the subgraphs G_e, corresponding to all the edges $e \in E$, all edges are incident upon one of the nodes corresponding to the subpaths of the

Hamiltonian circuit. These nodes therefore constitute a vertex cover consisting of k nodes.

It is left to the reader to check that (W, T) has a Hamiltonian circuit if $((V, E), k) \in$ *VERTEX-COVER*, and that the construction described above can be implemented by a $\lambda n[\log(n)]$-space-bounded Turing transducer. ■

The three reductions which remain to be proved are very simple.

Lemma 13.10

DHC, *D-SUB-ISO* and *TSP* are $\leqslant_L$-hard for **NP**.

Proof

UHC $\leqslant_L$ *DHC*. An instance (V, E) of *UHC* is transformed into the instance (V, T), where $T = \{(u, v), (v, u) | \{u, v\} \in E\}$.

DHC $\leqslant_L$ *D-SUB-ISO*. An instance (V, E) of *DHC* is transformed into the instance $((V, T), (V, E))$ where $T = \{(v, \sigma(v)) | v \in V\}$ and $\sigma: V \rightarrow V$ is some arbitrarily cyclic permutation of V – for instance, the cyclic permutation implied by the enumeration of V.

DHC $\leqslant_L$ *TSP*. An instance (V, E) of *DHC* is transformed into the instance $((V, E), w, k)$, where $w: E \rightarrow N$ assigns the weight 1 to every edge of E, and $k = card(V)$. ■

The preceding lemmas together prove the following theorem.

Theorem 13.3

All the problems given in Definition 13.5, *CLIQUE*, *VERTEX-COVER*, *U-SUB-ISO*, *D-SUB-ISO*, *UHC*, *DHC*, *TSP*, $\{0,1\}$-*LINEQ* and *SCHED* are $\leqslant_L$- and therefore $\leqslant_P^K$- and $\leqslant_P^C$-complete for **NP**. ■

The problems discussed above are but a small selection of all problems known to be **NP**-complete. **NP**-complete problems are abundant. Garey and Johnson (1979) give a long list of **NP**-complete problems from diverse fields such as graph theory, automata theory, the theory of formal languages, logic, mathematical programming, and various other fields which have optimization problems. New **NP**-complete problems are published almost daily.

13.3 Completeness in the strong sense

Some of the problems we have discussed in the preceding section contain in their instances numbers, next to other things such as nodes, edges and such. For example, an instance of the *CLIQUE* problem is a tuple $((V, E), k)$, where

k is a natural number. (We have used numbers as names for nodes and such; these are *not* counted as numbers in the present discussion.)

We know that if $\mathbf{NP} \neq \mathbf{P}$, then there can be no algorithm solving an **NP**-complete problem which is polynomial in the size of the problem instance. It might be, however, that there does exist an algorithm which is polynomial in the magnitude of the numbers involved; that is, an algorithm polynomial in the problem size provided we use *unary* notation for the numbers involved. We will consider this possibility for a moment.

For the rest of this section, we associate two measures with the instance I of a problem, namely

(1) $|I|$, is the size of the instance I. The size of I is identified with the length of the string representing the problem instance. This is the measure we have used till now and will continue to use.

(2) $mn(I)$, the *m*aximum *n*umber occurring in the instance I. If I does not contain any number, as for problem UHC for example, we set $mn(I) = 0$ by definition.

Definition 13.6

Let $L \subseteq \Sigma^*$ be a problem and p a polynomial in two variables.

(1) The problem is called a *number problem* if there is no polynomial q such that $mn(w) \leqslant q(|w|)$ for all $w \in L$.

(2) The problem L is called *solvable in pseudo-polynomial time* if there is a Turing acceptor which accepts L such that for every $w \in L$ there is an accepting computation of length no more than $p(|w|, mn(w))$.

(3) L is called *strongly $\leqslant$-complete for* **NP** if $L \in \mathbf{NP}$ and there exists a polynomial q such that the subproblem $L_q \triangleq \{w \in L \mid mn(w) \leqslant q(|w|)\}$ is $\leqslant$-complete for **NP** where $\leqslant_q$ is one of the reducibility relations $\leqslant_L$, $\leqslant_P^K$, or $\leqslant_P^C$. ■

For example, the instances of $CLIQUE$ $((V, E), k)$ contain a number k and there is no bound on the magnitude of k, and therefore $CLIQUE$ is a number problem. $CLIQUE$ is strongly $\leqslant_L$-complete for **NP**. This should be intuitively obvious, because all instances with $k > |(V, E)|$ are trivial, and therefore do not add to the complexity of the problem. Strong completeness also follows from the reduction $CNF\text{-}SAT \leqslant_L CLIQUE$ described in the proof of Lemma 13.4. This proof actually shows that the problem $CLIQUE_s \triangleq \{((V, E), k) \mid ((V, E), k) \in CLIQUE \text{ and } k \leqslant card(V) \leqslant |((V, E), k)|\}$ is $\leqslant_L$-hard for **NP**. Because it can be checked in linear time and logarithmic space whether $k \leqslant card(V)$, it follows that $CLIQUE_s$ is in **NP**, and thus $\leqslant_L$-complete for **NP**.

By similar reasoning we obtain the following lemma.

Lemma 13.11

TSP and $\{0,1\}$-*LINEQ* are strongly $\leqslant_L$-complete for **NP**. ■

Not all problems, however, which are $\leqslant_P^C$-complete for **NP**, are also strongly $\leqslant_P^C$-complete for **NP**. Examples are the following problems, *KNAPSACK* and *PARTITION*:

$$KNAPSACK \triangleq \{(V,k) \mid V \text{ is a set of positive integers and there is a subset } J \subseteq V \text{ such that } \sum_{x \in J} x = k\};$$

$$PARTITION \triangleq \{V \mid V \text{ is a set of positive integers and there is a subset } J \subseteq V \text{ such that } \sum_{x \in J} x = \sum_{x \in V-J} x\}.$$

In the following it will be shown that *KNAPSACK* is $\leqslant_P^K$-complete for **NP**, but not strongly $\leqslant_P^C$-complete. The proof to be given is somewhat involved, in that it makes a detour via the following problem 3-*DM* (three-dimensional matching):

$$3\text{-}DM \triangleq \{(U,V,W,X) \mid U, V \text{ and } W \text{ are non-empty sets of equal cardinality, } X \subseteq U \times V \times W, \text{ and there is a subset } M \subseteq X \text{ such that } card(M) = card(U)\,(= card(V) = card(W)), \text{ and if } (u_1,v_1,w_1) \text{ and } (u_2,v_2,w_2) \text{ are two triples in } M, \text{ then either } u_1 = u_2,\ v_1 = v_2 \text{ and } w_1 = w_2, \text{ or } u_1 \neq u_2,\ v_1 \neq v_2 \text{ and } w_1 \neq w_2\}.$$

In the above definition, the set M is called a matching because every element of U occurs in precisely one triple of M, every element of V occurs in precisely one triple of M, and every element of W also occurs in precisely one triple of M. Thus every element of U is matched with precisely one element of V and one element of W, and the same applies to the elements of V and W.

Lemma 13.12

3-*CNF-SAT* $\leqslant_L$ 3-*DM*, and therefore 3-*DM* is $\leqslant_L$-hard for **NP**.

Proof

From an instance $F = F_1 \wedge F_2 \wedge \cdots \wedge F_m$ with

$$\text{var}(F) = \{z_1, z_2, \ldots, z_n\}$$

of 3-*CNF-SAT* an instance (U, V, W, X) of 3-*DM* is constructed as follows:

(1) $U \triangleq \{u_{ij}, \bar{u}_{ij} \mid 1 \leqslant i \leqslant n \text{ and } 1 \leqslant j \leqslant m\}$.

(2) V consists of three kinds of elements.

$$V \triangleq \{a_{ij} | 1 \leqslant i \leqslant n \text{ and } 1 \leqslant j \leqslant m\} \cup$$
$$\{v_j | 1 \leqslant j \leqslant m\} \cup \{c_k | 1 \leqslant k \leqslant nm - m\}$$

(3) W is similar to V, and is defined as follows:

$$W \triangleq \{b_{ij} | 1 \leqslant i \leqslant n \text{ and } 1 \leqslant j \leqslant m\} \cup$$
$$\{w_j | 1 \leqslant j \leqslant m\} \cup \{d_k | 1 \leqslant k \leqslant nm - m\}$$

(4) The set X of triples is constructed as follows:

$$X \triangleq \left(\bigcup_{i=1}^{n} T_i\right) \cup \left(\bigcup_{j=1}^{m} C_j\right) \cup G$$

where T_i, C_j and G are the following sets of triples:

(a) $T_i \triangleq T_i^1 \cup T_i^0$ with

$$T_i^1 \triangleq \{(\bar{u}_{ij}, a_{ij}, b_{ij}) | 1 \leqslant j \leqslant m\}$$
$$T_i^0 \triangleq \{(u_{ij}, a_{ij+1}, b_{ij} | 1 \leqslant j \leqslant m-1\} \cup \{(u_{im}, a_{i1}, b_{im})\}$$

The elements a_{ij} and b_{ij} appear only in the triples of T. Thus, for any matching set M we must have $T_i^1 \subseteq M$ or $T_i^0 \subseteq M$, for all i, $1 \leqslant i \leqslant n$. Therefore we can associate with the matching M a variable assignment $\phi: \text{var}(F) \to \{0, 1\}$, namely $\phi(z_i) = [\text{if } (T_i^1 \subseteq M) \text{ then } 1 \text{ else } 0]$.

(b) $C_j \triangleq \{(u_{ij}, v_j, w_j) | z_i \text{ occurs as a literal in } F_j\} \cup$

$$\{(\bar{u}_{ij}, v_j, w_j) | \neg z_i \text{ occurs as a literal in } F_j\}$$

The v_j and w_j do not occur in triples of X outside C_j. Therefore any matching must contain exactly one triple from C_j for all j, $1 \leqslant j \leqslant m$. Assume for example that $(u_{45}, v_5, w_5) \in C_5$ occurs in a matching M. Then there is no other triple (u_{45}, x, y) in M. Consequently, *not* $T_4^0 \subseteq M$, and therefore $T_4^1 \subseteq M$. In other words, the variable assignment ϕ associated with M satisfies F_j, $\phi(F_j) = 1$.

(c) The set G provides the triples necessary to extend a partial matching which consists of triples in $\left(\bigcup_{i=1}^{n} T_i\right) \cup \left(\bigcup_{j=1}^{m} C_j\right)$ to a total matching. Such a partial match leaves uncovered $2(nm - m)$ elements of U, namely all those u_{ij} such that

T_i^0 is not included in the partial match, and neither is (u_{ij}, v_j, w_j), and similar elements $\bar{u}_{ij}$. G provides triples to cover all these:

$$G \triangleq \{(u_{ij}, c_k, d_k), (\bar{u}_{ij}, c_k, d_k) \mid 1 \leqslant i \leqslant n, 1 \leqslant j \leqslant m \text{ and } 1 \leqslant k \leqslant nm - m\}$$

It remains to verify that $F \in 3\text{-}CNF\text{-}SAT$ if and only if $(U, V, W, X) \in 3\text{-}DM$, and that the transformation described above can be performed by a $\lambda n[\log(n)]$-space-bounded Turing transducer. This is left to the reader. ∎

The remaining reductions, leading to the completeness result for *KNAPSACK*, are easy.

We begin by writing $3\text{-}DM$ as a special case of $\{0,1\}\text{-}LP$.

Definition 13.7

$\{0,1\}\text{-}LPN \triangleq \{(\mathbf{M}, \mathbf{b}) \mid \mathbf{M}$ is an $n \times m$ matrix over the natural numbers and $\mathbf{b}$ is an n-vector over the natural numbers, and there exists an m-vector $\mathbf{x}$ over $\{0,1\}$ such that $\mathbf{Mx} = \mathbf{b}\}$.

$\{0,1\}\text{-}LP\{0,1\} \triangleq \{(\mathbf{M}, \mathbf{b}) \mid \mathbf{M}$ is an $n \times m$ matrix over $\{0,1\}$ and $\mathbf{b}$ is an n-vector over $\{0,1\}$, and there exists an m-vector $\mathbf{x}$ over $\{0,1\}$ such that $\mathbf{Mx} = \mathbf{b}\}$. ∎

An instance (U, V, W, X) with $card(U) = a$ and $card(X) = b$ can be represented by a $3a \times b$ matrix over $\{0,1\}$ as depicted in Table 13.3.

We define a $3a \times b$ matrix $\mathbf{M}$ having a row for each element of $U \cup V \cup W$ and a column for each triple in X, and $m_{ij} =$ [if (element i from $U \cup V \cup W$ occurs in tripe j) then 1 else 0]. Thus $(U, V, W, X) \in 3\text{-}DM$ if and only if there is a vector $\mathbf{x} \in \{0,1\}^b$ such that $\mathbf{Mx} = \mathbf{1}$, where $\mathbf{1}$ is the unit vector of dimension $3a$, i.e. $\mathbf{1} = (1, 1, \ldots, 1)^{\mathrm{T}}$.

Corollary 13.1

$\{0,1\}\text{-}LPN$ and $\{0,1\}\text{-}LP\{0,1\}$ are strongly $\leqslant_L$-complete for **NP**. ∎

Lemma 13.13

KNAPSACK is $\leqslant_P^K$-complete for **NP**.

Proof

KNAPSACK clearly is in **NP** by Theorem 12.2. We show that $\{0,1\}\text{-}LP\{0,1\} \leqslant_P^K$ *KNAPSACK*, whence *KNAPSACK* is $\leqslant_P^K$-complete for **NP**.

Table 13.3 Representing an instance of 3-DM as an instance of $\{0,1\}$-*LP*.

	x_1	x_2	$\cdots$	x_b
u_1	1	1	$\cdots$	0
.	0	0	$\cdots$	0
.	.	.	$\cdots$	1
.	.	.	$\cdots$	0
.	.	.	$\cdots$	.
u_a	0	0	$\cdots 1 \cdots$	0
v_1	0	1	$\cdots$	0
.	0	0	$\cdots$	0
.	.	.	$\cdots$	1
.	.	.	$\cdots$	0
.	.	.	$\cdots$	.
v_a	1	0	$\cdots 1 \cdots$	0
w_1	1	0	$\cdots$	0
.	0	0	$\cdots$	0
.	.	.	$\cdots$	1
.	.	.	$\cdots$	0
.	.	.	$\cdots$	.
w_a	0	1	$\cdots 1 \cdots$	0

Let $(\mathbf{M}, \mathbf{b})$, where $\mathbf{M}$ is an $n \times m$ matrix, be an instance of $\{0,1\}$-$LP\{0,1\}$. This instance $(\mathbf{M}, \mathbf{b})$ is transformed into the instance (V, k) of *KNAPSACK* as follows:

(1) $V \triangleq \{v_i | 1 \leqslant i \leqslant m\}$, where $v_i \triangleq \sum_{i=1}^{n} m_{ij} r^{i-1}$, and $r = m + 1$.

In other words, we view the columns of $\mathbf{M}$ and the column vector $\mathbf{b}$ as representations of natural numbers, where the entry in the first row represents the least significant digit. The radix r of the number system is chosen equal to $m + 1$. This value is chosen to ensure that when adding at most m such numbers there will never occur an overflow from one digit to the next.

(2) $k \triangleq \sum_{i=1}^{n} b_i r^{i-1}$

If $(\mathbf{M}, \mathbf{b}) \in \{0,1\}$-$LP\{0,1\}$ then there is a vector $\mathbf{x} \in \{0,1\}^m$ such that $\mathbf{Mx} = \mathbf{b}$.

Let $J = \{i | x_i = 1\}$. Clearly, $\sum_{i \in J} v_i = k$, thus $(V, k) \in KNAPSACK$. The converse is also immediate.

It remains to verify that the transformation can be performed in polynomial time. This is left to the reader. ■

Table 13.4 A computation of a pseudo-polynomial-time algorithm for *KNAPSACK*.

$V=$		0	1	2	3	4	5	6	7	8	9	10
{7	1	1	0	0	0	0	0	0	1	0	0	0
4	2	1	0	0	0	1	0	0	1	0	0	0
5	3	1	0	0	0	1	1	0	1	0	1	0
2	4	1	0	1	0	1	1	1	1	0	1	1
1}	5	1	1	1	1	1	1	1	1	1	1	1

Having established that *KNAPSACK* is **NP**-complete, it remains to prove that *KNAPSACK* is not strongly $\leqslant_p^C$-complete for **NP**. This is done by using a pseudo-polynomial-time algorithm which solves *KNAPSACK*. The algorithm is a straightforward application of the technique of *dynamic programming*.

Let (V, k) be an instance of *KNAPSACK*. Number the elements of V. Assume that $V = \{a_1, a_2, \ldots, a_n\}$. We compute a table T having n rows and $k+1$ columns, numbered from 0 to k, such that

$$t(i,j) = [\text{if(there is a } J \subseteq \{a_s \mid 1 \leqslant s \leqslant i\} \text{ such that } \sum_{s \in J} a_s = j) \text{ then 1 else 0}].$$

We compute the table row by row using the following relations.

$$t(1,j) = [\text{if}(j = 0 \text{ or } j = a_1) \text{ then 1 else 0}]$$
$$t(i+1,j) = [\text{if}(t(i,j) = 1 \text{ or } t(i, j - a_{i+1}) = 1) \text{ then 1 else 0}]$$

Consider for example the instance $(\{7,4,5,2,1\}, 10)$. Table 13.4 corresponds to this instance.

The total time needed to compute the table is proportional to the size of the table, i.e. proportional to $k.card(V)$. Because $(V,k) \in KNAPSACK$ if and only if $t(card(V), k) = 1$, the above constitutes a pseudo-polynomial-time algorithm.

13.4 The structure of P, NP and *co*-NP

The answer to the crucial question whether or not **P** = **NP** is yet unknown. The general belief, however, is that **P** and **NP** are actually different. Many results have been established on the assumption **P** ≠ **NP**, some of which will be discussed in this section.

Assuming **P** ≠ **NP**, what can be said about **NP** − **P**? Obviously all **NP**-complete problems are in **NP** − **P**, otherwise **P** would be equal to **NP**.

It can be shown that **NP** − **P** must also contain languages which are not $\leqslant_P^C$-complete for **NP**. This is a corollary to the following theorem, whose proof gives a diagonalization procedure to find such a language in **NP** − **P**, given a problem which is $\leqslant_P^C$-complete for **NP**.

In this diagonalization procedure we need Gödel numberings of all polynomial-time-bounded Turing acceptors, all polynomial-time-bounded Turing transducers, and all polynomial-time-bounded Turing oracle acceptors.

In Section 11.5 we defined a Gödel numbering of Turing machines. Gödel numberings of Turing acceptors, Turing oracle acceptors and Turing transducers are constructed in the same way – that is, by writing down all the action items of a machine and then coding the resulting string as a number.

Since we only look at polynomial-time-bounded machines, by Theorem 11.4 and Corollary 11.3 it suffices to consider single-tape Turing machines, and by Theorem 11.3 it also suffices to consider Turing machines which have some fixed tape alphabet, say $V=\{0,1,\ldots,10\}$, and input alphabet, say $\{0,1\}$.

Using the Gödel numbering of Turing acceptors, we can construct a universal Turing acceptor U. On input x (= the Gödel number of a Turing acceptor) and w (= the string to be considered), U operates as follows:

(1) Compute m and k such that $(m,k)=\sigma_2^1(x)$.

(2) Simulate the Turing acceptor with Gödel number m for $|w|^k$ steps.

(3) If the computation has then not terminated, reject w. Otherwise accept w if and only if w is accepted by the mth Turing acceptor.

The machine described above implies a Gödel numbering of all polynomial-time-bounded Turing acceptors. From now on this Gödel numbering will be assumed, and the ith polynomial-time-bounded Turing acceptor will be denoted by P_i.

For polynomial-time-bounded Turing transducers and Turing oracle acceptors we proceed similarly. The ith polynomial-time-bounded Turing transducer will be denoted by T_i, the ith polynomial-time-bounded Turing oracle acceptor by M_i.

Let B be some language such that $B \in$ **NP** − **P**. The theorem below constructs a subset $A \subseteq B$ such that $A \leqslant_P^K B$, $A \notin$ **P**, and *not* $B \leqslant_P^C A$. By Lemma 12.3 it follows from $A \leqslant_P^K B$ and $B \in$ **NP** that $A \in$ **NP**. Consequently, the set A constructed is also in **NP** − **P**. Furthermore, A is not $\leqslant_P^K$-complete for **NP**. It is 'less difficult' than an **NP**-complete problem, but still not feasibly computable, i.e. not in **P**. We can apply the theorem again on this set A, which will produce a new set A' which is 'easier' to decide than A, but still not feasibly computable. Thus, if **P** ≠ **NP**, then there is an infinite hierarchy of complexity classes between **P** and **NP**.

Theorem 13.4

Every recursively decidable set B which is not in **P** contains a recursively decidable subset $A \subseteq B$ which is not in **P** such that $A \leqslant_P^K B$ but *not* $B \leqslant_P^C A$.

Proof

In the following a polynomial-time-bounded Turing transducer is constructed such that the set $A \triangleq \{w \in B \mid |T(w)| \text{is even}\}$ has the required properties. That is, $A \notin \mathbf{P}$, A is recursively decidable, $A \leqslant_P^K B$, and *not* $B \leqslant_P^C A$.

Transducer T is defined in such a way that an ever-growing initial part of the following list of conditions is satisfied.

$$
\begin{array}{ll}
0 & A \neq L(P_0), \\
1 & B \neq M_0(A), \\
2 & A \neq L(P_1), \\
3 & B \neq M_1(A), \\
\vdots & \\
2i & A \neq L(P_i), \\
2i+1 & B \neq M_i(A), \\
\vdots &
\end{array}
$$

where $L(P_i)$ denotes the language accepted by the ith polynomial-time-bounded Turing acceptor, and $M_i(A)$ denotes the language accepted by the ith polynomial-time-bounded Turing oracle acceptor assuming that the oracle decides membership of the set A.

Assuming this, it is clear that $A \notin \mathbf{P}$ and that *not* $B \leqslant_P^C A$. Also, $A \leqslant_P^K B$ via the function

$$\lambda w[\text{if}(|T(w)| \text{is even}) \text{ then } w \text{ else } w_0],$$

where w_0 is some word not in B. Because $B \notin \mathbf{P}$, B must be nontrivial, whence there must exist some word w_0 not in B.

The transducer T satisfies

(1) $T(\varepsilon) = \varepsilon$.

(2) $T(w) = T(0^{|w|})$.

(3) On input 0^n, $n > 0$, the transducer proceeds as follows. For n steps in all, it executes the steps of the computations of $T(\varepsilon)$, $T(0)$, $T(00), \ldots$. Let $T(0^m)$ be the last computation which has terminated within the allotted number of n computation steps in all. Because $n \geqslant 1$ and the computation of $T(\varepsilon)$ requires exactly one step, there does exist such a value m. Now T will

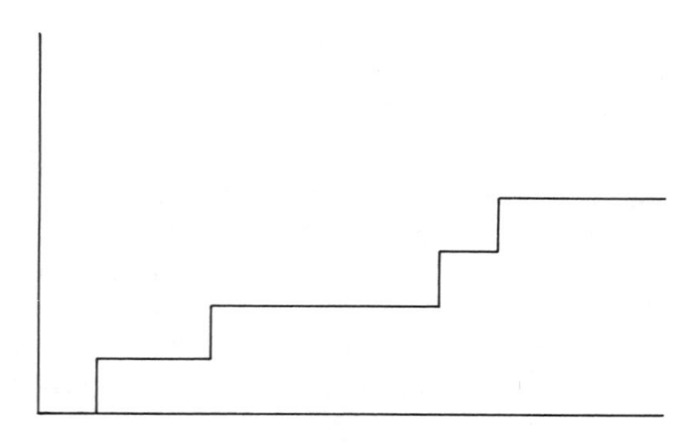

Figure 13.5 A staircase function.

spend another n steps trying to satisfy one of the requirements of the list given above.

Assume that $|T(0^m)|$ is even.

Let $i = |T(0^m)|/2$. For n steps, try to prove that $A \neq L(P_i)$. Do this by computing whether $w \in A$ and $w \in L(P_i)$ for all $w \in \{0, 1\}^*$ in the natural order $\varepsilon, 0, 1, 00, 01, 10, 11, 000, \ldots$, until a w is found such that $w \in (A - L(P_i)) \cup (L(P_i) - A)$ or time has run out, that is, the search has taken more than the allotted n steps. If no such w is found then $T(0^n) = 1^{2i}$, otherwise $T(0^n) = 1^{2i+1}$.

Thus, if no proof has been found that requirement i of the list above is indeed satisfied, then $T(0^n) = T(0^m)$. This proof will then be attempted again, using ever-growing values of n, the bound on the number of steps needed for the proof.

Assume that $|T(0^m)|$ is odd.

Let $i = (|T(0^m)| - 1)/2$. For n steps, try to prove that $B \neq M_i(A)$ by checking whether $w \in B$ and $w \in M_i(A)$ for $w = \varepsilon, 0, 1, 00, 01, \ldots$, until a proof has been found or time has run out. When checking whether $w \in M_i(A)$, the oracle acceptor M_i every now and then queries the oracle, to find out whether some word v belongs to A. The steps needed to determine the answer to this query must be counted among the n steps permitted.

If no proof has been found then $T(0^n) = 1^{2i+1}$, otherwise $T(0^n) = 1^{2i+2}$. Note once more that if no proof has been found, then $T(0^n) = T(0^m)$, and the proof will be attempted again.

Note that for all $n \geqslant 0$ there is some i such that $T(0^n) = 1^i$, and that $|T(0^n)| \leqslant |T(0^{n+1})| \leqslant |T(0^n)| + 1$ for all n. Thus T is a staircase function as depicted in Figure 13.5.

Furthermore, if $T(0^n) = 1^{i+1}$, then condition i of the list above is satisfied. Thus if we can prove that for all i there exists an n such that $T(0^n) = 1^i$, the theorem is proved.

We proceed by induction on i.

(1) $i = 0$. $T(\varepsilon) = \varepsilon = 1^0$ by definition, so we are done.

(2) $i = j + 1$. Let q be the least number such that $T(0^q) = 1^j$. Either there is an $n > q$ such that $T(0^n) = 1^{j+1}$, or $T(0^n) = T(0^q)$ for all $n > q$. The latter however, is impossible. Assume for the sake of argument that $T(0^n) = T(0^q) = 1^j$ for all $n > q$.

(a) Assume that j is even. Let $i = j/2$. Because $T(w) = T(0^q) = 1^j$ and j is even, we have $w \in A$ if and only if $w \in B$ for all w of length greater than or equal to q. Since $B \notin \mathbf{P}$, we must also have $A \notin \mathbf{P}$. In particular $A \neq L(P_i)$. Let z be the first string in the natural search order used such that

$$z \in (A - L(P_i)) \cup (L(P_i) - A).$$

There is some number of steps and thus a value n such that the procedure sketched above will find this z in its computation $T(0^n)$, thereby producing the output 1^{j+1}. Thus j cannot be even.

(b) Assume that j is odd, and let $i = (j-1)/2$. In this case, $w \notin A$ for all w of length greater than or equal to q. Therefore A and $M_i(A)$ are in $\mathbf{P}$. Because $B \notin \mathbf{P}$, we have that $B \neq M_i(A)$, whence there is a z such that $z \in (B - M_i(A)) \cup (M_i(A) - B)$, which will be found by the search procedure described above, making the transducer output the value 1^{j+1}. Thus j cannot be odd.

Since j is either odd or even, we have a contradiction.

Thus, for all i, there is some n such that $T(0^n) = 1^i$. ■

Assume now that $\mathbf{NP} \neq \mathbf{P}$, and let B be an **NP**-complete problem. The theorem above constructs a subproblem of B, which is also in $\mathbf{NP} - \mathbf{P}$, but is not in the same $\equiv_P^C$-class as B. It is therefore certainly not $\leqslant_P^C$-complete for **NP**. The theorem can also be applied to the newly constructed subproblem, which results in an infinite hierarchy $\cdots A'' \leqslant_P^K A' \leqslant_P^K A \leqslant_P^K B$ of problems $B, A, A', A'', \ldots$ in **NP**, all belonging to mutually different $\equiv_P^C$-classes.

Corollary 13.2

If $\mathbf{NP} \neq \mathbf{P}$ then every problem which is $\leqslant_P^C$-complete for **NP** has a subproblem which belongs to $\mathbf{NP} - \mathbf{P}$, but is not $\leqslant_P^C$-complete for **NP**. ■

Consider for instance the Hamiltonian circuit problem *UHC*. Then the preceding corollary tells us that there exists a set of graphs recognizable by a non-deterministic Turing machine in polynomial time (thus this set is a member of **NP**) such that the Hamiltonian circuit problem, when restricted to that class of graphs, is neither **NP**-complete nor in **P**. However, such a restricted version of an **NP**-complete problem is in general not a 'natural' problem in **NP** − **P**, since it is constructed by a complicated diagonalization procedure.

There are also known some 'natural' problems which are considered probable candidates for being neither in **P** nor **NP**-complete. These problems have withstood many attempts to prove that they are **NP**-complete or a member of **P**. Examples are the problems of testing primality of natural numbers, and of checking whether two given directed graphs are isomorphic.

Definition 13.8

$D\text{-}ISO \triangleq \{((V,E),(W,T)) \mid$ the directed graphs (V,E) and (W,T) are isomorphic, that is, there exist a bijection $\phi: V \to W$ such that $(u,v) \in E$ if and only if $(\phi(u),\phi(v)) \in T\}$.

$COMPOSITE \triangleq \{n \mid n$ is a natural number which is not a prime, i.e. $n=0$, or $n=1$, or there are natural numbers k and m such that $1 < m < n$, and $km = n\}$. ■

It is not obvious what makes *D-ISO* seemingly easier than *D-SUB-ISO*, an **NP**-complete problem (see Theorem 13.3). It seems that **NP**-complete problems have a certain robustness in their formulation, which makes them unsusceptible to minor changes. For example, if a propositional formula F belongs to 3-*CNF-SAT*, you can normally make various minor changes to F, such as adding a clause, or deleting a clause, or maybe changing a clause somewhat, without ruining the instance, i.e. such that the new formula F' still belongs to 3-*CNF-SAT*. This robustness is lacking in the problem *D-ISO*. Almost every change, however small, will ruin an instance of the isomorphism problem, e.g. adding a single edge destroys an isomorphism. The problem *D-SUB-ISO*, however, is relatively unsusceptible to such changes.

We have seen the problem *COMPOSITE* earlier. In Section 9.4, we discussed an ND-SAL program deciding whether a given number is prime. This problem has a property which is not common amongst **NP**-complete problems, namely that its complement is also in **NP**. As remarked earlier, showing that

$$PRIMES \triangleq \{n \mid n \text{ is a prime number}\}$$

is in **NP** is not trivial; a proof can be found in Pratt (1975). There are as yet no problems known such that both the problem and its complement are **NP**-complete.

Definition 13.9

(1) Let Σ be an alphabet and $A \subseteq \Sigma^*$ a problem. The *complement* of problem A is denoted by *co*-A and defined by *co*-$A \triangleq \Sigma^* - A$.

(2) The class *co*-**NP** consists of the complements of all problems in **NP**. That is, *co*-**NP** $\triangleq \{co\text{-}A \mid A \in \mathbf{NP}\}$. ■

There is an alternative definition which takes into account well-formedness. Consider for example the problem *CLIQUE*. Then, given a string w there are three possibilities:

- $w \in CLIQUE$
- w is not well-formed, i.e. w is not in the general format discussed in Chapter 13.
- w is an instance of the *CLIQUE* problem, but $w \notin CLIQUE$, i.e. the graph does not contain a clique of the specified number of vertices.

The problem *co-CLIQUE* is now restricted to the well-formed instances. In other words

co-CLIQUE $\triangleq \{((V, E), k) \mid$ where (V, E) is an undirected graph, k an integer, and (V, E) does *not* contain a k-clique as a subgraph, that is, there is no subset $W \subseteq V$ with k vertices and such that for all $u, v \in W$ there is an edge between u and $v\}$.

Since well-formedness is polynomially decidable, it does not matter which of these definitions is chosen.

Experience with many problems shows that if $A \in \mathbf{NP}$ then *co*-A seems to be not in **NP**. This leads to the following:

Conjecture: **NP** $\neq$ *co*-**NP**.

Since **P** $=$ *co*-**P**, this conjecture implies that **P** $\neq$ **NP**. On the other hand there is no evidence excluding the possibility that *co*-**NP** $=$ **NP**, but nevertheless **P** $\neq$ **NP**.

Theorem 13.5

NP is closed with respect to complementation if and only if there exists a problem A which is $\leqslant_P^K$-complete for **NP**, and such that *co*-$A \in$ **NP**.

Proof

- Assume that **NP** is closed with respect to complement. Then for any problem in **NP** and thus for a specific problem A which is $\leqslant_P^K$-complete for **NP**, we have that *co*-A is in **NP**.

- Let A be a problem which is $\leqslant_P^K$-complete for **NP** and such that $co\text{-}A \in \mathbf{NP}$. For any $B \in \mathbf{NP}$, $B \leqslant_P^K A$, and thus $co\text{-}B \leqslant_P^K co\text{-}A$. Since $co\text{-}A \in \mathbf{NP}$ and **NP** is closed with respect to Karp reducibility (Lemma 12.3), it follows that $co\text{-}B \in \mathbf{NP}$ and therefore **NP** is closed with respect to complement. ■

Thus a problem A with $co\text{-}A \in \mathbf{NP}$ can only be **NP**-complete if $\mathbf{NP} = co\text{-}\mathbf{NP}$.

The problem *COMPOSITE* clearly belongs to **NP**. It can also be shown that its co-problem *PRIMES* belongs to **NP**. So *COMPOSITE* cannot be **NP**-complete unless $\mathbf{NP} = co\text{-}\mathbf{NP}$. Thus it seems that the problem *COMPOSITE* is a likely candidate for a non-complete problem in $\mathbf{NP} - \mathbf{P}$.

Note, however, that the preceding discussion does not support the idea that *COMPOSITE* is not in **P**. On the contrary, there are algorithms which decide *COMPOSITE* and run in polynomial time if the extended Riemann hypothesis is true. Therefore it is far from evident that *COMPOSITE* is in **NP** but not in **P**.

13.5 Single-letter alphabets, or unary representation

The point has been stressed more than once (for example in Section 9.3) that problems should be properly coded as strings over a finite alphabet of at least two letters.

In this section it is explained why this is necessary. If we allow any serious amount of extraneous information, or some form of extremely inefficient coding, such as that obtained when using a single-letter alphabet, we get deceptive results. This is vividly demonstrated in the following theorem.

Theorem 13.6

If there exists a set A over a single-letter alphabet which is $\leqslant_P^K$-complete for **NP**, then $\mathbf{P} = \mathbf{NP}$. ■

This theorem is a relatively easily proved corollary to a more technical theorem. The formulation and proof of the latter are too complicated to be included here; details can be found in Berman (1978).

The above theorem also shows the need for sufficiently refined models of computation. We started our investigations of computability fixing a particularly simple programming language: SAL(**N**). We have seen overwhelming evidence that this programming language is sufficiently powerful, so that SAL-programmability may be identified with effectiveness in the intuitive sense. The arithmetical operations available were adding one ($+_1$) and subtracting one ($\dot{-}_1$).

Solely using these operations is equivalent to using unary notation. From a mathematical point of view there is nothing against using unary representation of numbers. The Peano axiom system of numbers uses as the basic operation the successor function, i.e. unary representation. So this is perfectly valid.

But in terms of complexity of computation the unary representation is ridiculously inefficient, and cannot be used to obtain sensible results, as is demonstrated in the above theorem.

This explains the need for a computational model such as the one introduced in Section 9.6, where the starting point is strings and operations on strings. There is no need for such a realistic model in recursion theory, where one is concerned with what can be done in principle, but it becomes unavoidable as soon as the cost of computing is taken into serious consideration.

13.6 Exercises

13.1 Prove that *FAL*, as defined below, is **NP**-complete:

$FAL \triangleq \{F \mid F$ is a propositional formula which is falsifiable, i.e. there is an assignment $v:\text{var}(F) \rightarrow \{0,1\}$, such that $v(F)=0\}$

Also consider the variants obtained by restricting F to be in conjunctive normal form and to be in disjunctive normal form.

13.2 The problem

$COL \triangleq \{((V,E),k) \mid (V,E)$ is an undirected graph and there is a labelling $L: V \rightarrow \{1,\ldots,k\}$ such that $L(u) \neq L(v)$ if $\{u,v\} \in E\}$

is called the 'colourability problem', the colours are the numbers 1 to k used as labels.

Prove that COL is **NP**-complete. Prove that the problem remains **NP**-complete, even if the graphs are restricted to having degree 4 and only colourability with five colours is considered.

13.3 Give a direct proof of the **NP**-completeness of *CLIQUE* by representing the computations of a non-deterministic Turing machine.

13.4 The *wandering salesman problem* (*WSP*) is a simple variant of the travelling salesman problem (*TSP*). The difference lies in the following facts:

(a) the salesman may start in any city he prefers;

(b) after visiting all cities the salesman need not return to the city where he started his tour.

Prove that WSP is **NP**-complete.

13.5 Prove that the following problem is **NP**-complete:

$$\{((V, E), k) \mid (V, E) \text{ is an undirected graph which has a spanning tree having } k \text{ leaves}\}$$

13.6 If there exists an **NP**-complete language L belonging to *co*-**NP**, then **NP** = *co*-**NP**. Prove this.

13.7 Let NAT_k be the set of all natural numbers which have decimal representations of length no more than k and let $f: NAT_k \to NAT_k$ be a surjective function such that $f(x)$ is computable in polynomial time, but $f^{-1}(x)$ is not. The set A is defined as follows:

$$A \triangleq \{(x, y) \mid f^{-1}(x) < y\}.$$

Prove that A belongs to (**NP** $\cup$ *co*-**NP**).

13.8 Prove that the function $f(n) = \lceil \log_2(n) \rceil$ is fully space constructible.

13.9 A Turing machine is called *oblivious* if at every moment the position of the read/write heads is solely dependent on the length of the input read so far. Prove that if a language can be accepted by a T-time-bounded, k-tape deterministic Turing machine, then this language can also be accepted by a $\lambda n[T(n).\log_2(T(n))]$-time-bounded, oblivious, k-tape deterministic Turing machine.

13.7 References

In the fundamental paper of Cook (1971) the classes **P** and **NP** were defined as classes of languages and the decision problems they encode. In this paper Cook also proved the theorem that satisfiability of propositional formulae is **NP**-complete.

Karp (1972) named the classes **P** and **NP** and observed that Cook's theorem remains valid if the notion of Turing reducibility is replaced by the simpler notion of polynomial transformability, which is called Karp reducibility.

The terminology **NP**-complete, **NP**-hard, polynomial transformability etc. is largely a result of the efforts of Knuth (1974). In 1974 Knuth circulated a private poll to a collection of researchers asking them for alternatives for the existing terminology. Knuth (1974) is an entertaining report of this poll.

Regarding the theory of **P** and **NP** an excellent reference is Garey and Johnson (1979). Wagner and Wechsung (1986) give a more complete survey of the vast literature on complexity of problems. Another valuable source is D. S. Johnson's 'The **NP**-completeness column: an ongoing guide', published more-or-less regularly in the Journal of Algorithms.

Chapter 14
The Classes NL, P and *PSPACE*

In this chapter we discuss problems $\leqslant_L$-complete for the classes **NL**, **P** and *PSPACE*. In Section 12.2 we saw (Corollary 12.1) that

$$\mathbf{L} \subseteq \mathbf{NL} \subseteq \mathbf{P} \subseteq \mathbf{NP} \subseteq PSPACE = NSPACE.$$

It is still unknown which of these inclusions are strict, although it is known that at least one of the inclusions must be strict, because **NL** is strictly included in *PSPACE* by Theorems 11.10 and 11.7.

In the following sections various practical problems, in the sense discussed in Chapter 13, will be shown to be $\leqslant_L$-complete for **NL**, **P**, or *PSPACE*. Intuitively, every problem that is $\leqslant_L$-complete for **NL** is just as difficult as any other problem in **NL**, and similarly for the other classes.

It is customary to speak of '**NL**-complete problems' instead of 'problems $\leqslant_L$-complete for **NL**', and similarly, of **P**-complete and *PSPACE*-complete problems. The reducibility relations $\leqslant_P^C$ and $\leqslant_P^K$ cannot be used to compare problems in **NL**, because the reduction is permitted to use more than logarithmic space. Similarly, $\leqslant_P^C$ and $\leqslant_P^K$ cannot be used to compare problems in **P**, because all nontrivial problems in **P** are equivalent with respect to $\equiv_P^C$ and $\equiv_P^K$. Thus, **NL**-completeness and **P**-completeness imply the use of $\leqslant_L$ as the reducibility relation. When speaking of *PSPACE*-complete problems, however, the reducibility relation used has to be stated explicitly.

14.1 NL-complete problems

The first problem which was shown to be **NL**-complete is the so-called accessibility problem for directed graphs.

Definition 14.1

$DGA \triangleq \{((V, E), n, O) \mid$ where (V, E) is a directed graph, $n \in V$ and $O \subseteq V$, and there is a vertex $m \in O$ such that there is a path from n to $m\}$

In the above, vertex n is called an *entry vertex* or entry node, and the vertices in O are called *exit vertices* or exit nodes. ■

We assume throughout this chapter that problems are represented as languages over some finite but sufficiently large alphabet, as discussed in Chapter 13.

The name 'accessibility problem' is due to Jones (1975), but the problem was first considered by Savitch (1970). Savitch called it 'the set of threadable mazes'.

Lemma 14.1

DGA is in **NL**.

Proof

A non-deterministic Turing machine can accept DGA in the following way. Let $((V, E), n, O)$ be the given instance of DGA.

(1) The machine keeps track of a *current vertex*; let x denote this current vertex. Initially, $x = n$.

(2) The machine checks whether the current vertex is an exit vertex, i.e. whether $x \in O$.

(3) If $x \in O$, then the Turing machine halts, accepting its input $((V, E), n, O)$.

(4) If $x \notin O$, the machine non-deterministically selects an edge $(x, y) \in E$; it then sets $x = y$, i.e. makes y the current vertex and then proceeds with step (2) above.

The above method can be implemented in a Turing machine having a work tape consisting of two tracks. The contents of these tracks are interpreted as pointers to the contents of the input tape. Only two pointers are needed, one to identify the current vertex, the other to check that $x \in O$ or to identify an edge of E.

Thus, the method can be implemented in a $\lambda n[\log(n)]$-space-bounded Turing machine. ■

Lemma 14.2

DGA is $\leqslant_L$-hard for **NL**.

Proof

Let $A \subseteq \Sigma^*$ be in **NL**. Then there is a non-deterministic, single-tape, $\lambda n[\log(n)]$-space-bounded Turing machine TM_A which accepts A.

Below we describe how to transform an element $w \in \Sigma^*$ into an instance $((V, E), n, O)$, such that $w \in A$ if and only if $((V, E), n, O) \in DGA$. Because the transformation can be carried out by a $\lambda n[\log(n)]$-space-bounded Turing transducer, it follows that $A \leqslant_L DGA$.

Because the construction works for any arbitrary language A, it follows that $A \leqslant_L DGA$ for all $A \in$ **NL**. Consequently DGA is $\leqslant_L$-hard for **NL**.

The instance $((V, E), n, O)$ is defined as follows:

(1) V has a vertex for each and every instantaneous description ID which can occur in a $\lambda n[\log(n)]$-space-bounded computation of TM_A on input w. An ID $((w_1, \alpha_1), q, (w_2, \alpha_2))$, where $w = w_1 w_2$ is represented as (i, j, α, q), where i is a number specifying the position of the read head on the input tape of TM_A and j is a number specifying the position of the read/write head on the work tape of TM_A.

(2) There is an edge from $x = (i_x, j_x, \alpha_x, q_x)$ to $y = (i_y, j_y, \alpha_y, q_y)$ if and only if $ID_x \vdash ID_y$, where ID_x denotes the ID corresponding to vertex x.

(3) The entry vertex n corresponds to the initial ID of TM_A.

(4) The set O consists of all vertices corresponding to final IDs of TM_A.

A $\lambda n[\log(n)]$-space-bounded Turing transducer can construct the instance $((V, E), n, O)$ as follows:

(1) It systematically generates the elements of V and writes them onto the output tape. The representation (i, j, α, q) of a vertex requires $O(\log(|w|))$ space. Thus, generating V can easily be achieved by a counting procedure.

(2) The set E is generated by once more generating V and for each element $x \in V$ writing all elements $(x, y) \in E$ to the output tape. There is a bound m_A, which depends on TM_A but not on w, on the number of elements y to be constructed for each x. Therefore, E can also be constructed and written to the output tape by a counting procedure.

(3) Finally, the elements of the set O can be written to the output

tape, by once more generating V and writing (i, j, α, q) to the output tape if q is an accepting state.

Consequently $A \leqslant_L DGA$, and therefore DGA is $\leqslant_L$-hard for **NL**. ■

Combining the two lemmas above, we have:

Theorem 14.1

DGA is **NL**-complete. ■

Savitch (1970) has shown that if $DGA \in \mathbf{L}$, then not only $\mathbf{L} = \mathbf{NL}$, but also $NDSPACE(t) = DSPACE(t)$ for every function t, such that $t(n) \geqslant \log(n)$ for all n.

The status of the accessibility problem for undirected graphs is unknown. It is unknown whether the problem is in **L**, and also unknown whether it is $\leqslant_L$-complete for **NL**.

Below a number of problems are defined. Some of them are **NL**-complete, some **P**-complete. Concepts related to propositional formulae, in particular *unit resolution* and *modus ponens* as rules of inference, have been explained in Section 13.1.

Definition 14.2

k-CNF-CONTRA $\triangleq$ $\{F \mid F$ is a propositional formula in conjunctive normal form, such that each clause contains at most k literals, and there is no assignment ϕ:var$(F) \rightarrow \{0, 1\}$ such that $v(F) = 1$; in other words F is *contradictory*$\}$

k-CNF-CONTRA-UR $\triangleq$ $\{F \mid F$ is a propositional formula in conjunctive normal form with at most k literals per clause and the empty clause can be deduced from F by *unit resolution*; in other words, F is contradictory, and this can be proved using unit resolution as a rule of inference$\}$

CNF-CONTRA-UR $\triangleq$ $\{F \mid F$ is a propositional formula in conjunctive normal form which can be proved contradictory by unit resolution$\}$

GEN $\triangleq$ $\{((X, .), S, z) \mid$ '.' is a binary operation on X, $S \subseteq X$, $z \in X$, and z can be generated from S; that is, there are elements $s_1, s_2, \ldots, s_n \in S$ such that $z = s_1 . s_2 . \ldots . s_n\}$

AGEN $\triangleq$ $\{((X, .), S, z) \mid$ '.' is an *associative* binary operation on X, $S \subseteq X$, $z \in X$, and $z = s_1 . s_2 . \ldots . s_n$ for some elements $s_1, \ldots, s_n \in S$. ('Associative' means that $a.(b.c) = (a.b).c$ for all a, b and c)$\}$. ■

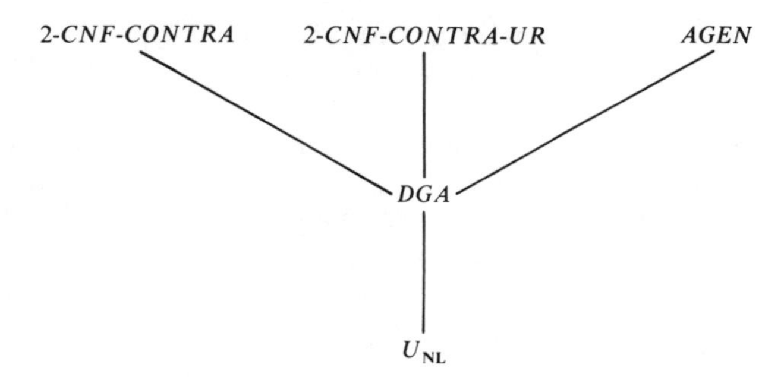

Figure 14.1 A scheme of reductions between problems.

Note that for propositional logic, unit resolution is almost the same as *modus ponens.*

(1) From A and $\neg A \vee B_1 \vee B_2 \vee \cdots \vee B_m$ the clause $B_1 \vee B_2 \vee \cdots \vee B_m$ may be deduced by unit resolution.

(2) From A and $A \rightarrow (B_1 \vee B_2 \vee \cdots \vee B_m)$ the formula $B_1 \vee B_2 \vee \cdots \vee B_m$ may be deduced by *modus ponens.*

(3) Because $\neg A \vee B_1 \vee B_2 \vee \cdots \vee B_m$ is logically equivalent to $A \rightarrow B_1 \vee B_2 \vee \cdots \vee B_m$, unit resolution is actually the same as *modus ponens* if $m \geqslant 1$. However, from the formulae A and $\neg A$ the empty clause $\square$ can be derived by unit resolution, but no other formulae can be deduced by the rule of *modus ponens.*

Note that $F \in CNF\text{-}CONTRA$ does not imply that $F \in CNF\text{-}CONTRA\text{-}UR$. Consider for example the following formula:

$$F_0 \triangleq (x \vee y) \wedge (x \vee \neg y) \wedge (\neg x \vee y) \wedge (\neg x \vee \neg y)$$

There is no unit clause, thus the rule of unit resolution is not applicable, but F_0 clearly is a contradiction.

In what follows it will be shown that *2-CNF-CONTRA*, *2-CNF-CONTRA-UR*, and *AGEN* are **NL**-complete. The structure of the reductions is depicted in Figure 14.1.

Lemma 14.3

Let F be a propositional formula in conjunctive normal form with at most two literals per clause. Write all clauses $x \vee y$ as a pairs $\neg x \rightarrow y$ and $\neg y \rightarrow x$. Let G denote the resulting formula.

(1) F is in *2-CNF-CONTRA-UR* if and only if there is a sequence of literals $x_1, x_2, \ldots, x_k$, such that statements (I) and (II) below are true.

(2) F is in *2-CNF-CONTRA* if and only if statement (I) and either statement (II) or (III) below are true.

(I) $x_i \to x_{i+1}$ is in G for each $1 \leqslant i < k$.
(II) x_1 is a clause of G and $x_k = \neg x_1$.
(III) $x_1 = \neg x_j = x_k$ for some j.

Proof

(1) If (I) and (II) hold, then x_1 is a clause of F and the clause $x_k = \neg x_1$ is deducible from F by unit resolution and also by *modus ponens*. Thus $F \in$ *2-CNF-CONTRA*. Since the empty clause $\square$ is deducible from x_1 and $\neg x_1$, F also belongs to *2-CNF-CONTRA-UR*.

(2) If (I) and (III) hold, then $x_1 \to \neg x_1$ and $\neg x_1 \to x_1$ are deducible from F by *modus ponens*. Therefore F is contradictory, i.e. $F \in$ *2-CNF-CONTRA*.

This is the case with formula F_0 above. The fact that F_0 is contradictory is not derivable by unit resolution, but it is derivable by resolution as follows:

(a) From $x \vee y$ and $\neg x \vee y$ we obtain y
(b) From $x \vee \neg y$ and $\neg x \vee \neg y$ we obtain $\neg y$
(c) From y and $\neg y$ we obtain $\square$.

(3) If $F \in$ *2-CNF-CONTRA-UR*, then $\square$ can be derived from F. The deduction can be written such that (I) and (II) hold, because there are at most two literals per clause.

(4) Assume that $F \in$ *2-CNF-CONTRA*. We show that there is a sequence of literals $x_1, x_2, \ldots, x_k$ such that (I) and (II) or (III) hold.

Assume there does not exist such a sequence of literals. Then the procedure sketched below constructs an assignment $v:\text{var}(F) \to \{0,1\}$ such that $v(F) = 1$. Therefore F is satisfiable, in contradiction with our assumption on F. Consequently, the assumption that there is no sequence of literals $x_1, x_2, \ldots, x_k$, such that (I) and (II) or (III) hold is wrong; there must exist such a sequence.

(a) $U \triangleq \{x \mid x \text{ is a unit clause of } F\} \cup$
$\{\neg x_1 \mid$ there is a sequence of literals $x_1, \ldots, x_n$ such that x_1 is in F and $x_n = \neg x_1$ and $x_i \to x_{i+1}$ is in G for all i, $1 \leqslant i < n\}$.

Assign the value 1 to every literal in U.

(b) **while** there is a literal without a value **do**

(c) $V \triangleq U \cup \{x_n \mid$ there is a sequence $x_1, \ldots, x_n$ such that $x_1 \in U$ and $x_i \to x_{i+1}$ is in G for all i, $1 \leqslant i < n\}$

(d) Assign the value 1 to every literal in V

(e) Let $U = V \cup \{u\}$, where u is the first variable which has not been assigned a value.

od

This procedure must eventually terminate since there are finitely many variables. Now assume there does not exist a sequence satisfying (I) and (II) or (III). Then neither U nor V can contain both a literal x and its negation, since otherwise x would satisfy (II) if x appeared in step (a), and would satisfy (III) if x appeared in step (c)

Therefore the above procedure constructs an assignment $v:\text{var}(F) \rightarrow \{0, 1\}$. Furthermore, $v(F) = 1$, and consequently F is satisfiable, contrary to our assumption on F.

Therefore, if $F \in$ 2-*CNF-CONTRA*, then there is a sequence of literals $x_1, x_2, \ldots, x_k$, such that (I) and (II) or (III) hold. ■

Theorem 14.2

2-*CNF-CONTRA* and 2-*CNF-CONTRA-UR* are **NL**-complete.

Proof

(1) To check whether $F \in$ 2-*CNF-CONTRA*, or $F \in$ 2-*CNF-CONTRA-UR* we non-deterministically choose a sequence $x_1, \ldots, x_n$ of literals and check that it satisfies (I) and (II) or (III), respectively (I) and (II). The literals can be chosen one by one. It is therefore only necessary to store the index of the initial literal and of the literal currently being considered, and this requires no more than $O(\log(|F|))$ space.

(2) Let $((V, E), n, O)$ be an instance of *DGA*. Without loss of generality, we assume that O contains but a single exit vertex; let m denote this exit vertex. From this instance of *DGA* we construct a propositional formula F using variables from the set $\{x_v | v \in V\}$. The formula F is defined as follows.

$$F \triangleq x_n \wedge \left(\bigwedge_{(u,v) \in E} (\neg x_u \vee x_v) \right) \wedge \neg x_m$$

There is a path from vertex n to vertex m in (V, E) if and only if the empty clause □ is deducible from F. Hence $((V, E), n, \{m\}) \in DGA$ if and only if $F \in$ 2-*CNF-CONTRA* if and only if $F \in$ 2-*CNF-CONTRA-UR*.

Thus 2-*CNF-CONTRA* and 2-*CNF-CONTRA-UR* are in **NL** and are also $\leqslant_L$-hard for **NL**. Consequently, 2-*CNF-CONTRA* and 2-*CNF-CONTRA-UR* are **NL**-complete. ■

It remains to consider the problem *AGEN*.

Lemma 14.4

$AGEN \in \mathbf{NL}$.

Proof

Let $((X,.), S, z)$ be an instance of *AGEN*. The element z can be generated from S if and only if there exist $x_1, \ldots, x_k \in S$ such that $z = x_1.x_2.\cdots.x_k$. Because the operation '.' is associative, there is no need for parentheses. If z can be generated from S, there is a shortest expression generating z. When evaluating this shortest expression from left to right, all partial results $x_1.\cdots.x_i$ obtained are mutually different.

Thus to see whether z can be generated from S nondeterministically, select a sequence of at most $card(X)$ elements from S and check that the result is equal to z ($card(X)$ denotes the number of elements of X).

Selecting the elements and computing the partial result can be done sequentially. Thus it is only necessary to store the indices of z and of the partial result. Thus $O(\log(|((X,.), S, z)|))$ space is sufficient. ∎

Lemma 14.5

$DGA \leqslant_L AGEN$, thus *AGEN* is **NL**-hard.

Proof

Let $((V, E), n, \{m\})$ be an instance of *DGA*, where without loss of generality we have assumed that there is but a single exit node. A corresponding instance $((X,.), S, z)$ of *AGEN* is defined as follows:

(1) $X \triangleq V \cup V \times V \cup \{\omega\}$, where ω is a new element, $\omega \notin V \cup V \times V$

(2) $S \triangleq E \cup \{n\}$

(3) $z \triangleq m$

(4) The binary operation '.' is defined as follows:

$$a.(a, b) \triangleq b$$

$$(a, b).(b, c) \triangleq (a, c)$$

for all $a, b, c \in V$, and $x.y \triangleq \omega$ for all other $x, y \in X$.

It is easily seen that this operation is associative, and that z can be generated from S if and only if there is a path from n to m in (V, E).

The instance $((X,.), S, z)$ can be constructed and written to the output tape by a $\lambda n[\log(n)]$-space-bounded Turing transducer using a counting procedure. To construct and write the table specifying the operation '.', four pointers to specify a pair of pairs $((a,b),(u,v))$ are needed; and this is the maximum that is ever needed. Thus the transducer requires $O(\log(|((V,E),n,O)|))$ space and is therefore $\lambda n[\log(n)]$-space-bounded. ■

14.2 P-complete problems

Problems which are $\leqslant_L$-complete for **P** particularize questions about the possibilities of time/space trade-off. If a problem is in **P** it is required to run in polynomial time regardless of the amount of space used. Obviously no more than a polynomial amount of space can be used, but it is not known whether it is ever necessary to use that much space. If some problem A, which is $\leqslant_L$-complete for **P**, can be solved using only a logarithmic amount of space, then this is true for all problems in **P**. To be specific, if A is **P**-complete and $A \in$ **NL**, then **P** = **NL**, and similarly for **L**.

Also note that a problem A *is* $\leqslant_L$-complete for **P** if and only if *co-A* is $\leqslant_L$-complete for **P**. As we have seen in Chapter 13, this is an open question for the class **NP**.

In this section, we will consider the problems *3-CNF-CONTRA-UR*, *CNF-CONTRA-UR* and *GEN*, defined in Definition 14.1. We will show that these problems are **P**-complete. There are many problems that are **P**-complete; we only discuss the above-mentioned ones because they fit in nicely between **NL**-complete and **NP**-complete problems.

Lemma 14.6

CNF-CONTRA-UR $\in$ **P**.

Proof

Let F be a propositional formula in conjunctive normal form, S a list of the unit clauses in F and T a list of the non-unit clauses in F. Let ST denote the list obtained by appending list T to list S.

Consider the following procedure.

while $S \neq \varnothing$ **do**
  let u be the first literal S;
  if u cannot be resolved with any clause of ST **then**

    delete u from S

  else

    let v be the first clause on ST with which u can be resolved and let w be the resolvent;

    delete v from the list ST;

```
        if w = □ then
            output ("F yields □");
            halt
        fi;
        if w is a unit clause then
            add w to the end of list S
        else
            add w to the end of list T
        fi
    fi
od;
output("F doesn't yield □")
```

The while loop in the above procedure is executed at most k^2 times, where k is the number of clauses in the given formula F. Thus the method works in polynomial time.

If $F \in$ *CNF-CONTRA-UR*, then the above method will yield the empty clause □, because the order in which the resolvents are determined is unimportant, and all possible unit resolutions are performed. Consequently, *CNF-CONTRA-UR*$\in$**P**. ■

It remains to show that *CNF-CONTRA-UR* is $\leqslant_L$-hard for **P**.

Lemma 14.7

CNF-CONTRA-UR is $\leqslant_L$-hard for **P**.

Proof

The proof is very similar to the proof of the **NP**-hardness of *CNF-SAT* given in Lemma 13.2. We also make use of the propositional formulae used in that proof.

Let $A \subseteq \Sigma^*$, $A \in \mathbf{P}$, TM_A a p-time-bounded deterministic Turing machine accepting A, p a polynomial and $w \in \Sigma^*$. We construct a propositional formula F_w such that $F_w \in$ *CNF-CONTRA-UR* if and only if $w \in A$. The formula consists of five subformulae

$$F_w \triangleq F_1 \wedge F_2 \wedge F_3 \wedge F_4 \wedge G$$

The construction proceeds exactly the same as in the proof of Lemma 13.2, and the formulae F_1 to F_4 are those defined in that proof. The formulae are somewhat simpler, because TM_A is deterministic.

Assume now that TM_A has a single accepting state q_f. Then $w \in A$ if and only if $S_{p(|w|),q_f}$ can be derived from $F_1 \wedge F_2 \wedge F_3 \wedge F_4$ by unit resolution. Thus, we define $G \triangleq \neg S_{p(|w|),q_f}$.

Consequently, $w \in A$ if and only if $\Box$ can be derived from F_w by unit resolution.

Thus the only thing that remains is to show that, without loss of generality, we may assume that TM_A has but a single accepting state.

For that purpose, let $T=(p, V, \Sigma, B, Q, q_0, F, M)$ be any Turing acceptor. Define the Turing acceptor $T_1=(p, V, \Sigma, B, Q \cup \{q_a\}, q'_0, F_1 M_1)$, where q_a is a new state not in Q, as follows:

(1) $q'_0 =$ if $q_0 \in F$ then q_a else q_0

(2) $F_1 = \{q_a\}$

(3) $M_1 = \{(p, x, y, r, q) \in M \mid p, q \notin F\} \cup$

$$\{(p, x, y, r, q_a) \mid (\exists q \in F)[(p, x, y, r, q) \in M]\} \cup$$

$$\{(q_a, x, x, r, q_a) \mid x \in V^p, r=(S, S, \ldots, S)\}$$

Then $L(T)=L(T_1)$ and furthermore T_1 has a single accepting state.

Thus $A \leqslant_L$ *CNF-CONTRA-UR* for every $A \in$ **P** and consequently *CNF-CONTRA-UR* is **P**-hard. ■

Combining the preceding two lemmas we have

Theorem 14.3

CNF-CONTRA-UR is **P**-complete. ■

The number of literals per clause can be restricted to three by the technique explained in the proof of Theorem 13.2. However, if the formulae are restricted to have at most two literals per clause, then the problem is $\leqslant_L$-complete for **NL**, as we have seen in Theorem 14.2.

Corollary 14.1

3-*CNF-CONTRA-UR* is **P**-complete. ■

We turn now to the problem *GEN*. Consider an instance $((X, .), S, z)$. To see whether z can be generated from S an expression $x_1.x_2.\cdots.x_k$ is selected and it is checked that this expression evaluates to z. If the operation '.' is associative, as in the problem *AGEN*, the order of evaluation is unimportant; in the non-associative case all possible ways of parenthesizing must be taken into account.

Lemma 14.8

$GEN \in$ **P**.

Proof

Let $((X,.), S, z)$ be an instance of *GEN*. In the algorithm below $W.W$ is defined as follows:

$$W.W \triangleq \{u.v \mid u, v \in W\}$$

The following algorithm determines whether or not z can be generated by S.

$W := S$;

while $W.W \not\subseteq W \wedge z \notin W$ **do**

$W := W.W$

od;

if $z \in W$ **then** return ("yes") **else** return ("no") **fi**

The while loop in the above program is executed at most $card(X)$ times, because each time the loop is executed, at least one new element must be added to W. The evaluation of $W.W$ requires no more than $(card(X))^2$ evaluations of the elementary operation '.'. Therefore the algorithm runs in polynomial time and consequently, $GEN \in \mathbf{P}$. ■

Lemma 14.9

$3\text{-}CNF\text{-}CONTRA\text{-}UR \leqslant_L GEN$, thus *GEN* is **P**-hard.

Proof

Let F be an instance of *3-CNF-CONTRA-UR*. Define an instance $((X,.), S, z)$ of *GEN* as follows:

(1) $X \triangleq \{\omega\} \cup \{C \mid C \text{ is a subclause of a clause in } F\}$,

where a subclause of a clause $x_1 \vee x_2 \vee \cdots \vee x_k$ is a clause $y_1 \vee y_2 \vee \cdots \vee y_r$ such that for every i, $1 \leqslant i \leqslant r$, there exists a j, $1 \leqslant j \leqslant k$, such that $y_i = x_j$.

(2) $S \triangleq \{C \mid C \text{ is a clause in } F\}$

(3) $z \triangleq \square$

(4) The operation '.' is defined as follows:

$x.y \triangleq$ if (x and y are subclauses of clauses of F, which can be unit-resolved and have u as the resolvent) then u else ω.

It is obvious that $z = \square$ is generated by $S =$ the set of all clauses of F if and only if $F \in 3\text{-}CNF\text{-}CONTRA\text{-}UR$.

Let n_F denote the number of clauses of F. Because a clause in F has at most three literals, $card(X) \leqslant 1 + 7n_F$. Hence, this instance of GEN can be constructed and written to the output by a $\lambda n[\log(n)]$-space-bounded Turing transducer. ■

Combining the preceding two lemmas we have:

Theorem 14.4

GEN is **P**-complete. ■

14.3 Polynomial space completeness

Obviously, all problems in **P** can be solved in polynomial space, but it is still unknown whether there exist problems solvable in polynomial space which cannot be solved in polynomial time. It is generally believed that $\mathbf{P} \neq \mathbf{NP}$, thus the conjecture $\mathbf{P} \neq PSPACE$ is even more plausible, since $\mathbf{P} \subseteq \mathbf{NP} \subseteq PSPACE$. In addition, many problems that can be solved in polynomial space appear to be harder than problems in **NP**.

Problems in $PSPACE$ can be compared with respect to $\leqslant_L$, $\leqslant_P^K$, or $\leqslant_P^C$. All problems known to be $\leqslant_P^C$-complete for $PSPACE$ are also $\leqslant_L$-complete for $PSPACE$. In this section we use the phrase '$PSPACE$-complete' to denote $\leqslant_P^C$-completeness for $PSPACE$.

In Section 12.5, we have seen that the resource-bounded Halting Problem, as represented in the set U_{PSPACE} is $\leqslant_L$-complete for $PSPACE$. There also are many more or less practical problems known to be $PSPACE$-complete. The following problem 'Quantified Boolean formulae' was shown to be $PSPACE$-complete by Stockmeyer and Meyer (1973).

$$QBF \triangleq \{(Q_1x_1)(Q_2x_2)\cdots(Q_nx_n)F \mid F \text{ is a propositional (or Boolean) formula with } \mathrm{var}(F) = \{x_1, x_2, \ldots, x_n\}, Q_i \text{ is either } \forall \text{ or } \exists \text{ and the formula } (Q_1x_1)(Q_2x_2)\cdots(Q_nx_n)F \text{ is true}\}$$

It is easy to prove that QBF is in $PSPACE$. The value of F can be computed for all possible assignments $v: \mathrm{var}(F) \to \{0, 1\}$ by a simple counting procedure. Although polynomial space is clearly sufficient, exponential time is needed to enumerate the 2^n possible assignments.

Theorem 14.5

QBF is $PSPACE$-complete. ■

Proofs of this theorem can be found in Stockmeyer and Meyer (1973) and Wrathall (1977). The restricted versions obtained by requiring that F be in conjunctive normal form (QB-CNF) and furthermore requiring that there are at most three literals per clause (*QB*-3-*CNF*), are also *PSPACE*-complete.

A particularly rich source of *PSPACE*-complete problems is the area of combinatorial games. Of interest are two-person games that can be specified concisely.

Consider a game played by two players, call them Ben and Wilma. Let $P_0, P_1, \ldots$ denote positions of the game. Here a 'position' completely specifies a game situation, in draughts (checkers), for example, the 'position' specifies for each square whether it is occupied, and if so, the colour of the piece occupying the square. Let P_0 denote the initial position. The fact that Wilma, who is the first to move, has a forced win in n moves can be expressed as follows:

> There is a move for Wilma from P_0 to P_1 such that
>
> for all possible moves for Ben from P_1 to P_2
>
> there is a move for Wilma from P_2 to P_3 such that
>
> for all possible moves for Ben from P_3 to P_4
>
> ⋮
>
> there is a move for Wilma from P_{n-2} to P_{n-1} such that
>
> for all possible moves for Ben from P_{n-1} to P_n
>
> the position P_n is a win for Wilma.

This statement displays a similar alternation of quantifiers as *QBF*. This makes it plausible that decision problems of the type 'Has Wilma a forced win from a given position in a particular game?' should be *PSPACE*-complete.

Consider for example the game of *HEX*. The game is played on a honeycomb structure as sketched in Figure 14.2. Players in turn select a cell and fill it with a white or black peg. The game is won by the player who succeeds in constructing a path connecting two opposite sides. At the beginning of the game, the players decide who uses the white pegs and who tries to connect which opposite sides.

A generalization, which was shown to be *PSPACE*-complete by Even and Tarjan (1976), is as follows. *GENERALIZED HEX* is played on an undirected graph (V, E) with two marked vertices, s and t. Positions of the game are triples (X, Y, Z) such that $X \cap Y = X \cap Z = Y \cap Z = \varnothing$ and $X \cup Y \cup Z = V - \{s, t\}$. The first player owns the set X, the second player owns the set Y. The initial position is $(\varnothing, \varnothing, V - \{s, t\})$. The players alternatively select a vertex from Z and move it to their own set. A final position $(V_1, V_2, \varnothing)$ is a win for player i if there is a path from s to t in (V, E), which uses only vertices from V_i and s and t.

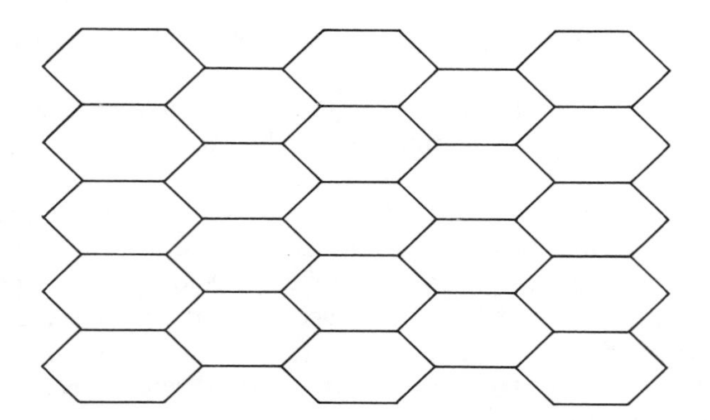

Figure 14.2 Board for the game of HEX.

The problem is to determine whether the first player has a forced win. More formally:

$GENERALIZED\ HEX \triangleq \{((V, E), s, t) \mid (V, E)$ is an undirected graph, $s, t \in V$, and the first player has a forced win starting from the position $(\emptyset, \emptyset, V - \{s, t\})$ according to the rules of the game as explained above$\}$.

Many other games have been proved to be *PSPACE*-complete. Examples are various generalizations of draughts and Go.

Also many problems associated with automata and languages have been shown to be *PSPACE*-complete. A list of such problems can be found in Garey and Johnson (1979) and in Wagner and Wechsung (1986).

14.4 Exercises

Below we define a number of problems concerning formal languages. A *grammar* $G = (V, \Sigma, P, S)$ consists of a *rewriting system*, a subset $\Sigma \subseteq V$ of so-called *terminal symbols*, and an element $S \in (V - \Sigma)$, called the *sentence symbol*, or the *start symbol*. (Rewriting systems have been defined in Section 10.5.)

Definition 14.3

(1) The language generated by G is denoted by $L(G)$ and is defined as follows:

$$L(G) \triangleq \{w \in \Sigma^* \mid S \Rightarrow w\}$$

(2) The grammar is *context-free* if all rewrite rules are of the form $A \rightarrow \omega$, where $A \in (V - \Sigma)$ and $\omega \in \mathbf{V}^*$.

(3) The grammar is *linear* if all rewrite rules are of the form

- $A \to \omega$, where $A \in (V - \Sigma)$ and $\omega \in \Sigma^*$, or
- $A \to \omega_1 B \omega_2$, where $A, B \in (V - \Sigma)$ and $\omega_1, \omega_2 \in \Sigma^*$.

(4) The grammar is *regular* if all rewrite rules are of the form

- $A \to \omega B$, where $A, B \in (V - \Sigma)$ and $\omega \in \Sigma^*$, or
- $A \to \omega$, where $A \in (V - \Sigma)$ and $\omega \in \Sigma^*$.

(5)
- $R\varnothing \triangleq \{G \mid G$ is a regular grammar and $L(G) = \varnothing\}$
- $CF\varnothing \triangleq \{G \mid G$ is a context-free grammar and $L(G) = \varnothing\}$
- $R\infty \triangleq \{G \mid G$ is a regular grammar and $L(G)$ is infinite$\}$
- $CF\infty \triangleq \{G \mid G$ is a context-free grammar and $L(G)$ is infinite$\}$
- $LCF \triangleq \{(G, w) \mid G$ is a linear context-free grammar and $w \in L(G)\}$
- $CF \triangleq \{(G, w) \mid G$ is a context-free grammar and $w \in L(G)\}$
- $\{0,1\}$-$LINEQ$-2 $\triangleq \{(\mathbf{M}, \mathbf{b}) \mid \mathbf{M}$ is an $n \times n$ matrix over the rationals Q having at most two non-zero entries per row, and $\mathbf{b} \in Q^n$, and there *does not exist* a $\{0,1\}$ vector $\mathbf{x}$ such that $\mathbf{Mx} \geqslant \mathbf{b}\}$

■

14.1 Prove that the problems $R\varnothing$, $R\infty$, LCF and $\{0,1\}$-$LINEQ$-2 are $\leqslant_L$-complete for **NL**.

14.2 Prove that the problems $CF\varnothing$, $CF\infty$, and CF are $\leqslant_L$-complete for **P**.

14.3 If $DSPACE(\lambda n[n^k]) \subseteq \mathbf{P}$ for some $k \geqslant 1$, then $PSPACE = \mathbf{P}$. Prove this.

14.4 If $DSPACE(\lambda n[n^k]) \subseteq \mathbf{NP}$ for some $k \geqslant 1$, then $PSPACE = \mathbf{NP}$. Prove this.

14.5 References

For literature concerning the various relations between the complexity classes defined in this chapter and to complete problems we refer to Wagner and Wechsung (1986).

For problems complete for **NL** and **P**, we refer to Jones (1975), Jones and Laaser (1976, 1977) and for *PSPACE*-complete problems to Stockmeyer and Meyer (1973).

Chapter 15
Final Remarks

In this final chapter a number of topics will be introduced, a thorough discussion of which is outside the scope of this book. We will discuss – without undue formalism – the theory of approximation algorithms, the theory of probabilistic algorithms, and the theory of parallel algorithms.

A good overview of the theory of algorithms, including these topics, was given by Cook (1983) in his Turing Lecture.

15.1 Approximation algorithms

In the preceding chapters we saw strong evidence supporting the conjecture that **NP**-hard problems cannot be solved by deterministic polynomial-time algorithms. Among these **NP**-hard problems are many which are derived from optimization problems and which are of great practical importance.

The best algorithms for **NP**-hard problems known, have an exponential worst-case time complexity. Since we cannot expect polynomial-time algorithms for these problems, we may try to find algorithms with subexponential time complexity, such as $\lambda n[2^{n/c}]$, or $\lambda n[2^{\sqrt{n}}]$ for example.

The invention of subexponential algorithms for **NP**-hard problems will increase the maximum of the sizes of problem instances whose solutions are feasibly computable.

Another possible way to decrease the time complexity of **NP**-hard problems is the use of heuristics. However, heuristics do not improve the worst-case execution time. Thus exponential-time algorithms equipped with heuristics will still require exponential time for some of their inputs.

In this section, we will follow a different track. The idea is to abandon the goal of finding optimal solutions, and instead settle for solutions which are suboptimal, but not too far from the optimal solution. Algorithms producing such suboptimal solutions are called *approximation algorithms*. An approximation algorithm for an optimization problem P is supposed to compute '*near optimal*' solutions. A 'near optimal' solution is a solution which is sufficiently close, in the sense defined below, to the optimal solution.

Especially for **NP**-hard optimization problems, where optimal solutions cannot be obtained within a feasible time, it may be advantageous to have an approximation algorithm computing 'near optimal' solutions in polynomial time.

Let X be an optimization problem, I an instance of **X** and $S^*(I)$ the value of an optimal solution of **X**. An approximation algorithm A computes a feasible solution for I whose value $S(I)$ is 'near optimal'. Thus, for a maximization problem $S(I) \leqslant S^*(I)$, and for a minimization problem $S(I) \geqslant S^*(I)$.

Whether a feasible solution of an instance I is 'near optimal' is a matter of definition. In the following, several possibilities are considered.

Definition 15.1

(1) Let c be a natural number. A is an *absolute c-approximation algorithm* for problem **X** if for every instance I of **X**

$$|S^*(I) - S(I)| \leqslant c.$$

(2) Let f be a function from N to the set of positive real numbers. A is a *relative $f(n)$-approximation algorithm* for problem **X** if for every problem instance I of **X**

$$\frac{|S^*(I) - S(I)|}{S^*(I)} \leqslant f(|I|),$$

where $|I|$ denotes the size of I. (We assume that $S^*(I) > 0$ for all instances I.)

(3) A is a *relative ε-approximation scheme* for problem **X** if for every $\varepsilon > 0$ A_ε is a relative $\lambda n[\varepsilon]$-approximation algorithm for problem **X**, that is, if for every instance I,

$$\frac{|S^*(I) - S(I)|}{S^*(I)} \leqslant \varepsilon.$$

(We assume that $S^*(I) > 0$ for all instances I.) Thus a relative ε-approximation scheme is an algorithm which has two inputs, namely an instance of problem **X**, and a value for the bound ε.

(4) A relative ε-approximation scheme A is called *polynomial-time-bounded*, if A_ε is polynomial-time-bounded for every $\varepsilon > 0$.

(5) A relative ε-approximation scheme is called *fully polynomial-time-bounded* if the time required by the execution of A is bounded by a polynomial in $|I|$ and $1/\varepsilon$. ■

If **X** is a maximization problem, then $(|S^*(I) - S(I)|/S^*(I)) \leqslant 1$ for every instance I of **X**. Thus every algorithm which returns some feasible solution is already a relative $\lambda n[1]$-approximation algorithm. To make relative approximation useful, we must consider functions f such that $f(n)$ is small compared to 1.

Obviously the most desirable approximation algorithm is an absolute approximation algorithm. However, for many **NP**-hard problems **X** it can be proved that efficient absolute approximation algorithms exist if and only if **P** = **NP**. There are but a few **NP**-hard optimization problems for which polynomial-time-bounded absolute c-approximation algorithms are known, a trivial example concerning the colouring of planar graphs is given below.

EXAMPLE 15.1

It can be proved that colouring a planar graph requires no more than four colours. Thus the following is an absolute 1-approximation algorithm solving the problem to determine the minimum number of colours required to colour a given planar graph $g = (V, E)$.

input: $G = (V, E)$
output: an approximation of the number of colours required to colour the input graph G, such that $|S^*(I) - S(I)| \leqslant 1$.
method:

if $E = \emptyset$ **then** return (1);
if G is bipartite **then** return(2);
return(4)

In the following example, a relative ε-approximation algorithm is described, which solves a scheduling problem. The problem is as follows:

Given is a computer system with $m > 2$ processors, which is to process n independent computing tasks, each requiring a computing time t_i, $1 \leqslant i \leqslant n$. The problem is to find a schedule for these tasks, such that the total time is minimal.

The problem differs from the problem *SCHED* as given in Definition 13.5, in that there are no precedence relations. Let *I-SCHED* denote the language version of the scheduling problem described above:

$I\text{-}SCHED \triangleq \{(V, m, t) \mid V$ is a sequence $(a_1, a_2, \ldots, a_n)$ of not necessarily distinct natural numbers, specifying the time required by the n computing tasks, m and t are natural numbers, $m \geqslant 2$ and $t \geqslant 1$, and there is a function $s: \{1, 2, \ldots, n\} \to \{1, 2, \ldots, m\}$ such that $\sum_{s(j)=i} a_j \leqslant t$, for all i, $1 \leqslant i \leqslant m\}$.

Clearly, $I\text{-}SCHED \in \mathbf{NP}$. It can be shown that $PARTITION \leqslant_P^K I\text{-}SCHED_2$, where $I\text{-}SCHED_2$ is the problem *I-SCHED* restricted to the case where $m = 2$, and *PARTITION* is the problem to determine whether a given set V of positive integers can be divided into two subsets J and $V - J$ such that $\sum_{x \in J} x = \sum_{x \in V - J} x$ (see Section 13.3). Thus *I-SCHED* is **NP**-complete.

To find a schedule for a given instance (V, m) of the optimization version of the problem *I-SCHED*, various approximating strategies are known, for example the Longest Processing Time (LPT) method.

EXAMPLE 15.2

The LPT scheduling method assigns to a free processor a task not yet scheduled, which has maximal required processing time. In case more than one such maximum time task exists, or more than one processor is free, one of those tasks is chosen at random and assigned to an arbitrarily chosen free processor.

It can be proved that at most $O(n \log(n))$ time is required to generate an LPT schedule for n tasks and m processors such that

$$\frac{|S^*(I) - S(I)|}{S^*(I)} \leqslant \left(\frac{1}{3} - \frac{1}{3m}\right)$$

where $S(I)$, the value of a feasible solution, in this case denotes the total time needed by the schedule produced.

Thus the LPT algorithm is a relative $\frac{1}{3}$-approximation algorithm.

Consider the instance $((5, 5, 4, 4, 3, 3, 3,), 3)$. In other words, there are seven tasks and 3 processors. The LPT algorithm produces the following schedule.

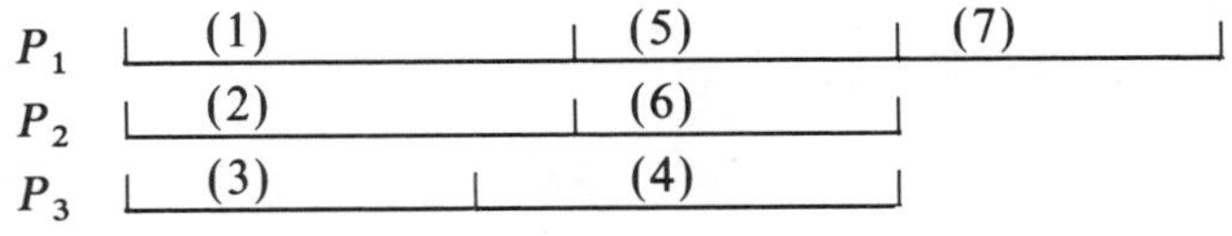

```
P1  |____(1)_____________|____(5)________|____(7)_____________|
P2  |____(2)_____________|____(6)________|
P3  |____(3)________|________(4)_________|
```

Thus, the total time required by this schedule is 11 units. The following is an optimal schedule.

```
P1  |_(1)____________|____(3)__________|
P2  |_(2)____________|____(4)__________|
P3  |_(5)________|_(6)___|_(7)_________|
```

The total time required by this schedule is 9 units. Thus, in this case we have $S(I) = 11$ and $S^*(I) = 9$. Therefore

$$\frac{|S^*(I) - S(I)|}{S^*(I)} = \frac{11 - 9}{9} = \frac{2}{9}$$

Effective relative ε-approximation algorithms do exist for many scheduling problems.

The LPT method can also be used to obtain a polynomial-time-bounded relative ε-approximation scheme. The scheme is not fully polynomial-time-bounded.

EXAMPLE 15.3

Let k be a natural number. Consider the following method to solve the optimization version of the *I-SCHED* problem.

(1) determine an optimal schedule for the k longest tasks;

(2) schedule the remaining $n - k$ tasks using the LPT method.

Graham (1969) proved that using this method we have

$$\frac{|S^*(I) - S(I)|}{S^*(I)} \leqslant \frac{(1 - 1/m)}{(1 + \lfloor k/m \rfloor)}.$$

Using this result, we can obtain the following polynomial-time-bounded relative ε-approximation scheme.

input: $((V, m), \varepsilon)$
output: an approximation of the time required by an optimal schedule of the tasks in V
method:

compute k such that $\varepsilon \geqslant \dfrac{1 - 1/m}{1 + \lfloor k/m \rfloor}$;

find an optimal schedule for the k longest tasks;

construct a schedule for the remaining $n-k$ tasks using the LPT method.

The time required by the above algorithm largely depends on the time needed to find an optimal schedule for k tasks on m processors.

(1) The value $k=(m-1)/\varepsilon$ satisfies the requirements, and can be easily computed.

(2) Having found k, we sort the tasks, which takes $O(n\log(n))$ comparisons, determining in this way the k longest tasks.

(3) Determining an optimal schedule for the k longest tasks can be achieved in time $O(m^k)$.

(4) Finally, determining a suboptimal schedule using the LPT method takes $O((n-k)\log(n-k))$ time.

In all, the time required by the above algorithm is $O(n\log(n)+m^{(m-1)/\varepsilon})$. Thus for any fixed m, and therefore also for the optimization version of the **NP**-hard problem $I\text{-}SCHED_2$, the above algorithm implements a polynomial-time-bounded relative ε-approximation scheme. The scheme is not fully polynomial-time-bounded, because the running time is not polynomial in $1/\varepsilon$.

As already mentioned, for most **NP**-hard problems there exists a polynomial-time-bounded absolute approximation algorithm if and only if the optimization problem is solvable by a polynomial-time-bounded algorithm. The optimization version of the *CLIQUE* problem is an example. This optimization version *MAX-CLIQUE* is defined as follows.

Given is an undirected graph $G=(V,E)$ and the problem is to find the largest value k such that (V,E) contains a k-clique.

Theorem 15.1

The *MAX-CLIQUE* problem is solvable by a polynomial-time-bounded algorithm if and only if there exists an absolute c-approximation algorithm for *MAX-CLIQUE* for some natural number $c>0$.

Proof

- If *MAX-CLIQUE* is solvable by a polynomial-time-bounded algorithm, then this algorithm is also a polynomial-time-bounded c-approximation algorithm for any $c \geqslant 0$.
- Let c be a natural number, and assume that A is a polynomial-time-bounded absolute c-approximation algorithm for *MAX-*

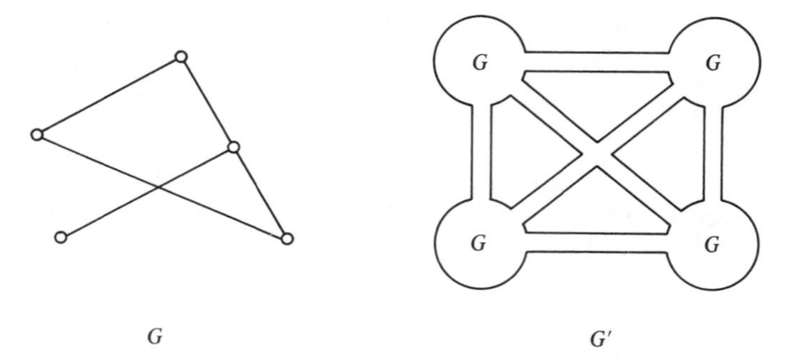

Figure 15.1 Gluing four copies of G together resulting in G'.

CLIQUE. We show how to use this approximation algorithm to obtain an exact solution.

Given is an undirected graph $G=(V,E)$. Construct a new graph G', which consists of $c+1$ copies of G glued together by adding all possible edges between vertices in mutually different copies, as sketched in Figure 15.1 for $c=3$.

If G contains a k-clique, then G' contains a $(c+1)k$-clique. Let m be the size of the maximal clique in G' and let s be the size as returned by the absolute c-approximation algorithm A. Then

$$(c+1)m-k \leqslant s \leqslant (c+1)m.$$

Thus to find the size of the maximal clique in $G=(V,E)$, proceed as follows:

(1) Construct G' as described above.

(2) Call the approximation algorithm A on input G' and let s be the value returned.

(3) Return($\lceil s/(c+1)\rceil$).

Clearly, the above procedure is polynomial-time-bounded if the absolute c-approximation algorithm is polynomial-time-bounded. ■

Thus, finding a polynomial-time-bounded absolute c-approximation algorithm can be as hard as finding a polynomial-time-bounded algorithm for an **NP**-complete problem. This is also true for the problem of finding polynomial-time-bounded relative ε-approximation schemes.

Theorem 15.2

If there exists a polynomial-time-bounded relative ε-approximation algorithm for the optimization version of the Travelling Salesman Problem, then the Directed Hamiltonian Circuit Problem can be solved in polynomial-time; that is, $DHC \in \mathbf{P}$.

Proof

Assume that there is a polynomial-time-bounded relative ε-approximation algorithm for the Travelling Salesman Problem. This algorithm returns the length of a tour, which is not necessarily minimal. We show how to use this algorithm to solve DHC.

Let $G=(V,E)$ be an instance of DHC, and let $|(V,E)|=n$. An instance of the optimization version of the Travelling Salesman Problem is a pair (G',w), where $G'=(W,T)$ is a directed graph, and $w: W \to N$ is a function which assigns distances to the edges. This instance (G',w) is constructed as follows:

(1) $W=V$

(2) $T=\{(u,v|u,v \in W \text{ and } u \neq v\}$

(3) $w((u,v))=\begin{cases} 1 & \text{if } (u,v) \in E \\ 2+\varepsilon.card(V) & \text{otherwise.} \end{cases}$

Call the relative ε-approximation algorithm on the instance (G',w), and let k be the length returned by this algorithm.

(1) If G contains a Hamiltonian circuit, then (G',w) has a tour of length $card(V)$. Therefore $|k-card(V)| \leqslant \varepsilon.card(V)$, and thus $k \leqslant (1+\varepsilon)card(V)$.

(2) If G does not contain a Hamiltonian circuit, every tour of (G',w) has length at least $card(V)-1+(2+\varepsilon.card(V))$. Thus

$$k \geqslant card(V)+1+\varepsilon.card(V) > (1+\varepsilon)card(V).$$

Consequently, $G \in DHC$ if and only if the approximation algorithm returns a length k such that $k \leqslant (1+\varepsilon)card(V)$. Hence $DHC \in \mathbf{P}$ if the relative ε-approximation algorithm is polynomial-time-bounded. ■

More information about approximation algorithms can be found in Hochbaum and Shmoys (1986), where a simple technique is described for devising approximation algorithms for a wide variety of **NP**-complete problems in routing, location and communication network design.

15.2 Probabilistic algorithms

Many computer algorithms use random numbers, in particular algorithms which are used to simulate complex processes. This use of random numbers is mathematically well established. There also exist algorithms, intended to solve combinatorial problems, which use random numbers to determine which of several alternative operations to perform. It is not obvious that such 'coin-tossing' algorithms can be useful at all. However, there are coin-tossing algorithms which are very efficient, and, for all practical purposes, are as good as deterministic algorithms.

The first such algorithm was invented by Berlekamp (1968). It can be used to factor a polynomial f over a field $GF(p)$ of p elements. The running time of this algorithm is bounded by a polynomial in the degree of f and in $\log(p)$. Applying the algorithm to a polynomial f produces one of two possible results:

- the algorithm reports its failure to find a factorization, or
- the algorithm produces a factorization of f.

If the polynomial *can* be factored, the probability of getting 'failure' for a result is less than $\frac{1}{2}$. The failure events are independent, so in practice factors can always be found within a feasible amount of time by calling the algorithm a number of times.

To illustrate the methods used, we describe a probabilistic algorithm to determine whether a given number is prime, devised by Rabin (1976).

Miller (1975) proved that a natural number n is composite if and only if there exists a natural number b $(1 < b < n)$ satisfying $W(b)$, where $W(b)$ is the following condition:

$$W(b) = [b^{n-1} \neq 1 (\mathrm{mod}\, n)]$$
$$\text{or } (\exists i > 0)[(n-1)/2^i \in Z \text{ and } 1 < \gcd(b^n - 1, n) < n]$$

An integer b for which the condition $W(b)$ holds is called a *witness*. It can be proved that for every composite number n at least half of the numbers $b < n$ are witnesses to the compositeness of n. That is,

$$\frac{n-1}{2} \leqslant card(\{(b \mid 1 < b < n \text{ and } W(b)\}).$$

For a given input n Rabin's primality testing algorithm proceeds by randomly selecting m numbers $b_1, \ldots, b_m$ and testing whether or not $W(b_i)$ holds, $1 \leqslant i \leqslant m$. If $W(b_i)$ holds for a certain i, then n is a composite number, otherwise the algorithm declares (maybe mistakenly) m to be a prime number. Thus if n is a prime the output is 'prime' and if n is composite, then the output may be 'prime', or 'composite'. The probability that 'prime' is produced for an n which is composite is less than $(\frac{1}{2})^m$.

The condition $W(b)$ is quickly testable; the value of b_i^{n-1} (mod m) can be computed using $O(\log(n))$ multiplications of $\log(n)$-bit numbers, and thus proceeds in polynomial time; finding the greatest common divisor of b_i^{n-1} and n can also be achieved in polynomial time.

Rabin (1976) reports some experiments with this algorithm. Using the value $m = 10$, and therefore allowing $(\frac{1}{2})^{10} \simeq 0.001$ error probability, the primality of the Mersenne numbers $2^p - 1$ with $p \leqslant 500$ was tested. The whole sequence was tested within minutes and without a single error. The largest number less than 2^{400} declared prime by the algorithm was $2^{400} - 593$. This number was also found within minutes. Subsequently this number was tested 100 times, with no change in the conclusion. Thus there is good evidence that the number n is prime; there is a probability of 2^{-100} that the algorithm produces the wrong result.

An interesting application of probabilistic prime testers is described by Rivest *et al.* In their crypto systems they frequently need to find primes of some 100 digits. They propose to find these by testing randomly selected 100-digit numbers using a probabilistic algorithm, until a prime is found.

Let **R** denote the class of sets recognizable by polynomial-time-bounded probabilistic algorithms in the sense described above. More precisely, the set L is in **R** if and only if there exists a polynomial-time-bounded probabilistic recognition algorithm that always halts and in addition never announces a wrong result for inputs belonging to L. Hence the set of prime numbers is in **R**. The following inclusion holds.

$$\mathbf{P} \subseteq \mathbf{R} \subseteq co\text{-}\mathbf{NP}$$

It is an open question whether $\mathbf{P} = \mathbf{R}$. A related question is whether a probabilistic algorithm is as good as a deterministic algorithm for all practical purposes. In actual practice, a probabilistic algorithm is deterministic, because deterministic algorithms are used to generate the (pseudo-) random numbers employed by the probabilistic algorithm. The error probability can be made extremely low. Therefore, in many cases a probabilistic algorithm is as good as any other algorithm, because it is efficient and almost always correct.

There are also circumstances in which probabilistic algorithms are of little or no value. For example, in number-theoretic research, probabilistic algorithms testing primality are utterly worthless, because the algorithms are essentially incorrect. In this type of research any error probability greater than zero makes the results useless.

15.3 Parallel algorithms and parallel computers

In the near future the theory of parallel algorithms will probably become more and more important, since many sequential algorithms can be modified into parallel algorithms which run substantially faster on parallel computers.

Examples of such problems are computations in relation to weather forecasting and computations necessary to solve certain combinatorial problems in nuclear physics. Other examples are problems in numerical mathematics, partial differential equations and numerical linear algebra. Some of these problems can be very efficiently solved on special-purpose parallel computers or on supercomputers.

In this section we present an introduction to parallel machine models and examples of parallel algorithms; also theoretical questions regarding parallel computation models are discussed.

A parallel computer consists of a number of processing elements (PEs) interconnected by a network, or sharing a global memory. Since the introduction of VLSI technology, people think of parallel computers containing many thousands of PEs. Each PE is able to execute, more or less independently from others, certain computational tasks. Parallel algorithms and programming languages have to be designed in order to make better use of these hardware facilities.

Problems to be solved by such parallel computers have to be partitioned into subtasks which can be executed independently by different PEs. The results of the subtasks must subsequently be combined into the final result of the parallel computation.

Parallel computers, with parallel systems software and parallel programming languages, are commercially available at the moment. The usefulness of these computers strongly depends on the possibility of dividing the computational problem at hand into more or less independent subproblems, and also on the particular hardware configuration, and the available software tools. In many cases a parallel system may have a lack of resources such as processing power, and local or global memory. The system may also have an undesirable connection network topology with respect to the problem at hand. Efficiently using a parallel computer is far from easy.

But apart from these important practical questions, there are also a number of fundamental questions to be answered, e.g. 'What is the inherent parallelism of a problem?', 'Is it possible to construct a general-purpose parallel computer?', and 'Which problems can be solved substantially faster on a parallel computer than on a sequential computer?'

15.3.1 Parallel machine models

Many architecture models of parallel machines have been proposed in the literature. Some of these models have actually been built in hardware; others are used in the theoretical analysis of parallel algorithms.

One of the best-known classifications of parallel and sequential computer models has been devised by Flynn (1972). In this classification four computer models are distinguished:

(1) *SISD* (= *Single-Instruction, Single-Data streams*)
In every time unit one instruction is executed on one data item. This class contains the well-known sequential computer model, the so-called von Neumann computer model.

(2) *SIMD* (= *Single-Instruction, Multiple-Data streams*)
In every time unit an instruction is executed. The computer contains an unspecified number of processing elements, each of which is either idle or executes this same instruction, using possibly mutually different data items. The active PEs are selected using a mask operation. The ICL DAP (= Distributed Array Processor) belongs to this class.

(3) *MISD* (= *Multiple-Instruction, Single-Data streams*)
In every time unit different PEs execute different instructions on the same data item.

(4) *MIMD* (= *Multiple-Instruction, Multiple-Data streams*)
In every time unit different PEs can execute different instructions on different data items. The PEs in a synchronous MIMD machine perform the instructions in a lock-step fashion; the processors of an asynchronous MIMD machine run independently of each other and may need to wait if information from other PEs is needed. The Denelcor HEP (= Heterogeneous Element Processor) and the DPP (= Delft Parallel Processor) are examples of asynchronous MIMD machines.

Flynn's (1972) classification of computer architectures is rather coarse. Many additional distinctions could be made, for example with respect to the amount of local memory available for each PE, the way the global memory is shared among the PEs (if global memory is at all available), and the particular connection network used for the PEs and the memories. For details see for example Hockney and Jesshope (1981).

The development of parallel computers and distributed systems has direct implications for the theory of algorithms (practical and theoretical). This is illustrated for example in the excellent annotated bibliography on parallel combinatorial algorithms by Kindervater and Lenstra (1983), and the wealth of references in van Leeuwen and Lenstra (1985). In fact parallelism has given a new dimension to the design and analysis of algorithms.

One may distinguish the following four sorts of parallel algorithms.

(1) *Vectorized algorithms*
These are algorithms formulated in terms of operations on (long) vectors of numbers, digits, or whatever.

(2) *Systolic algorithms*
These are algorithms designed for highly parallel computers with a regular connection network with many PEs (VLSI chips, for example)

and which are characterized by a constant stream of data between the processors.

(3) *Parallel processing algorithms*
These are algorithms devised for computers with a particular connection network topology.

(4) *Distributed algorithms*
These are algorithms designed to be implemented as communicating asynchronous processes.

Parallel processing algorithms can be executed on SIMD machines consisting of PEs which are interconnected by a network. In each time unit all PEs may communicate with or transfer data to their neighbours in the connection network. The following are some frequently occurring connection networks.

(1) *Mesh Connected Computer (MCC)*
In this model each PE has an address which is an n-tuple $(x_1,\ldots,x_n)$ with $x_i \in \{0,1,\ldots,p-1\}$ for some $p>1$ and all i, $1 \leqslant i \leqslant n$. The PE with address $(x_1,\ldots,x_n)$ is connected to the PE with address $(y_1,\ldots,y_n)$ if and only if for some i, $1 \leqslant i \leqslant n$, $|y_i - x_i| \leqslant 1$ and $y_j = x_j$ for all j, $1 \leqslant j \leqslant n$ and $j \neq i$.

(2) *Cube Connected Computer (CCC)*
Every PE has an address written as a string $x_0 \cdots x_{n-1}$ over the alphabet $\{0,1\}$. Two PEs are connected if their addresses differ in precisely one bit. Thus one can think of the PEs as being located at the corners of an n-dimensional hypercube.

(3) *Perfect Shuffle Computer (PSC)*
As in the CCC model each PE has an address represented by a string over the alphabet $\{0,1\}$. Define the operations *shuffle* and *unshuffle* by

(a) $\mathit{shuffle}(x_0,\ldots,x_{n-1}) = x_1,\ldots,x_{n-1},x_0$
(b) $\mathit{unshuffle}(x_0,\ldots,x_{n-1}) = x_{n-1},x_0,\ldots,x_{n-2}$.

$PE(w)$ is connected to $PE(\mathit{shuffle}(w))$, $PE(\mathit{unshuffle}(w))$, and $PE(w_0)$, where $w_0 = xw_1,\ldots,w_{n-1}$ if $w = yw_1,\ldots,w_{n-1}, x,y \in \{0,1\}$ and $x \neq y$.

Siegel (1979) proved that each of these networks can simulate the others efficiently.

The efficiency of the parallelization of an algorithm may be judged by the speed-up s_p (= the running time of the best sequential implementation of the algorithm divided by the running time of the parallel implementation using p processors), and also by the processor utilization $u_p = s_p/p$. The best one can hope to achieve is a speed-up of p and a processor utilization of 1.

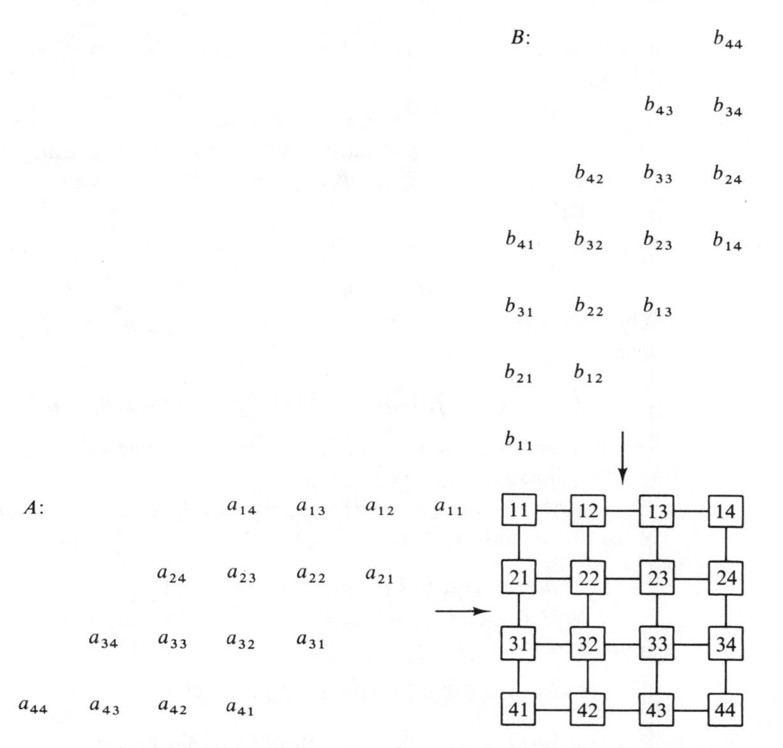

Figure 15.2 A systolic algorithm for matrix multiplication.

15.3.2 Examples of parallel algorithms

EXAMPLE 15.4

(1) *Matrix multiplication*
Two $n \times n$ matrices $A = (a_{ij})$ and $B = (b_{ij})$ can be multiplied in $O(n)$ time on an MCC with $p = n$. The basic idea is sketched in Figure 15.2.

At each time unit the matrix A makes one step to the right, the matrix B goes one step down and each $PE(i,j)$ multiplies its current values a_{ij} and b_{ij} and adds the result in its accumulator. After $2n - 1$ time units $PE(i,j)$ contains the required value $\sum_{t=1}^{n} a_{it}.b_{tj}$.

This algorithm is a typical example of a systolic algorithm suitable for VLSI implementation.

In Dekel *et al.* (1981) it is proved that the product of two $n \times n$ matrices can be computed on a CCC in $O(\log(n))$ parallel time. They also proved that this algorithm can be adapted to a PSC such that the computing time is also $O(\log(n))$.

(2) *The transitive closure of a direct graph*
Let $G = (V, E)$ be a directed graph, and assume that $V = \{v_1, v_2, \ldots, v_n\}$. The *adjacency matrix* of G is an $n \times n$ matrix A defined as follows:

$$a_{ij} = \begin{cases} 1 & \text{if } (v_i, v_j) \in E \\ 0 & \text{otherwise} \end{cases}$$

The *transitive closure* of G is a directed graph $G^* = (V, E^*)$, where

$$E^* \triangleq \{(v_i, v_j) \mid \text{there is a directed path from } v_i \text{ to } v_j \text{ in } G\}.$$

The adjacency matrix A^* of G^* is called the transitive closure of the adjacency matrix A of G.

The following algorithm computes the transitive closure of an $n \times n$ matrix A over $\{0, 1\}$ in $O(n \log(n))$ parallel time.

input: A {an $n \times n$ matrix over $\{0, 1\}$}
output: C {$C = A^*$, the transitive closure of A}
method:

for $1 \leqslant i \leqslant n$ **in-parallel-do** $c_{i,i} = 1$ **od**;

for $1 \leqslant i, j \leqslant n \wedge i \neq j$ **in-parallel-do** $c_{i,j} := a_{i,j}$ **od**;

for $1 \leqslant k \leqslant \log(n)$ **do**

  for $1 \leqslant i, j \leqslant n$ **in-parallel-do**

    $c_{i,j} =$ **if** $(\exists m)[c_{i,m} = 1 \wedge c_{m,j} = 1]$ **then** 1 **else** 0

  od

od

(3) *Membership testing*
Given a set S consisting of n elements and an object x to be tested for membership in S, one can test whether $x \in S$ in $O(\log(n))$ time on a binary tree network with n leaves. The elements $s_1, \ldots, s_n$ of S are stored in the n PEs in the leaves of the tree. It takes $\log(n)$ steps to send the element x from the root to all leaves, and one step to compare each element s_i with x. After this parallel comparison each leaf sends the message 1, meaning '$x = s_i$', or 0, meaning '$x \neq s_i$' to its father; each father computes the Boolean sum $(0 + 0 = 0, 0 + 1 = 1 + 0 = 1 + 1 = 1)$ of the messages of his sons and sends the result to his own father, etc.

Obviously the whole procedure takes $O(\log(n))$ parallel time.

If the processors in the vertices of the tree can compare the elements of S, the same model can be used to find the maximum (or the minimum) of n elements in $O(\log(n))$ time.

The major resources of a sequential algorithm are time and space. The resources for parallel computation are also time and space. Parallel computation time is the sum of the following components:

(1) The time needed to execute all operations involved.

(2) The idle time spent during the synchronization of the different subcomputations, during which no operations can be exeucted.

(3) The time overhead introduced by the execution of the synchronization operations during which no other operation can be executed.

In many discussions the components mentioned under (2) and (3) are ignored; thus in this case parallel time is equal to the total execution time of the longest sequence of operations performed. Instead of execution time, the number of operations performed is commonly used to measure the parallel computation time. Thus the time complexity t of a certain parallel computation might be equal to n multiplication steps and m addition steps. For example, Kung (1974) proved that the parallel computation of a rational function of degree n requires at least $O(\log(n))$ parallel steps, if the operations provided by the underlying data type are the four arithmetical operations $+, -, \cdot,$ and $/$.

Analogous results can be proved for the evaluation of a polynomial in one variable of degree n. With the well-known Horner algorithm this value can be computed sequentially using n additions and n multiplications; thus sequential time is $O(n)$. This computation takes at least $O(\log(n))$ parallel steps. Brent (1974) proved that every 'general arithmetical' expression in n variables can be computed in $O(\log(n))$ parallel steps if the four arithmetical operations, $+, -, \cdot,$ and $/$, are provided by the underlying data type.

To give an idea of how these results are obtained we consider a simple case, namely computing $\sum_{i=1}^{n} a_i$, where $n = 2^k$, for some $k > 0$. The sum is computed by the so-called cascade method, also called the method of recursive doubling. The parallel algorithm for this computation is as follows:

input: $a_1, \ldots, a_n$ $\{n = 2^k\}$
output: S $\{S$ is sum of the n terms a_1 to $a_n\}$
method:

for $1 \leqslant i \leqslant n$ **in-parallel-do** $s_i := a_i$ **od**;

for $1 \leqslant i \leqslant k$ **in-parallel-do**

$s_1 := s_1 + s_2;$

$s_2 := s_3 + s_4;$

$s_3 := s_5 + s_6;$

$\vdots$

$s_{n/2} := s_{n-1} + s_n;$

od;

$S := s_1$

This method can be extended so as to handle arbitrary arithmetical expressions instead of expressions with addition only.

Another interesting general result of Munro and Paterson (1973) states that a computation Q consisting of q binary arithmetical operations requires at least $O(\log(q))$ parallel computational steps. This result is proved under the assumption that there is an unbounded number of PEs available and that the PEs can communicate with each other in a 'suitable way'.

Let t be the parallel computation time for some algorithm that computes Q. At time t at most one PE can be usefully employed; at time $t-1$ at most two PEs are usefully employed and at time $t-r$ at most 2^r PEs can be usefully employed in the computation of Q. Thus,

$$q \leqslant 1 + 2 + 2^2 + \cdots + 2^{t-1} = 2^t - 1,$$

which implies that $t \geqslant \lceil \log(q+1) \rceil$ and proves the statement.

The argument used in the proof is called a *fan-in argument*.

EXAMPLE 15.5

Matrix multiplication revisited

From the theorem of Munro and Patterson it follows that the inner product $\sum_{i=1}^{n} a_i . b_i$ of two vectors $\mathbf{a} = (a_1, \ldots, a_n)$ and $\mathbf{b} = (b_1, \ldots, b_n)$ can be computed in $1 + \lceil \log n \rceil$ steps, assuming unbounded parallelism. The product can be computed as follows:

(1) In the first parallel step, compute all the products $c_i \triangleq a_i . b_i$, $1 \leqslant i \leqslant n$.

(2) In the second step compute $\sum_{i=1}^{n} c_i$.

This requires $1 + \log(n)$ parallel steps. It can be proved that this is optimal.

The above method can be used to compute the product of two $n \times n$ matrices in $1 + \lceil \log(n) \rceil$ parallel steps. This is also optimal.

If only a bounded number, say k, of PEs is available, then the obvious sequential algorithm to compute the inner product of two vectors is better than the sequentialization of the above parallel one.

15.3.3 Parallel complexity theory

One of the basic questions in the field of parallel complexity theory is: 'Which problems can be solved substantially faster using many processors rather than one processor?'

The question was raised by Nicholas Pippinger (1979) and resulted in the definition of a class of problems known nowadays as the class NC (Nick's Class). The class NC is defined as the class of problems solvable 'ultra-fast' ($T(n) = (\log n)^{O(1)}$) on a parallel computer with a feasible amount of hardware ($H(n) = n^{O(1)}$). The definition of the class NC proved to be independent of the particular parallel computation model chosen.

Among the problems known to be in NC are the evaluation of arithmetic expressions using the operations $+$, $-$, $\odot$, and $/$ on binary represented numbers. Also in Nick's class are sorting, graph connectivity, matrix operations (multiplication, inverse, determinant, rank), finding the greatest common divisors of polynomials, and finding the maximum spanning forest in a graph.

In the complexity theory of sequential algorithms there are a large number of computation models available, such as variants of Turing machines, random-access machines, unlimited-register machines etc.; definitions of these machines can be found in Chapter 10.

Each of these models has (or can be extended with) a reasonable definition of the time and space complexity of a computation. All these measures have the nice property that for each pair C_1 and C_2 of such models, any computation in C_1 using time t and space s can be simulated in C_2 with a polynomially bounded overhead in time and a constant-factor overhead in space.

The time and space bounds of an algorithm solving a specific problem depend on the model of computation used. However, the fundamental complexity classes are closed with respect to this (polynomial) overhead and thus independent of the particular computation model used. These classes, by Corollary 12.1, form the hierarchy

$$\mathbf{L} \subseteq \mathbf{NL} \subseteq \mathbf{P} \subseteq \mathbf{NP} \subseteq PSPACE = NPSPACE.$$

All computation models for which this hierarchy holds may be considered as ‘reasonable’ computation models. Examples of such reasonable computation models are SAL(**V***) and Turing machines.

For parallel computations there also exists a large variety of models, but their mathematical theory is not very satisfactory, since too much of their detailed specification is arbitrary. The parallelism in these models is also defined in rather different ways. In some models the parallelism has been made very explicit, for example in the acyclic Boolean circuits of Borodin (1977) or the SIMDAG model of Goldschlager (1982); parallelism may also be hidden in some tree of possible computations, as for example in the alternating Turing machine of Chandra *et al.* (1981). In some other models, parallelism is disguised as a sequential computation in which operations can be performed on unreasonably large objects in a single step, as for example in the MRAMs (Random Access Machines with Multiplication) of Hartmanis and Simon (1974).

Most of these parallel computation models are polynomially related, that is, there exist polynomials p_{AB} and q_{AB} such that a computation according to model A, which is t-time-bounded and s-space-bounded, can be simulated by a computation according to model B, which is $p_{AB}(t)$-time-bounded and $q_{AB}(s)$-space-bounded.

For all of the above models of parallel computation, it can be shown that a language can be accepted in polynomial parallel time if and only if L can be accepted in polynomial sequential space, i.e.

$$//\mathbf{P} = PSPACE.$$

In expressions such as the one above, $//\mathbf{X}$ denotes the set of all languages which can be accepted by a machine from the class of parallel machines considered, such that the resources needed, i.e. parallel time or space, satisfy the bounds as given in the definition of the sequential class $\mathbf{X}$. For example, $//\mathbf{P}$ is the set of all languages which can be accepted by polynomial-time-bounded parallel machines. The above result $//\mathbf{P} = PSPACE$ was first proved for the vector machine model of Pratt and Stockmeyer (1976), but later for all parallel models.

An interesting question is whether our familiar hierarchy $\mathbf{L} \subseteq \mathbf{NL} \subseteq \mathbf{P} \subseteq \mathbf{NP} \subseteq PSPACE = NPSPACE$ also holds in the parallel case. That is, whether

$$//\mathbf{L} \subseteq //\mathbf{NL} \subseteq //\mathbf{P} \subseteq //\mathbf{NP} \subseteq //PSPACE = //NPSPACE$$

This proves to be the case.

If this hierarchy is valid for a particular model C of parallel computation, then the model is said to satisfy the *parallel computation thesis.* With some minor exceptions all parallel computation models for which the parallel computation thesis holds also satisfy the following equalities:

$$//\mathbf{L} = \mathbf{P}, //\mathbf{NL} = \mathbf{NP}, //\mathbf{P} = PSPACE, //\mathbf{NP} = NPSPACE (= PSPACE).$$

By Savitch's theorem (Theorem 11.7 on page 281) we have $NPSPACE = PSPACE$. Consequently, $//\mathbf{P} = //\mathbf{NP}$. There is no $\mathbf{P} \stackrel{?}{=} \mathbf{NP}$ problem in parallel computation models satisfying the parallel computation thesis.

Van Emde Boas (1985) collects into a single class all computation models for which the parallel computation thesis holds, called the *second machine class*. Placing a machine in this second machine class indicates that the complexity classes determined by this parallel model have properties similar to those of the corresponding sequential complexity classes.

Machines for which the parallel computation thesis does not hold are collected in the so-called *third machine class*.

Van Emde Boas (1985) shows that for parallel computation models belonging to the second machine class Savitch's theorem holds, i.e. $//PSPACE = //NPSPACE$. In the same paper a rather extensive list of references is given.

The machine models in the second machine class can be divided into four categories.

(1) machines which manipulate large data items in unit time;

(2) machine models with parallel recursion;

(3) random access machines with unbounded parallelism;

(4) alternating Turing machines.

Well-known representatives of machines manipulating large data items in unit time are the vector machines of Pratt and Stockmeyer (1976) and the MRAMs of Hartmanis and Simon (1976). Both types of machines can be programmed using a very simple programming language.

Both types of machines have bit-parallel manipulation instructions (i.e. *and*, *or*, *exclusive-or*, and *not*, each performed bitwise, on large bit strings), and instructions which can be used to construct large numbers in polynomial time, for example 2^x, can be constructed in time bounded by a polynomial in $|x|$. In the vector machine these large numbers are obtained by shift instructions. The number of positions to shift is given by a number in a scalar register. For the scalar registers only the standard arithmetical operations are included in the instruction repertory. The MRAM of Hartmanis and Simon produces these large numbers by multiplication and division instead of shifting. Hartmanis and Simon proved for their class of machines, used as acceptors, that $//\mathbf{P} = //\mathbf{NP}$.

Recursion can be implemented on traditional RAMs or Turing machines by using a stack mechanism, but there is hardly any profit in time compared with a sequential computation without recursion. The reason for this is that the recursive calls are executed sequentially. After a recursive call the machine has to wait until this call has terminated before it can compute any further. Machines based upon parallel recursion remedy this by executing a recursive call on a newly created copy of the computing device. In this way the machine can proceed with its computation while the recursive call is being processed. Using a tree structure of recursive calls an exponential number of subordinate machines all working in parallel can be activated in linear time.

An example of such a model is the k-PRAM model of Savitch and Stimson (1979). A k-PRAM is a RAM which can make k recursive calls in parallel. To make a recursive call the machine puts (a finite set of) parameters in the first registers of the newly created machine. This machine executes its procedure taking its inputs from these registers. On its turn it can execute recursive calls, which results in the creation of further new devices. After terminating the recursive call, the newly created device places the result of the computation in a special register which can be read by the parent machine. Eventually a parent machine can inspect this register and see whether the computation of the recursive call is already completed. This is the only way a machine can inform a parent that its computation is completed. Besides the instructions needed for a recursive call, reading the contents of the communication register, and testing for termination of the recursive calls, the k-PRAM can execute the addition, the subtraction, direct and indirect loading and storing, and the conditional jump.

Savitch and Stimson proved for the deterministic version of the k-PRAM (with $k > 1$) that $PSPACE = POLYNOMIAL\text{-}PRAM\text{-}TIME$ (i.e. parallel time).

EXAMPLE 15.6

The following recursive algorithm computes the maximum of a sequence of consecutive array elements $A_p, A_{p+1}, \ldots, A_q$. The algorithm can be implemented on a k-PRAM with $k \geqslant 2$ under the assumption that the array elements $A_p, \ldots, A_q$ are stored in global memory. The parallel time complexity of this algorithm is $O(\log(q-p))$.

input: $A_p, A_{p+1}, \ldots, A_q$ {We assume that there are 2^k elements, i.e. $q-p+1=2^k$, $q \geqslant p$}
output: *maximum*$\{A_p, \ldots, A_q\}$
method:

procedure $MAX(p, q)$

if $p = q$ **then** *return*(A_p)

else

$t := (q-p+1)/2;$

$m_1 := MAX(p, t);\ m_2 := MAX(t+1, q);$

$return(max\{m_1, m_2\})$

fi

end-of-procedure

Goldschlager (1982) described a RAM with unbounded parallelism, which he called a SIMDAG. This model has a main RAM called a CPU and a possibly infinite sequence of subordinate RAM processors $PE_0, PE_1, \ldots, PE_i, \ldots$. The CPU is a RAM with the usual instruction set containing load instructions (direct and indirect), store instructions (direct and indirect), arithmetical instructions (addition, subtraction), parallel bitwise Boolean operations, shifting-one-position, conditional jump, accept, and reject. What makes the CPU powerful is its ability to broadcast instructions to all the PEs.

There is a main memory which can be accessed by the CPU and all PEs. Every PE has some local memory, and an instruction repertory containing direct and indirect load instructions from the global memory to the PE memory, direct and indirect store instructions from local to global memory and the same arithmetical instructions as the CPU. The instruction to write in the global memory is a conditional instruction; there are no other conditional instructions and there is no accept and reject instruction.

At every moment only a finite number of PEs are active, otherwise in one computation step an infinite amount of work could be done. This is achieved by adding a parameter to the broadcast instruction. Only PEs with an index smaller than this parameter are allowed to execute this instruction; the other ones remain inactive. Write conflicts in the global memory are resolved by allowing the PE with the least index to execute its write operation.

As an example of the use of this machine, we take the implementation of the algorithm to compute the transitive closure of an $n \times n$ matrix A given as algorithm (2) of Example 15.4.

The elements of A are stored in the global memory, and the matrix C, which in the end will be equal to A^*, is kept in the global memory as well.

The algorithm performs the following steps:

(1) The elements $c_{i,i}$ are set to 1 and the elements $c_{i,j}$, $i \neq j$, are set to the values of the corresponding elements of the adjacency matrix $(a_{i,j})$.

(2) $PE_{\langle i, \langle j,k \rangle \rangle}$ inspects the matrix elements c_{ik} and c_{kj} and puts a 1 in c_{ij} if both elements are equal to 1.

The time complexity of this parallel algorithm is $O(\log(n))$, whereas the sequential algorithm of Warshall for the computation of the transitive closure is $O(n^3)$.

Another model which satisfies the parallel computation thesis is the alternating Turing machine of Stockmeyer *et al.* (1976). An alternating Turing machine M is in fact a non-deterministic Turing machine M with a modified mode of acceptance. All the states of M are labelled *universal* or *existential.*

Given some input x, the machine M determines the unique computation tree T_x consisting of all possible computations on input x. The nodes of this tree are labelled by IDs of the Turing machine. In T_x, each path represents a possible computation of M on input x.

A computation tree T_x is an *accepting computation tree* if and only if it satisfies the following requirements:

(1) It is finite, i.e. there are no non-terminating computations.

(2) The root corresponds to the initial ID.

(3) The route is an accepting node.

A node of T_x is called accepting if:

(1) It is a leaf and the corresponding ID is accepting.

(2) The node is labelled by an ID whose state is *universal* and all its sons are accepting.

(3) The node is labelled by an ID whose state is existential and at least one of its sons is accepting.

The language $L(M)$ accepted by an alternating Turing machine M is defined as follows:

$$L(M) = \{x \mid T_x \text{ is an accepting computation tree}\}.$$

The computation time $T(|x|)$ of an alternating Turing machine M on an input x is equal to the maximal length of a path in an accepting computation tree T_x. The space complexity $S(n)$ of such a computation is the maximum of the space used by the IDs in the tree T_x.

Consider QBF, the satisfiability problem for quantified Boolean formulae, which is defined as follows:

$$QBF = \{Q_1x_1Q_2x_2 \cdots Q_nx_n[F] \mid F \text{ is a propositional formula with } \mathrm{var}(F) \subseteq \{x_1 \cdots x_n\},\ Q_i \text{ is either } \exists \text{ or } \forall,\ 1 \leqslant i \leqslant n, \text{ and } Q_1x_1 \cdots Q_nx_n[F] \text{ is true with respect to the standard interpretation}\}.$$

QBF can be accepted by an alternating Turing machine as follows:

(1) The machine has two special *guessing states*, $q_\exists$, and $q_\forall$, labelled *existential* and *universal.* All other states of the machine are labelled *existential.*

(2) The machine constructs a variable assignment $\phi:\{x_1,\dots,x_n\}\rightarrow\{0,1\}$, represented by a list $(x_1,i_1),(x_2,i_2),\dots,(x_n,i_n)$ by making a non-deterministic choice for each variable. While making the choice, the machine passes through $q_\exists$ or $q_\forall$ depending on whether Q_i is $\exists$ or $\forall$. This can be achieved using action items of the following type:

$$(q,\ (\forall,\dots), (\forall,\dots), (R,S,\dots),\ q_\forall)$$

$$(q,\ (\exists,\dots), (\exists,\dots), (R,S,\dots),\ q_\exists)$$

$$(q_\forall, (\dots),\quad (\dots),\quad (S,\dots,S),\ q_0)$$

$$(q_\forall, (\dots),\quad (\dots),\quad (S,\dots,S),\ q_1)$$

$$(q_\exists, (\dots),\quad (\dots),\quad (S,\dots,S),\ q_0)$$

$$(q_\exists, (\dots),\quad (\dots),\quad (S,\dots,S),\ q_1)$$

From q_i, $i=0,1$, the machine stores the chosen value i in the list representing ϕ and proceeds with choosing a value for the next variable.

(3) Finally, the machine evaluates $\phi(F)$. If $\phi(F)=1$, then the machine enters a final state and terminates. Similarly, if $\phi(F)=0$, then the machine enters a non-final state, after which the computation terminates.

Thus *QBF* can be accepted by a $\lambda n[n^2]$-parallel-time-bounded alternating Turing machine.

QBF is $\leqslant_P^C$-complete for *PSPACE*, therefore $PSPACE \subseteq //_A\mathbf{P}$, where $//_A\mathbf{P}$ is the set of all languages which can be accepted by polynomial-parallel-time-bounded alternating Turing machines. As stated earlier, it can also be shown that $//_A\mathbf{P} = PSPACE$.

15.4 References

This chapter shows that the theory of algorithms does not end with the theory of sequential algorithms. It also includes the study of the computational possibilities of parallelism. As mentioned in the introduction to this chapter, besides the theory of sequential and parallel algorithms one must also consider the theory of approximate algorithms, probabilistic algorithms, etc. A wealth of references can be found in Cook's Turing Lecture (1983), Wagner and Wechsung (1986) and Parberry (1987). For a general introduction to the field of parallel computing we refer to van Leeuwen and Lenstra (1985) in which the reader can find a lot of references to the literature on parallel computing. For practical information on parallel computers and the design of parallel algorithms, see Quinn (1987).

Hints for the Solution of Selected Exercises

Chapter 3

3.3 Consider the composition $h \triangleq f \cdot (g_1, g_2, \ldots, g_r)$, where $f: N^m \to N^k$ and $g_i: N^n \to N^{m_i}$ are functions computable by SAL(**N**) programs whose computation times are bounded by polynomials p and p_i, $1 \leqslant i \leqslant r$. The function h is computed by the following macroprogram:

input: $x_1, \ldots, x_n$
output: $y_1, \ldots, y_k$
method:

$(z_1^1, \ldots, z_{m_1}^1) := g_1(x_1, \ldots, x_n);$

$(z_1^2, \ldots, z_{m_2}^2) := g_2(x_1, \ldots, x_n);$

$\vdots$

$(z_1^r, \ldots, z_{m_r}^r) := g_r(x_1, \ldots, x_n);$

$(y_1, \ldots, y_k) := f(z_1^1, \ldots, z_{m_r}^r)$

Executing this program on input x takes no more than $p_i(x_1, \ldots, x_n)$ steps to compute z_1^i to $z_{m_i}^i$ and no more than $p(z_1^1, \ldots, z_{m_r}^r)$ steps to compute y_1 to y_k.

Note that $z_j^i \leqslant \max\{x_1, \ldots, x_n\} + p_i(x_1, \ldots, x_n) \leqslant \bar{p}_i(x_1, \ldots, x_n)$ for some suitable polynomial $\bar{p}_i$, because at every step a variable can increase by at most one.

Thus the total time required by the above program on input $x_1, \ldots, x_n$ is less than

$$\sum_{i=1}^{r} p_i(x_1, \ldots, x_n) + p(\bar{p}_1(x_1, \ldots, x_n), \ldots, \bar{p}_r(x_1, \ldots, x_n)) + c$$
$$\leqslant \bar{p}(x_1, \ldots, x_n)$$

for some suitable polynomial $\bar{p}$ (c is the number of computation steps introduced by expanding the macros). Closure with respect to Cartesian product is shown similarly.

To see that *POL* is not closed with respect to exponentiation and iteration consider the function $f \triangleq \lambda x[2x]$. Then $f \in POL$, but $f^* = \lambda xy[2^y.x]$ does not belong to *POL* because the output is too large. Non-closure with respect to iteration is shown similarly. The fact that *POL* is not closed with respect to exponentiation is not due to the fact we have chosen a particularly simple underlying data type (see Section 4.5).

3.12 The set of all SAL(**N**) programs $(1, 1, p, P)$ is countably infinite and therefore the set $\{f:N \to N \mid f \text{ is SAL}(\mathbf{N}) \text{ computable}\}$ is countably infinite too. Because the set of all functions from N to N is uncountable (Theorem 1.1 on page 14) there is a function $\zeta: N \to N$ which is not SAL(**N**) computable.

We can extend the compound data type **N** to include ζ as a base function

$$^{\zeta}\mathbf{N} \triangleq (\mathbf{N}, \mathbf{B}; 0, +_1, \dot{-}_1, \zeta, \neq_0).$$

Then $F(\mathrm{SAL}(^{\zeta}\mathbf{N})) - F(\mathrm{SAL}(\mathbf{N}))$ is countably infinite. However, $F(\mathrm{SAL}(^{\zeta}\mathbf{N}))$ is also countably infinite. Thus, once more there remain uncomputable functions. The model SAL($^{\zeta}$**N**) is worse in the sense that Church's thesis does not hold for this model.

Chapter 4

Most of the exercises of Section 4.6 are straightforward. As regards Exercise 4.6, if the time complexity function of a SAL(**N**) program Q can be bounded by a *total* recursive function, then for this program the question 'Does the computation determined by Q on input x terminate?' is algorithmically solvable: just simulate this computation until it terminates, or the number of steps exceeds the limit as given by the total recursive function bounding the time complexity function. Hence the Halting Problem would be solvable. In Section 6.1 a particular computable function can be found for which the time complexity function cannot be bounded by any total recursive function – namely $\lambda x[\phi_x(x)]$.

Exercise 4.7 asks for the proof of Theorem 4.4 (page 88). It proceeds by induction as discussed in Section 4.5.

The remaining exercises are straightforward. To show that a function is primitive (or elementary) you can either give a (loop) program computing the function, or give an expression which shows how the function can be obtained from the base functions. An example of both methods is given below.

4.5 From the definition it is immediate that f is total. By Church's thesis,

f is algorithmically computable. The function f is computed by the following macro program.

```
input: x
output: m
method:
  y:= 0; m:= 0;
  do x times
    y:= y + 1;
    if g(y) < g(m) then m:= y fi
  od
```

Since $g \in PRIM$ this macro program can be expanded into a loop program. Consequently $f \in PRIM$ (and f is therefore recursive and total).

4.8 The characteristic function χ_Q satisfies

$$\chi_Q = \lambda x_1 x_2 \cdots x_n [\chi_P(x_1, x_2, \ldots, x_{n-1}, 0) \cdot \chi_P(x_1, x_2, \ldots, x_{n-1}, 1) \cdots \chi_P(x_1, x_2, \ldots, x_{n-1}, x_n)]$$

Define the function ζ as follows:

$$\zeta \triangleq \lambda x_1 x_2 \cdots x_{n-1} yz [(x_1, x_2, \ldots, x_{n-1}, y + 1, z \cdot \chi_P(x_1, x_2, \ldots, x_{n-1}, y))]$$

Then

$$\chi_Q = \lambda x_1 x_2 \cdots x_n [\zeta^*(x_1, x_2, \ldots, x_{n-1}, 0, 1, x_n + 1)]$$

Therefore, from $\chi_P \in PRIM$, it follows that $\chi_Q \in PRIM$.

To show that R is a primitive recursive predicate, you can proceed similarly.

Chapter 5

5.4 An algorithm is a *finite description* of a method of computation. Thus the class A must be countable. On the other hand there are uncountably many number-theoretic functions (see Section 5.4).

5.5 The proof proceeds by diagonalization as outlined in Section 5.4. The only problem is to define the function f. Letting $f = \lambda x[f_x(x) + 1]$ as in Section 5.4 will not do, because this function is not necessarily monotonically increasing. A possibility is

$$f(0) \triangleq 1 + f_0(0)$$
$$f(n+1) \triangleq 1 + \max\{f(n), f_{n+1}(n+1)\} \qquad \text{for } n = 0, 1, 2, \ldots$$

5.7 The function $p \triangleq \lambda x[p_x]$ is primitive recursive (see Section 4.3.3). Show, for example by writing loop programs computing them, that the functions

$$e \triangleq \lambda xy[\text{the largest } z \text{ such that } p_x^z \text{ divides } y] \qquad \text{and}$$
$$r \triangleq \lambda xy[x^y]$$

are primitive recursive.

(a) The following function ψ is universal:

$$\psi \triangleq \lambda ix\left[\sum_{j=0}^{x} e(j, i+1)\cdot r(x, j)\right]$$

It is easy to construct an expression for ψ using e, r and exponentiation along the lines of the solution of Exercise 4.8.

(b) The coefficient of x^i in the function $f_{g(u,v)}$ is the sum of the coefficients of x^i in the functions f_u and f_v. Therefore $g = \lambda uv[(u+1)(v+1) \dot{-} 1]$. Case (ii) is similar but more complicated.

The exercises about acceptable indexings can be solved using Theorem 5.3, stating that there are total recursive functions to go from a π-index to a π_0-index and vice versa. Exercise 5.12 is most easily proved using the *recursion theorem* (see Example 5.5(2)).

Chapter 6

6.2 (h) Let P denote the predicate $\lambda x[dom(\phi_x) \subseteq \{p \mid p \text{ is prime}\}]$ and let f be a total recursive function such that $\phi_{f(x)} = \lambda y[\phi_x(x) + y]$. Then $\neg K \leqslant P$ via f. Therefore the predicate P is not recursively enumerable.

On the other hand $\lambda x[\neg(dom(\phi_x \subseteq \{p \mid p \text{ is prime}\})]$ is equivalent to

$$(\exists t)[T(x,0,t) \vee T(x,1,t)] \vee$$
$$(\exists y)(\exists z)(\exists t)[y > 1 \wedge z > 1 \wedge T(x, y.z, t)].$$

Therefore, the predicate $\neg P$ is recursively enumerable by the closure properties of the class of all recursively enumerable predicates.

6.7 (b) If $h: N \to N$ is a monotonically increasing total recursive function, i.e. $h(n+1) > h(n)$ for all $n \in N$, then the set $\{h(n) | n \in N\}$ is recursively decidable; it is easy to write a program computing the characteristic function of this set.

Let an infinite recursively enumerable set be given. By Theorem 6.6 (page 140) there is a total recursive function f such that $\{f(n) | n \in N\} = A$. Define the function g as follows:

$$g(0) = f(0)$$
$$g(n+1) = f(\mu z[f(z) > g(n)]) \qquad \text{for } n = 0, 1, 2, \ldots$$

Then g is a monotonically increasing, recursive, total function, the range of which is an infinite recursively decidable subset of A.

6.11 An indexing of the recursively enumerable sets is given in Definition 6.3 (page 143). Let D denote the set of all recursively decidable subsets of N. Then D is countable; therefore there exists an indexing of D. This indexing cannot be effective, because D is not recursively enumerable (see Example 6.3 items (1) and (3) on page 146).

Chapter 7

7.5 Consider for example sets (d) and (e). Let

$$A \triangleq \{x | W_x \text{ is infinite}\} \qquad \text{and} \qquad B \triangleq \{x | \phi_x \text{ is total}\}.$$

We will show that $A \leqslant_m B$ and $B \leqslant_m A$.

- Define the function ψ_1 such that $\psi_1(x, 0)$ is some element of the domain W_x of ϕ_x and $\psi_1(x, y)$ is an element of W_x different from all the elements $\psi_1(x, j), j < y$, already generated, if at least one such element exists. Otherwise $\psi_1(x, y)$ is undefined ($\uparrow$). More formally,

$$\psi_1(x, 0) \triangleq \sigma^1_{2,1}(\mu t[T(x, \sigma^1_{2,1}(t), \sigma^1_{2,2}(t))])$$
$$\psi_1(x, y+1) \triangleq \sigma^1_{2,1}(\mu t[T(x, \sigma^1_{2,1}(t), \sigma^1_{2,2}(t)) \wedge (\forall j)[\text{if } 0 \leqslant j \leqslant y \text{ then } \sigma^1_{2,1}(t) \neq \psi_1(x, j)]])$$

Then $A \leqslant_m B$ via the function f such that $\phi_{f(x)} = \lambda y[\psi_1(x, y)]$, as can easily be checked.

- Define the function ψ_2 as follows:

$$\psi_2 \triangleq \phi_x(y)$$
$$\psi_2(x, y+1) \triangleq \psi_2(x, y).\phi_x(y+1)$$

Then $B \leqslant_m A$ via the function h such that $\phi_{h(x)} = \lambda y[\psi_2(x, y)]$, as can easily be checked.

7.7 Using the s–n–m theorem (Theorem 5.2) construct a total recursive function g such that

$$\phi_{g(z,x)} = \lambda y[sg(1 + \phi_z(z)).\phi_x(y)].$$

Let $f: N \to K$ be a surjective total recursive function. Such a total surjection exists because K is recursively enumerable.

Define a function h which is computed as follows. To determine $h(x)$ simultaneously compute $\psi(x), \psi(f(0), x), \psi(f(1), x), \ldots$ and assign to $h(x)$ the output of the first computation which terminates. Use Kleene's T predicate to formally define h. The function h is recursive, total and productive for A.

7.8 This is most easily proved using the recursion theorem; for details see Rogers (1967), Chapter 11.

7.10 Otherwise, pairs of recursive functions can be listed such that in every pair precisely one function would be total. This listing can then be used to diagonalize over the set of total recursive functions, yielding a contradiction.

7.14 Let $A \simeq B$ mean that A and B are recursively isomorphic, i.e. there is a recursive permutation p such that $A = p(B)$. Thus, if A is a cylinder, then $A \simeq C \times N$ for some set B (see Definition 7.5). The following properties hold:

- $A \times B \simeq B \times A$
  because $B \times A = p(A \times B)$, where $p = \sigma_1^2 \cdot (\sigma_{2,2}^1, \sigma_{2,1}^1)$.
- $A \times (B \times C) \simeq (A \times B) \times C$
  because $A \times (B \times C) = q((A \times B) \times C)$, where the permutation q is defined as follows:

$$q = \sigma_1^2 \cdot (\sigma_{2,1}^1 \cdot \sigma_{2,1}^1, \sigma_1^2 \cdot (\sigma_{2,2}^1 \cdot \sigma_{2,1}^1, \sigma_{2,2}^1)).$$

Thus, if B is a cylinder, then there is some set C such that

$$A \times B \simeq A \times (C \times N) \simeq (A \times C) \times N$$

and consequently $A \times B$ is a cylinder. If A is a cylinder a similar argument applies. For part (b) show that $(A \times N) \sqcup (B \times N) \simeq (A \sqcup B) \times N$.

Chapter 8

8.2 If $A \leqslant_T B$ then there is a SAL($^{\chi_B}$**N**) program P such that $f_P = \chi_A$. Therefore $[B]_{\equiv_T}$ is countable. Also if $[B]_{\equiv_T}$ contains a non-empty set A, then it also contains all sets $p(A)$, where p is a recursive permutation, and is therefore infinite. Finally $\{\varnothing\}$ is not a Turing degree.

8.4 Let $X, Y \in \{A \mid A \leqslant_T B\}$. Then $\chi_{X \cap Y} = \lambda x[\chi_X(x).\chi_Y(x)]$. There are SAL($^{\chi_B}$**N**) programs computing χ_X and χ_Y. Consequently, there is a SAL($^{\chi_B}$**N**) program that computes $\chi_{X \cap Y}$. The other cases are similar.

8.5 By the $s\text{–}n\text{–}m$ theorem (Theorem 5.2, page 110) there are total recursive functions f and h such that

$$\phi_{f(x)} = \lambda y[sg(1 + \phi_x(x)).y] \text{ and}$$

$$\phi_{h(x)} = \lambda y[\text{if } (\exists z)[\phi_x(z) = x] \text{ then } y \text{ else } \uparrow].$$

Then $A \leqslant_m B$ via f and $B \leqslant_m A$ via h, thus $A \equiv_m B$.

The $s\text{–}n\text{–}m$ theorem can be strengthened by requiring that s^n_m must be injective. Such an injective function can be constructed using the padding lemma (Lemma 7.1, page 158). Thus $A \equiv_1 B$ and $A \equiv_m B$ and $A \equiv_T B$.

Chapter 9

9.1 This is a version of the partition problem, discussed as item (7) in Example 9.8 (page 223). In this exercise the program can be somewhat simpler, because unary representation of numbers is assumed. Note that this is not a concise representation of the partition problem and is therefore not acceptable from a complexity-theoretic point of view.

9.3 This if far from easy; consult Pratt (1975).

9.8 Every SAL(**V***) computable function clearly is ND-SAL(**V***) computable too. Proving the converse is the goal of Exercise 9.7. To show that an ND-SAL(**V***) computable function can also be computed by a deterministic program, the technique shown in the proof of Lemma 9.1 (page 216) can be used. Thus we use a string s consisting of zeros and ones of length $T_P(n)$ to code the possible computations of the non-deterministic program P.

By systematically trying all possible strings, a terminating computation is found if one exists; then the output is also known since we assume that P computes a function. The total time required by this process is less than $2^{T_P(n)} \cdot T_P(n) \leqslant c^{T_P(n)}$ for some suitable constant c.

Chapter 10

10.1 Proving that every partial recursive function can be computed on a URM requires a formal definition of the URM and the function computed by a URM. The same applies to the other exercises of this type. All these proofs are similar to the proof of the computational equivalence between SAM(V*) machines and SAL(V*) programs given in Section 10.2.

The importance of these equivalences is that they add credibility to Church's thesis and give us the opportunity to select that model of computation which is best suited to the task at hand.

10.9 An accepting computation on an input w of size n consists of a sequence of *ID*s which use no more than $S(n)$ space. This sequence can be chosen such that it does not contain repetitions. Thus the length of an accepting computation is at most equal to the number of mutually different *ID*s using no more than $S(n)$ space.

There are $r^{S(n)}$ ways to fill a tape segment of $S(n)$ cells, $S(n)$ ways to position the read/write head in this segment and the machine can be in any of s states. Thus the length of an $S(n)$-space-bounded accepting computation is at most $s \cdot S(n) \cdot r^{S(n)}$.

Chapter 11

11.2 Lower bounds such as this can be proved using so-called crossing sequences. See Hennie (1965) and Hopcroft and Ullman (1969a, b).

11.3 Part (a) is easy; just add an extra tape and have the read/write head move to the right at each computation step.

For (b) we might use a counter of $S(n)$ positions which counts in base c for some sufficiently large constant c. The problem with this scheme is that the counter does not give the number of computation steps, because updating the counter may take more than a single step due to carries. For example, it takes 1217 steps to count from 0 to 1000 using decimal representation.

We can make a Turing machine execute 2^r steps by a process of repeated doubling. Assume that we have a string $ab^{n-2}a$ of length n on a tape A. Then a string $ab^{2n-2}a$ can be written on tape B in precisely $2n-1$ steps by moving from left to right and back over the string on tape A, meanwhile constructing the string $ab^{2n-2}a$ on tape B. One extra move initiates a repetition of this process with the roles of tapes A and B interchanged. Proceeding in this way a machine can be constructed that performs exactly 2^r steps for an input of length r.

To show that $\lambda n[c^{S(n)}]$ is time constructible for some c, the above process can be used if we choose c sufficiently large and a power of 2. We use a Turing machine M to mark off $S(n)$ cells, which are used as the input for the doubling machine. Such a machine M exists, because S is space constructible. During the marking of the $S(n)$ cells, the doubling process is running in parallel. We must choose c sufficiently large to ensure that the number of steps executed by M is less than $c^{S(n)}$.

11.4 This is rather complicated. A proof can be found in Hennie and Stearns (1966).

11.5 This is actually Theorem 11.12 (page 293). The proof by diagonalizing over two-tape deterministic Turing machines is similar to the proof of Theorem 11.11 (page 292).

Chapter 12

12.3 A non-deterministic Turing machine accepting *QBF-SAT* can operate as follows:

Non-deterministically select values for the variables in $\text{var}(F) - \{x_1, x_2, \ldots, x_k\}$ and also for those variables from $\{x_1, x_2, \ldots, x_k\}$ which are existentially quantified. Enumerate all possible assignments of the remaining variables and check that $v(F) = 1$ for all the valuations of $\text{var}(F)$ determined in this way. If they all evaluate to 1, then accept the input and otherwise enter a non-terminating computation.

This procedure can be implemented by a polynomial-space-bounded non-deterministic Turing machine. Thus $QBF\text{-}SAT \in NPSPACE = PSPACE$.

12.4 Consult Garey and Johnson (1979), Section 7.3, and Valiant (1979).

12.8 Note that $\mathbf{NP}^{\Sigma_k^p} = \mathbf{NP}^{\Pi_k^p}$ because the same oracles are involved in both cases. Similarly, $\mathbf{P}^{\Sigma_k^p} = \mathbf{P}^{\Pi_k^p}$. Also, if $A \in \mathbf{P}^X$ then $co\text{-}A \in \mathbf{P}^X$ because we are dealing with deterministic oracle acceptors.

We turn now to the inclusions we are to prove. The case $k = 0$

is immediate by definition and for $k > 0$ we proceed as follows:

- $\Delta_k^p = \mathbf{P}^{\Sigma_{k-1}^p} \subseteq \mathbf{NP}^{\Sigma_{k-1}^p} = \Sigma_k^p$.
- If $A \in \Delta_k^p = \mathbf{P}^{\Sigma_{k-1}^p}$ then $co\text{-}A \in \mathbf{P}^{\Sigma_{k-1}^p}$. Therefore $co\text{-}A \in \Sigma_k^p$ and consequently $A \in \Pi_k^p$.

In conclusion, $\Delta_k^p \subseteq (\Sigma_k^p \cap \Pi_k^p)$

- $\Sigma_k^p \subseteq \mathbf{P}^{\Sigma_k^p} = \Delta_{k+1}^p$
- $\Pi_k^p = co\text{-}\Sigma_k^p \subseteq \mathbf{P}^{\Sigma_k^p} = \Delta_{k+1}^n$

Consequently, $(\Pi_k^p \cup \Sigma_k^p) \subseteq \Delta_{k+1}^p$.

Chapter 13

13.1 It is obvious that $FAL \in \mathbf{NP}$. Also, FAL is **NP**-complete, because $SAT \leqslant_P^K FAL$ via the function which transforms any propositional formula F into the formula $\neg(F)$.

We turn now to the normal forms. Conjunctive normal form has been defined in Definition 13.3 (page 340). A propositional formula is in *disjunctive normal form*, abbreviated DNF, if it is of the format $D_1 \vee D_2 \vee \cdots \vee D_n$ for some n, and the so-called *disjuncts* D_i are of the format $L_1 \wedge L_2 \wedge \cdots \wedge L_{m_i}$, where each L_j is a literal.

- *CNF-SAT* is **NP**-complete.
- *DNF-SAT* is in **P**; a satisfying assignment always exists unless the formula in disjunctive normal form is empty.
- *CNF-FAL* is also in **P**.
- *DNF-FAL* is **NP**-complete, *CNF-SAT* $\leqslant_P^K$ *DNF-FAL*.

13.2 $COL \in \mathbf{NP}$ is obvious.
3-*CNF-SAT* $\leqslant_P^K COL$ via the following transformation.

Let $F = C_1 \wedge C_2 \wedge \cdots \wedge C_n$ be an instance of 3-*CNF-SAT* and assume that $\text{var}(F) = \{x_1, \ldots, x_p\}$. Construct an instance $((V, E), k)$ of COL as follows:

- $V \triangleq \{x_i, \bar{x}_i, v_i \mid 1 \leqslant i \leqslant p\} \cup \{c_i \mid 1 \leqslant i \leqslant n\}$
- $E \triangleq \{\{v_i, v_j\} \mid i \neq j\} \cup \{\{x_i, \bar{x}_i\} \mid 1 \leqslant i \leqslant p\} \cup \{\{v_i, x_j\}, \{v_i, \bar{x}_j\} \mid i \neq j\} \cup$
  $\{\{c_i, x_j\} \mid x_j$ does not occur as a literal in $C_i\} \cup$
  $\{\{c_i, \bar{x}_j\} \mid \bar{x}_j$ does not occur as a literal in $C_i\}$.
- $k \triangleq p + 1$

The subgraph with nodes $\{v_i | 1 \leqslant i \leqslant p\}$ is complete and requires p colours. x_i and $\bar{x}_i$ are adjacent and also adjacent to $v_j, j \neq i$. Therefore at least one extra colour is necessary, call it RED.

Assume that the graph can be coloured using $p+1$ colours. Then either x_i and v_i, or $\bar{x}_i$ and v_i have the same colour. Also, if z_i denotes that node of x_i and $\bar{x}_i$ that is not RED, then the nodes z_1 to z_p all have mutually different colours.

Consider the colours of the nodes c_i. Each c_i is adjacent to at least $2p-3$ nodes of $\{x_i, \bar{x}_i | 1 \leqslant i \leqslant p\}$. Without loss of generality, we assume that $p \geqslant 4$. Then $2p-3 \geqslant p+1$. Therefore each c_i is adjacent to x_j and $\bar{x}_j$ for some j. Consequently, c_i is not RED.

For each i there is a literal y occurring in C_i such that y is not RED. Otherwise c_i is adjacent to all nodes z_1 to z_p that have mutually different colours, and also adjacent to RED nodes. Thus, more than $p+1$ colours would be required.

The assignment $v \triangleq \lambda x[\text{if (node } x \text{ is RED) then } 0 \text{ else } 1]$ satisfies F, i.e. $v(F)=1$. Convince yourself that the converse also holds: if a satisfying assignment exists, then the graph can be coloured using $p+1$ colours.

For more information about the colourability problem, also called 'chromatic number', consult Garey and Johnson (1979). Johnson's '**NP**-completeness column: an ongoing guide' which appears more-or-less regularly in the *Journal of Algorithms* is also a valuable source.

Chapter 14

14.1 and **14.2**. The reductions proving these completeness results can be found in the papers of Jones (1975), and Jones and Laaser (1976, 1977).

14.3 Let $L \in PSPACE$, $L \subseteq \Sigma^*$, and assume that L is accepted by a p-space-bounded deterministic Turing machine M, where p is some polynomial. Let $L_1 \triangleq \{wa^{p(|w|)} | w \in L\}$, $a \notin \Sigma$.

Then $L_1 \in DSPACE(\lambda n[n])$. A Turing machine M_1 accepting L_1 can simply simulate M on input w. It will need no more than $p(|w|)$ space, which is linear in the size of the input $wa^{p(|w|)}$ of M_1. By assumption $DSPACE(\lambda n[n]) \subseteq \mathbf{P}$, thus there is a q-time-bounded deterministic Turing machine M_2 which accepts L_1, where q is some polynomial.

Construct a deterministic Turing machine M_3 which accepts L. On input w it computes $wa^{p(|w|)}$. This takes $O(p(|w|))$ steps. The machine accepts w if and only if M_2 accepts $wa^{p(|w|)}$. Testing whether or not M_2 accepts $wa^{p(|w|)}$ takes $O(q(|w|+p(|w|)))$ steps. Therefore M_3 is polynomial-time-bounded.

Consequently, $PSPACE \subseteq \mathbf{P}$ if $DSPACE(\lambda n[n]) \subseteq \mathbf{P}$.

References

Ackermann W. (1928), Zum Hilbertschen Aufbau der reelen Zahlen. *Math. Ann.* **99**, 118–133.

Asser G. (1960), Rekursive Wortfunctionen. *Z. Math. Log. Grundlag. Math.* **6**, 258–278.

Berlekamp E. (1968), *Algebraic Coding Theory*. McGraw-Hill.

Berman P. (1978), Relationship between density and deterministic complexity of NP-complete languages. In G. Ausiello, C. Bohm (eds) *Automata, Languages and Programming*, fifth ICALP Coll Udine, Springer LNCS-62.

Blum M. (1967), A machine independent theory of the complexity of recursive functions, *J. ACM* **14**, 322–336.

Blum M. (1983), A note on the parallel computation thesis. *Inf. Proc. Letters* **17**, 203–205.

Book R. V. (1970), Relationships between non-deterministic and deterministic tape complexity classes. *J. Comp. & Syst. Sc.*, **4**, 177–192.

Book R. V. (1976), Translational lemmas, polynomial time and $(\log n)^j$ space. *Theor. Comp. Sc.*, **1**, 215–226.

Borodin A. B. (1972), Computational complexity and the existence of complexity gaps. *J. ACM*, **19**, 158–174.

Borodin A. B. (1977), On relating time and space to size and depth. *SIAM J. Comp.* **6**, 733–744.

Brainerd W. S., Landweber L. H. (1974), *Theory of Computation*, Wiley.

Brent R. P. (1974), The parallel evaluation of general arithmetic expressions. *J. ACM* **21**, 201–206.

Chandra A. K., Kozen D. C., Stockmeyer L. J. (1981), Alternations. *J. ACM* **28**, 114–133.

Church A. (1936), A note on the Entscheidungsproblem. *J. Symbolic Logic* **58**, 345–363 (also in Davis (1965), 89–107).

Cobham A. (1964), The intrinsic computational difficulty of functions. *Proc. Congress. Logic, Methodology and Phil. of Science.* Haifa Israel North Holland Amsterdam, 24–30.

Constable R. L., Borodin A. B. (1972), Subrecursive programming languages part 1: efficiency and program structure. *J. ACM* **19**, 526–568.

Cook S. A. (1971), The complexity of theorem proving procedures. *Proc. Third Ann. ACM Symp. on Theory of Computing.* 151–158.

Cook S. A. (1973), A hierarchy for nondeterministic time complexity. *J. Comp. Syst. Sc.*, **7**, 343–353.

Cook S. A. (1974), An observation on time-storage trade-off. *J. Comp. & Syst. Sci.* **9**, 308–316.

Cook S. A. (1983), An overview of computational complexity. *CACM* **26**, 401–408.

Coppersmith D., Winograd S. (1982), On the asymptotic complexity of matrix multiplication. *SIAM J. Comput.* **11**, 472–492.

Davis M. (ed.) (1965), *The Undecidable.* New York Raven Press.

Dedekind R. (1930). *Was sind und sollen die Zahlen.* Braunschweig. 6th edition 1930.

Dekel E., Nassimi D., Sahni S. (1981), Parallel matrix and graph algorithms. *SIAM J. Comp.*, 657–675.

Dekker J. C. E., Myhill J. (1958), Some theorems on classes of recursively enumerable sets. *Trans AMS* **89**, 25–29.

Eilenberg S., Elgot C. C. (1970), *Recursiveness.* Academic Press, New York.

Eilenberg S. (1974), *Automata, Languages and Machines*, Vol. A, Academic Press.

Emde Boas P. van (1985), The second machine class: Models of parallelism. In van Leeuwen and Lenstra (1985), 133–163.

Engeler E. (1968), *Formal Languages.* Lectures in advanced mathematics. Murkham Publ. Company 1968.

Even S., Tarjan R. E. (1976), A combinatorial problem which is complete in polynomial space. *J ACM* **23**, 710–719.

Flynn F. J. (1972), Some computer organizations and their effectiveness. *IEEE Trans. Comp.* **C-21**, 948–960.

Friedberg R. M. (1957), Two recursively enumerable sets of incomparable degrees of unsolvability. *Proc. Nat. Acad. Sc.* **43**, 236–238.

Garey M. R., Johnson D. S. (1979), *Computers and Intractability, A Guide to the Theory of NP-completeness.* Freeman.

Gödel K. (1965), On undecidable propositions of formal mathematical systems. Lecture Notes Institute for Advanced Study Princeton N.Y. (published in Davis (1965), 5–38).

Goldschlager L. M. (1982), A universal interconnection pattern for parallel computers. *J. ACM* **29**, 1073–1086.

Grätzer G. (1968), *Universal Algebra.* Van Nostrand.

Graham R. (1969), Bounds on multi-processor timing anomalies. *SIAM J.* **17**, 416–429.

Grzegorczyk A. (1953), Some classes of recursive functions. *Rozpraqy Mathematyczny* **4**, 1–45.

Hardy G. H., Wright E. M. (1954), *An Introduction to the Theory of Numbers* (*3rd edn.*), Oxford.

Hartmanis J. (1987), *The Structural Complexity Column.* EATCS Bulletin 33, October, 26–39.

Hartmanis J., Hunt H. B. (III) (1974), *The LBA Problem and its Importance in the Theory of Computing.* Proc. SIAM-AMS, (**7**).

Hartmanis J., Lewis P. M., Stearns R. E. (1965), Classification of computations by time and memory requirements. *Proc. IFIP*, 31–35.

Hartmanis J., Simon J. (1974), On the power of multiplication in random access machines. *Proc. SWAT* **15**, 13–23.

Hartmanis J., Simon J. (1976), On the structure of feasible computations. In M. Rubinoff, M. C. Yovits (eds), *Advances in Computers* **14**, 1–43. Academic Press.

Hartmanis J., Stearns R. E. (1964), Computational complexity of recursive sequences. *Proc. Fifth Ann. Symp. on Switching Theory and Logical Design*, Princeton, 82–90.

Hartmanis J., Stearns R. E. (1965), On the complexity of algorithms. *Trans. AMS* **117**, 285–306.

Hennie F. C. (1965), One-tape off-line Turing machine computations, *Inf. and Control.* **8**, 553–578.

Hennie F. C. (1976), *Introduction to Computability*. Addison-Wesley, Mass.

Hennie F. C., Stearns R. E. (1966), Two-tape simulation of multi-tape Turing machine. *J. ACM* **13**, 533–546.

Hermes H. (1961), *Aufzählbarkeit, Entscheidbarkeit, Berechenbarkeit*. Springer, Berlin.

Heyenoort J. van (1981), *From Frege to Gödel* (4th printing). Harvard University Press.

Hilbert D. (1901), Mathematical problems. *Bull. AMS* **8**, 161–190. See also Heyenoort (1981), 478–479.

Hilbert D., Barnays P. (1934/1939), *Grundlagen der Mathematik* Vol. I (1934), Vol II (1939) Springer.

Hochbaum D. S., Shmoys D. B. (1986), A unified approach to approximation algorithms for bottleneck problems. *J. ACM* **33**, 533–550.

Hockney R. W., Jesshope C. R. (1981), *Parallel Computers*. Adam Hilger, Bristol.

Hopcroft J. E., Ullman J. D. (1969a), Some results on tape bounded Turing machines, *J. ACM* **16**, 168–177.

Hopcroft J. E., Ullman J. D. (1969b), *Formal Languages and Their Relation to Automata*, Addison-Wesley.

Horowitz E., Sahni S. (1978), *Algorithms: Design and Analysis*. Computer Science Press, Potomac.

Ibarra O. H. (1972), A note concerning non-deterministic tape complexities. *J. ACM* **19**, 608–612.

Immerman N. (1987), Nondeterministic space is closed under complement. Yale University Technical Report, August.

Jones N. D. (1975), Space-bounded reducibility among combinatorial problems. *J. Comp. & Syst. Sci.* **11**, 68–85.

Jones N. D., Laaser W. T. (1976), New problems complete for nondeterministic log space. *Math. Syst. Th.* **10**, 1–17.

Jones N. D., Laaser W. T. (1977), Complete problems for deterministic polynomial time. *Theor. Comp. Sci.* **3**, 105–117.

Kalmar L. (1943), Ein einfaches Beispiel für ein unentscheidbares arithmetisches Problem. *Mathematikai és Fizikai Lapok* **50**, 1–23 (Hungarian with a German abstract).

Karmarkar N. (1984), A new polynomial time algorithm for linear programming. *Combinatorica* **4**, 373–395.

Karp R. M. (1972), Reducibility among combinatorial problems. In R. E. Miller and J. W. Thatcher (eds) *Complexity of Computer Computations*, Plenum Press.

Khachian L. G. (1979), A polynomial algorithm for linear programming. *Doklady Akad. Nauk USSR* **244**, 1093–1096. Translation in *Soviet Math. Doklady* **20**, 191–194.

Kindervater G. A. P., Lenstra J. K. (1983), *Parallel Algorithms in Combinatorial Optimization: an Annotated Bibliography*. Techn. Rep. Math. Centre, Amsterdam.

Kleene S. C. (1936), General recursive functions of natural numbers. *Math. Ann.* **112**, 727–742. See also Davis (1965), 237–253.

Kleene S. C. (1943), Recursive predicates and quantifiers. *Trans. AMS* **53**, 41–74. (Also in Davis (1965), 255–287).

Kleene S. C. (1952), *Introduction to Meta-mathematics*. North-Holland.

Knuth D. E. (1973), *The Art of Computer Programming*, Addison-Wesley.

Knuth D. E. (1974), A terminological proposal. *SIGACT News* **6**, 12–18.

Kung H. T. (1974), Some complexity bounds for parallel computation. *Proc. 6th. Annual ACM Symp. Theory of Comp.*, 323–333.

Ladner R. E., Lynch N. A., Selman A. L. (1975), A comparison of polynomial time reducibilities. *Theor. Comp. Sci.* **1**, 103–123.

Ladner R. E. (1975), On the structure of polynomial time reducibility. *J. ACM* **22**, 155–171.

Lawler E. L., Lenstra J. K., Rinnooy Kan A. H. G., Shmoys D. B. (1985). *The Travelling Salesman Problem*, Wiley.

Leeuwen J. van, Lenstra J. K. (eds) (1985). *Parallel Computers and Computations*, CWI Amsterdam.

Liskov B., Guttag J. (1986), *Abstraction and Specification in Program Development*, MIT Press.

McCall E. H. (1980), *A study of the Khachian Algorithm for real world linear programming problems.* Report of the Computer Science Dept., Univ. of Minnesota.

Markov A. A. (1962), *Theory of Algorithms* (translated from Russian). Published by the Israel Program for Scientific Translations Jerusalem.

Meyer A. R., Ritchie D. M. (1967a), *Computational Complexity and Program Structure.* IBM research paper RC-1817. Yorktown Heights IBM Watson Research Center.

Meyer A. R., Ritchie D. M. (1967b), The complexity of loop programs. *Proc. ACM Nat. Meeting*, 465–469.

Miller G. L. (1975), Riemann's hypothesis and tests for primality. *Proc. Seventh Ann. ACM Symp. on Theory of Comput.*, 234–239.

Mostowski A. (1966), *Thirty Years of Foundational Studies.* Basil Blackwell, Oxford.

Muchnik A. A. (1956), On the unsolvability of the problem of reducibility in the theory of algorithms (Russian). *Doklady Akad. Nauk. USSR* **108**, 194–197.

Munro I., Paterson M. (1973), Optimal algorithms for parallel polynomial evaluation. *J. Comp. Sys. Sc.* **7**, 189–198.

Naur P. (1963), Revised report on the algorithmic language ALGOL 60. *Numerische Mathematik* **4**, 420–453 (Sonderdruck).

Papadimitriou C. H., Steiglitz K. (1982), *Combinatorial Optimization, Algorithms and Complexity*, Prentice-Hall.

Parberry I. (1987), *Parallel Complexity Theory.* Pitman and Wiley.

Peter R. (1936), Über die mehrfache Rekursion. *Math. Ann.* **113**, 489–527.

Pippinger N. (1979), On simultaneous resource bounds. *Proc. 20th IEEE Symp. on Foundations of Computer Science. IEEE Comp. Soc.*, 307–311.

Post E. L. (1936), Finite combinatory processes. *J. S. Logic* **1**, 103–105. See also Davis (1965), 289–291.

Post E. L. (1943), Formal reductions of the general combinatorial decision problem. *Am. J. Math.*, **65**, 197–215.

Post E. L. (1944), Recursively enumerable sets of positive natural numbers and their decision problems. *Bull. AMS* **50**, 284–316.

Post E. L. (1948), Degrees of unsolvability, preliminary report. *Bull. AMS* **54**, 641–642.

Pratt V. (1975), Every prime has a succinct certificate. *SIAM J. Comput.* **4**, 214–220.

Pratt V. R., Stockmeyer L. J. (1976), A characterization of the power of vector machines. *S. Comp. Syst. Sc.* **12**, 198–221.

Quinn M. J. (1987), *Designing efficient algorithms for parallel computers.* McGraw-Hill.

Rabin M. O. (1976), Probabilistic algorithms. In J. F. Taub (ed.) *Algorithms and Complexity, Recent Results and New Directions.* Academic Press.

Rice H. G. (1953), Classes of recursively enumerable sets and their decision problems. *Trans. AMS* **89**, 25–59.

Ritchie R. W. (1963), Classes of predictably computable functions. *Trans. AMS* **106**, 139–173.

Rivest R. L., Shamir A. and Adleman L. (1977), A method for obtaining digital signatures and public-key cryptosystems, MIT technical memo LCS/T1982.

Robson J. M. (1984), *N* by *N* checkers is exponential time complete w.r.t. log space reducibility. *SIAM J. Comput.* **13**(2), 252–267.

Rogers H. Jr. (1967), *Theory of Recursive Functions and Effective Computability.* McGraw-Hill.

Savitch, W. J. (1970), Relationships between non-deterministic and deterministic tape complexity classes. *J. Comp. & Syst. Sci.* **4**, 177–192.

Savitch W. J., Stimson M. J. (1979), Time bounded random access machines with parallel processing. *J. ACM* **26**, 103–118.

Schaefer T. J. (1978), Complexity of some two-person perfect information games. *J. Comp. & Syst. Sci.* **16**, 185–229.

Schönage A. (1980), Storage modification machines. *SIAM J. Comput.* **9**, 490–508.

Shepherdson J. C., Sturgis H. E. (1963), Computability of recursive functions. *J. ACM* **10**, 217–255.

Shoenfield J. R. (1971), *Degrees of Unsolvability*, Amsterdam.

Siegel H. J. (1979), A model of SIMD machines and a comparison of various interconnection networks and the effects of processor address masks. *IEEE Trans. Comp.* **C-26**, 907–917.

Soare R. I. (1982), Computational complexity of recursively enumerable sets. *Inf. Control* **52**, 19–35.

Stockmeyer L. J. (1977), The polynomial time hierarchy. *J. Theor. Comp. Sc.* **3**, 1–22.

Stockmeyer L. J., Meyer A. R. (1973), Word problems requiring exponential time. *Proc. Fifth Ann. Symp. on the Theory of Comput.* 1–9, ACM New York.

Strassen V. (1969), Gaussian elimination is not optimal. *Numerische Mathematik* **13**, 354–356.

Szelepcsenyi R. (1987), The method of forcing for nondeterministic automata. (Extended abstract) *EATCS Bulletin* **33**, 96–100.

Tsichritzis D. (1970), The equivalence problem of simple programs. *J. ACM* **17**, 729–738.

Turing A. M. (1936/1937), On computable numbers with an application to the Entscheidungsproblem. *Proc. London Math. Soc. series 2*, **42** (1936), 230–265 and **43** (1937), 544–546. See also Davis (1965), 116–151.

Valiant L. G. (1979), The complexity of computing the permanent. *Theor. Comp. Sci.*

Vitany P. M. B. (1984), *One queue or two pushdown stores take square time on a one-head tape-unit.* Report CS-R8406, March 1984, CWI dept. Comp. Sc.

Wagner K., Wechsung G. (1986), *Computational Complexity.* Reidel Publ. Comp. Dordrecht, Boston, Lancaster, Tokyo.

Wang H. (1957), A variant to Turing's theory of computing machines. *J. ACM* **4**, 63–92.

Wrathall C. (1977), Complete sets and the polynomial time hierarchy. *J. Theor. Comp. Science* **3**, 23–33.

Index